Transactions on Computer Systems and Networks

Transactions on Computer Systems and Networks is a unique series that aims to capture advances in evolution of computer hardware and software systems and progress in computer networks. Computing Systems in present world span from miniature IoT nodes and embedded computing systems to large-scale cloud infrastructures, which necessitates developing systems architecture, storage infrastructure and process management to work at various scales. Present day networking technologies provide pervasive global coverage on a scale and enable multitude of transformative technologies. The new landscape of computing comprises of self-aware autonomous systems, which are built upon a software-hardware collaborative framework. These systems are designed to execute critical and non-critical tasks involving a variety of processing resources like multi-core CPUs, reconfigurable hardware, GPUs and TPUs which are managed through virtualisation, real-time process management and fault-tolerance. While AI, Machine Learning and Deep Learning tasks are predominantly increasing in the application space the computing system research aim towards efficient means of data processing, memory management, real-time task scheduling, scalable, secured and energy aware computing. The paradigm of computer networks also extends it support to this evolving application scenario through various advanced protocols, architectures and services. This series aims to present leading works on advances in theory, design, behaviour and applications in computing systems and networks. The Series accepts research monographs, introductory and advanced textbooks, professional books, reference works, and select conference proceedings.

Amrita Rai · Dinesh Kumar Singh · Rupali Singh ·
Korhan Cengiz
Editors

Convergence of Artificial Intelligence, Machine Learning, and the Internet of Things in Industry 4.0 Applications

Springer

Editors
Amrita Rai
Department of Electronics
and Communication Engineering
Lloyd Institute of Engineering
and Technology
Greater Noida, Uttar Pradesh, India

Dinesh Kumar Singh
Department of Electronics
and Communication
G. L. Bajaj Institute of Technology
and Management
Greater Noida, Uttar Pradesh, India

Rupali Singh
Department of Electronics
and Communication Engineering
SRM Institute of Science and Technology
Delhi-NCR Campus
Ghaziabad, Uttar Pradesh, India

Korhan Cengiz
Department of Electrical and Electronics
Engineering
Biruni University
Zeytinburnu, Istanbul, Türkiye

ISSN 2730-7484 ISSN 2730-7492 (electronic)
Transactions on Computer Systems and Networks
ISBN 978-981-96-5268-6 ISBN 978-981-96-5269-3 (eBook)
https://doi.org/10.1007/978-981-96-5269-3

This Springer imprint is published by the registered company Springer Nature Singapore Pte Ltd.
The registered company address is: 152 Beach Road, #21-01/04 Gateway East, Singapore 189721, Singapore

Contents

Editors and Contributors

About the Editors

Amrita Rai received her B.Tech. (Electronics) degree from the College of Engineering Chandpur, affiliated with Nagpur University, in 1998. She earned her M.Tech. degree from Maharishi Dayanand University, Rohtak, where she was the university topper, in 2007, and her Ph.D. in Signal Processing from Thapar University, Patiala (previously Thapar Institute of Engineering and Technology), in 2015. With over 21 years of experience in teaching, research, and administration, Dr. Rai has published several papers in reputed international journals and has attended various national and international IEEE conferences. She has guided more than 20 postgraduate (PG) and 100 undergraduate (UG) theses. Additionally, she is the author of the book *Principles of VLSI and CMOS Integrated Circuits* and has written six book chapters. Dr. Rai serves as a reviewer for several prestigious journals and conferences. She has filed six patents and delivered expert lectures on topics such as Smart Sensors, IoT and sensors, and Adaptive Signal Processing at various universities and colleges. She has also chaired sessions at numerous national and international conferences. Dr. Rai was a senior member of the IEEE Delhi Section Society and has contributed to many international/SCI journals and conferences. She has published 13 national and international patents and received grants for two of her patents.

Dinesh Kumar Singh earned his B.E. in Electronics and Communication Engineering from Kumaon Engineering College in Dwarahat, Almora, in 2003. He holds an M.Tech. in Digital Communication from RGPV University, Bhopal, India, and a Ph.D. from the Indian Institute of Technology (ISM), Dhanbad, Jharkhand. His research interests include microwave engineering, particularly designing high-gain, compact, reconfigurable, fractal-shaped, circularly polarized microstrip antennas, substrate integrated waveguides (SIW), and magneto-electric (ME) dipole antennas for modern communication systems. Currently, he is a Professor in the Department of Electronics and Communication Engineering at G. L. Bajaj Institute of Technology and Management, Greater Noida, Uttar Pradesh, India. He has published more than

20 papers in various reputed international journals and conferences and serves as a reviewer for the *AEU International Journal of Electronics and Communication* and *Electronics Letters*, among others.

Rupali Singh obtained her B.E. degree from Nagpur University in 2001, her M.Tech. degree from Shobhit University, Meerut, in 2009, and her Ph.D. from the SRM Institute of Science and Technology, Ghaziabad, in 2021. Her research focuses on reversible logic circuits, quantum dot cellular automata, and nano-circuits. Dr. Singh possesses extensive teaching and research experience and has a good number of national and international publications to her name. She currently works at the School of Electronics and Communication Engineering at the SRM Institute of Science and Technology, Ghaziabad, India. She has published three patents and has received a grant for one of her patents.

Korhan Cengiz is an Assistant Professor who completed his Ph.D. in the Department of Electronics Engineering at Kadir Has University. He is the author of more than 40 papers in the field of communications and has participated as a keynote speaker at several conferences. Dr. Cengiz holds various referee and editor positions in several IEEE journals and conferences and is a Senior Member of IEEE. He serves as an Associate Editor for *IET Electronics Letters*.

Contributors

Reshu Agarwal Amity Institute of Information Technology, Amity University Noida, Noida, India

P. Amsini Department of Computer Science, Padmavani Arts and Science College for Women, Salem, India

Harsh Arora Banarsidas Chandiwala Institute of Information Technology, Kalkaji, New Delhi, India

Sanya Arora Amity School of Engineering and Technology, Amity University Uttar Pradesh, Noida, Uttar Pradesh, India

Shilpa Choudhary Department of Electronics and Communication Engineering, G. L. Bajaj Institute of Technology and Management, Greater Noida, Uttar Pradesh, India

Abhishek Das School of Engineering and Technology, Centurion University of Technology and Management, Paralakhemundi, Odisha, India;
Department of Computer Science and Engineering, Centurian University of Technology and Management Paralakhemundi, Jatani, India

Shradha Gupta Department of Physics, ABES Institute of Technology, Ghaziabad, Uttar Pradesh, India

Sindhu Hak Gupta Amity School of Engineering and Technology, Amity University Uttar Pradesh, Noida, Uttar Pradesh, India

Karabi Kalita Department of Electronics and Communication Engineering, Lloyd Institute of Engineering and Technology, Greater Noida, India

Balaram Yadav Kasula Department of Information Technology, University of the Cumberlands, Williamsburg, KY, USA

Imran Ahmed Khan Department of Electronics and Communication, Jamia Millia Islamia, New Delhi,
India

Dipti Kiran Department of Commerce, Guru Ghasidas University, Bilaspur, C.G, India

Naveen Kumar University School of Management, Gautam Buddha University, Greater Noida, Uttar Pradesh, India

Rohit Kumar University School of Management, Gautam Buddha University, Greater Noida, Uttar Pradesh, India

Md RashId Mahmood Department of ECE, Guru Nanak Institutions Technical, Campus, Hyderabad, Hyderabad, Uttar Pradesh, India

Asha Rani Mishra Department of Artificial Intelligence and Machine Learning, Galgotias College of Engineering and Technology, Greater Noida, Uttar Pradesh, India

Ashok Kumar Mishra Dean Department of Commerce, Guru Ghasidas Vishwavidyalaya, Bilaspur, C.G, India

Mihir Narayan Mohanty Department of Electronics and Communication Engineering, ITER, Siksha 'O' Anusandhan (Deemed to be University), Bhubaneswar, Odisha, India

Saumendra Kumar Mohapatra Faculty of Engineering and Technology, Sri Sri University, Cuttack, Odisha, India

Om Prakash Department of Electronics and Communication Engineering, SVCET, Chittoor, India

Hemant Purohit Department of Computer Engineering and Science, Florida Institute of Technology, Melbourn, FL, USA

Amrita Rai Department of Electronics and Communication Engineering, Lloyd Institute of Engineering and Technology, Greater Noida, Uttar Pradesh, India

Visakshi Rajput Department of Electronics and Communication Engineering, G. L. Bajaj Institute of Technology and Management, Greater Noida, Uttar Pradesh, India

Agarwal Reshu Amity Institute of Information Technology, Amity University, Noida, India

Owais Ahmad Shah Department of Electronics and Communication Engineering, Dayananda Sagar University, Bengaluru, Karnataka, India

V. Shanmugasundaram Department of Electrical and Electronics Engineering, Sonna College of Technology, Salem, India

Abhishek Sharma Department of Electronics and Communication Engineering, G. L. Bajaj Institute of Technology and Management, Greater Noida, Uttar Pradesh, India

Sanjeev Sharma Department of Applied Science and Humanities, IMS Engineering College, Ghaziabad, Uttar Pradesh, India

Dinesh Kumar Singh Department of Electronics and Communication Engineering, G. L. Bajaj Institute of Technology and Management, Greater Noida, Uttar Pradesh, India

Neha Singh Department of Electronics and Communication Engineering, GCRG Group of Institution, Lucknow, Uttar Pradesh, India

Ritesh Pratap Singh School of Electrical and Computer Engineering, Haramaya Institute of Technology, Haramaya University, Dire Dawa, Ethiopia

Sri Krishna Singh Department of Physics, ABES Institute of Technology, Ghaziabad, Uttar Pradesh, India;
Faculty of Department of Physics, Government College Lamta Balaghat, Bhondwa, Madhya Pradesh, India

Ayush Thakur Amity Institute of Information Technology, Amity University Noida, Noida, India

Vinita Amity Institute of Information Technology, Amity University, Noida, India

Pawan Whig Vivekananda Institute of Professional Studies, New Delhi, India

Kosheen Wighmal Amity School of Engineering and Technology, Amity University Uttar Pradesh, Noida, Uttar Pradesh, India

Nikhitha Yathiraju University of the Cumberlands, Williamsburg, KY, USA

Chapter 1
Artificial Intelligence—An Overview

Abhishek Sharma, Shilpa Choudhary, Dinesh Kumar Singh, and Hemant Purohit

Abstract This chapter serves as an introduction to AI, providing a clear definition of the concept and explaining its importance. It also provides an overview of the foundations and modifications of AI, highlighting significant developments in the field. The chapter covers various ways of AI, including deep learning, and examines the advantages and limitations of each approach. It then delves into machine learning, explaining the concept and its significance, along with its different types, how they are used in various applications. Furthermore, the chapter discusses NLP and provides examples of their applications in different fields. It also examines the functions of robotics in AI and provides examples of how robots are being used in various applications. The chapter then shifts its focus to the applications of AI in many industries like healthcare, finance, manufacturing, and transportation. It also discusses the benefits and challenges of AI in these areas. In addition, the chapter examines the ethical and societal impacts of AI, including concerns such as bias, privacy, accountability, and employment. It emphasizes the need for ethical guidelines and regulations for the development and deployment of AI to mitigate these issues. The chapter also provides an overview of emerging trends and areas of research in AI and discusses the potential future impact of AI on society. Finally, the chapter concludes by summarizing the key points covered and highlighting the importance of understanding the fundamentals of AI.

Keywords Artificial intelligence · Machine learning · Deep learning · NLP · Computer vision · Robotics

A. Sharma · S. Choudhary (✉) · D. K. Singh
Department of Electronics and Communication Engineering, G L Bajaj Institute of Technology and Management, Greater Noida, UP, India
e-mail: shilpadchoudhary@gmail.com

A. Sharma
e-mail: ec20092@glbitm.ac.in

H. Purohit
Department of Computer Engineering and Science, Florida Institute of Technology, Melbourn, FL, USA

A. Rai et al. (eds.), *Convergence of Artificial Intelligence, Machine Learning, and the Internet of Things in Industry 4.0 Applications*, Transactions on Computer Systems and Networks, https://doi.org/10.1007/978-981-96-5269-3_1

1.1 Introduction to Artificial Intelligence

AI encompasses the design and development of computer system that can perform certain tasks as well or better than human educational purposes. Such tasks include reasoning, learning (the ability to determine what is happening in the environment at present time), problem-solving, and natural language comprehension. It is rapidly growing field, and AI applications can potentially have a ground-breaking impact on regulating industries as well as improving using daily life procedures in sectors such as healthcare transportation and entertainment.

1.2 Outline of AI's History and Advancement

Artificial intelligence (AI) first originated when researchers began discussing the potential of machines so far back as in the mid-twentieth century, tasks such that only human learning and intelligence could perform. One of the first major achievements in the field was accomplished with a computer capable of playing chess, named Checkers. In the 1980s, a number of rules-based systems emerged, using sets of if–then rules to perform key tasks such as pattern matching and decision-making.

After emerging from the initial AI winter, in which funding had dried up for a period following high expectations during its early years as researchers believed that machines with near-human cognitive abilities were just around the corner on accounts of optimistic results obtained by successful research, artificial intelligence (AI) research went into an extended hiatus. Still, De Juvigny said progress stagnated by the 1970s and 1980s—a time that would eventually represent an artificial intelligence winter marked by diminished funding and interest.

Machine learning fell out of favor as an academic area in the 1980s, and then when machine learning started to rise, we saw a resurgence of interest toward its approach with data rather than knowledge during 1990s. Arising neural network-enabled systems to cater large amounts of data and, as a result, have improved predictive accuracy. Between 2000 and 2019, AI grew very quickly because we had more data than ever and better, faster computers to use that data. These days, the advancement of deep learning has brought great change in specializations (Fig. 1.1).

1.3 Approaches to AI

Rule-based systems, expert systems, neural networks, and deep learning are all different approaches to machine learning and artificial intelligence. Each approach has its own strengths and weaknesses, making them appropriate for various types of problems (Table 1.1).

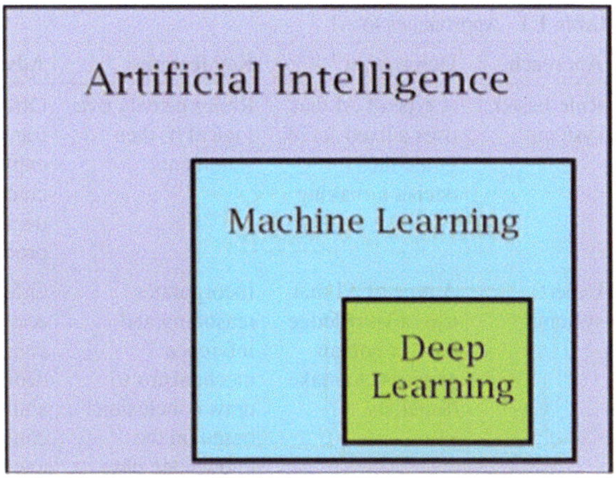

Fig. 1.1 Relation between AI, machine learning, and deep learning

- Rule-Based Systems: It depends on a set of rules if–then that are specially designed to take some decision or prediction. For example, rule-based systems are commonly used in areas like medical diagnosis, fraud detection, and credit scoring to name but a few since they require some human expertise. One of the benefits of these systems is their transparency and interpretability, since they outline rules that are easy to follow by humans. Nonetheless, they are not as versatile in dealing with tasks involving a high level of decision-making and actually understanding data (Russell 2010).
- Expert Systems: Though similar to rule-based systems, expert system is more advanced and can handle additional knowledge for better reasoning. They aim to replicate how experts think in various fields such as medicine, engineering, or finance. The power of the expert systems is in processing hard decision and reasoning tasks. Nonetheless, their principal drawback is that they require the intervention of human specialists to specify the rules and knowledge base; this fact may limit its scalability (Approach 2011).
- Neural Networks: These are made up of connected nodes (also known as neurons) that take in data, perform operations on it, and then pass info along to the next node. Neural networks are commonly employed in many applications. The main advantage of neural networks is that they can learn from data and increase their performance over time, which make them really good candidates for pattern recognition or classification type tasks. The lack of interpretability in neural networks is a critical issue, considering there are applications where comprehending the reasoning behind decision-making can be crucial (Artificial neural networks in medical diagnosis 2013).
- Deep Learning: This subfield of neural networks, deep learning, trains and operates AI simulations on 1000sK+ layers—including some like this one with billions of neurons. Deep learning is good at big data and can find very hard patterns, so it also works well on, e.g., computer vision or voice recognition tasks. The

Table 1.1 Approaches to AI

Approach	Description	Key features	Advantages	Disadvantages
Rule-based systems	A type of AI that uses a fixed set of guidelines for decision-making	Relies heavily over logical if–then statements	Offers transparency, explain ability, and control over decision-making processes	Limited ability to handle complex and uncertain data, difficult to scale and maintain
Expert systems	A type of AI that uses a knowledge base of human expertise to make decisions	Incorporates reasoning and inference mechanisms to draw conclusions based on the knowledge base	Offers high accuracy and consistency in decision-making, ability to handle complex and uncertain data	Limited ability to learn and adapt to new situations, high development and maintenance costs
Neural networks	A type of AI that uses a layered structure of interconnected nodes to process data and make decisions	Emphasizes pattern recognition and statistical analysis	Offers high accuracy and predictive power for large and complex datasets	Limited transparency and explainability
Deep learning	A subset of neural networks that uses multiple layers of processing to learn representations of data	Emphasizes automatic feature extraction and hierarchical learning	Offers high accuracy and predictive power for unstructured data	Limited transparency and explainability, high computational and data requirements

main advantage of this is that it lets you accommodate the data more easily than traditional machine learning algorithms for complex methods (Miotto et al. 2017).

1.3.1 Some Other Important Approaches to AI

- **Symbolic Logic and Reasoning****: This method employs symbols and formal rules to represent and manipulate knowledge systematically. It enables the system to draw inferences and make decisions based on logical reasoning. Often utilized to develop expert systems, this approach aims to replicate the decision-making processes of human specialists within a particular field.
- **Connectionism**: Motivated by the organization and workings of the brain, this approach leverages artificial neural networks to understand intricate relationships between inputs and outputs. It facilitates tasks such as object recognition, natural language processing, and interaction between humans and computers. Typically

used to build deep learning models, this method allows systems to learn from extensive datasets and achieve high levels of accuracy and effectiveness.

- **Evolutionary algorithms**: This approach is based on the principles of natural selection and genetic variation and uses a population of candidate solutions that evolve over time to find the optimal or near-optimal solution. It enables the system to adjust to evolving conditions and solve problems that are difficult to define or optimize. This approach is often used to create optimization and search systems that can find solutions to complex and dynamic problems.

However, the main limitation of deep learning is its requirement for large amounts of data and computational resources. Training deep learning models can be time-consuming and expensive, which can be a barrier to their adoption in certain applications.

1.4 Machine Learning

It is a domain within AI dedicated to design algorithms that can analyze and interpret data to make predictions and generate outcomes without requiring explicit manual programming by humans. These algorithms have the ability to autonomously identify patterns in data, enabling them to execute tasks such as categorization, prediction, grouping, and forecasting. Machine learning has become a pivotal area of research and application, offering significant practical applications and potential advantages for society.

A key benefit of machine learning is its capability to enhance over time. By examining data, recognizing patterns, and adjusting to new information, machine learning models can refine their accuracy and effectiveness in making predictions. This characteristic renders machine learning particularly useful in fields like healthcare, finance, marketing, and manufacturing, where extensive data and complex patterns are prevalent. In summary, machine learning is a wide and fast expanding field with many potential applications and societal benefits. Ongoing research works continue to explore innovative techniques and approaches, and there is still much to uncover about how machine learning can contribute to advancing our world.

1.4.1 Kinds of ML

ML is mainly divided into the following three of the types (Table 1.2).

Table 1.2 Types of ML

Type of learning	Supervised learning	Unsupervised learning	Reinforcement learning
Type of input data	Labeled data—both the input as well as the output variables should be known	Unlabeled data—only input variables are known	Feedback from environment/experience
Learning objectives	To develop a function that reliably translates inputs into outputs	To find patterns and structure in input data	To develop a strategy that optimizes the total reward
Examples of applications	Image classification, speech recognition, regression	Clustering, anomaly detection, dimensionality reduction	Game playing, robotics, autonomous driving

1.4.1.1 Supervised Learning

Supervised learning involves training a model using labeled data to make predictions on new, unlabeled data. Some of common tasks that supervised learning can perform are:

- Classification: This is the task of assigning a discrete label to an input. For example, spam detection, face recognition, sentiment analysis, etc.
- Regression: This is the task of predicting a continuous value for an input. For example, house price prediction, stock market forecasting, etc.
- Object detection: This is the task of locating and identifying objects in an image or video. For example, self-driving cars, face detection, etc.

Some of the popular algorithms that are used for supervised learning are:

- Linear regression: This is a straightforward algorithm that represents the relation among input variables and output variables by simply using a linear function. It can be used for regression tasks.
- Support Vector Machine: This robust algorithm identifies the optimal hyperplane that divides different classes in the input space. It is applicable for both classification and regression tasks.
- Decision Tree: This approach uses a tree-like model to partition the input space into distinct regions based on certain criteria. It can be employed for classification as well as regression tasks.
- K-Nearest Neighbor: The straightforward algorithm divides the new input by referencing the labels of its closest neighboring inputs. It is versatile, useful for both classification and regression problems.

Some examples of supervised learning applications are:

- Image classification: identifying the objects or scenes in an image.
- Spam detection: filtering out unwanted or malicious emails.

- Sentiment analysis: determining the emotional tone of a text.
- Speech recognition: converting spoken words into text.
- Credit scoring: assessing the risk of lending money to a customer.

Supervised learning is a powerful technique that enables computers to learn from data and make intelligent decisions. However, it also has some limitations, such as.

- It may not generalize well to new or unseen data, especially if the data is noisy or biased.
- It may overfit the data, which means that it learns the specific details of the data rather than the general patterns.

1.4.1.2 Unsupervised Learning

Unsupervised learning involves analyzing unlabeled data to uncover patterns and relationships within the dataset. In this approach, the algorithm works with data that lacks predefined output labels. The aim of unsupervised learning is to reveal hidden structures or patterns in the data, which can aid in future decision-making. Common algorithms in unsupervised learning include clustering, dimensionality reduction, and association rule mining.

Common tasks performed includes:

- Dimensionality Reduction: This process reduces the number of features or dimensions in the dataset while retaining the core information. Techniques such as principal component analysis and singular value decomposition are commonly used.
- Anomaly Detection: This task focuses on identifying data points that substantially deviate from the anticipated patterns or distributions. Applications include fraud detection and outliner identification.

Some of the popular algorithms that are used for unsupervised learning are:

- K-means clustering: This is a simple algorithm that partitions the data into k clusters based on the distance to the cluster centroids. It can be used for clustering tasks.
- Hierarchical clustering: This is an algorithm that constructs a hierarchy of clusters based on the similarity or differences between the data points. It can be used for clustering tasks.
- Neural networks: This is a complex algorithm that consists of multiple layers of interconnected nodes that can learn nonlinear relationships between inputs and outputs. It can be used for various tasks such as clustering, dimensionality reduction, and anomaly detection.

Some examples of unsupervised learning applications are:

- Customer segmentation: dividing customers into groups based on their behavior, preferences, or demographics.

- Topic modeling: discovering the main themes or topics in a collection of text documents.
- Image compression: reducing the size of an image by removing redundant or irrelevant pixels.
- Face recognition: identifying and verifying the identity of a person from their face image.
- Recommender systems: suggesting products or services to users based on their preferences or past interactions.

Unsupervised learning is a challenging and exciting technique that enables machines to discover hidden knowledge from data. However, it also has some limitations, such as.

- It may be difficult to evaluate the performance or accuracy of unsupervised learning models, as there is no ground truth or correct answer to compare with.
- It may be hard to interpret or explain the results or outputs of unsupervised learning models, as they may not match human intuition or expectations.
- It may be sensitive to the choice of parameters, such as the number of clusters or the distance metric, which may affect the quality or consistency of the results.

1.4.1.3 Reinforcement Learning

The algorithm improves through trial-and-error feedback, aiming to make choices that optimize long-term rewards. The goal of reinforcement learning is to develop an agent capable of making the best possible decisions within a given environment (Wang 2001).

Some of the common tasks that reinforcement learning can perform are:

- Game playing: This is the task of learning to play and win games such as chess, Go, or Atari. For example, AlphaGo is a famous RL system that defeated the world champion of Go.
- Robotics: This is the task of learning to control and manipulate robots in various scenarios. For example, OpenAI's Dactyl is a RL system that learned to manipulate a robotic hand.
- Recommendation systems: This is the task of learning to recommend products or services to users based on their preferences and behavior. For example, Netflix uses RL to personalize its recommendations to users.

Some of the popular algorithms that are used for reinforcement learning are:

- Q-learning: This algorithm learns a value function that estimates the expected future reward for each action in each state. It is suitable for discrete action spaces and deterministic environments.
- Policy Gradient: This algorithm learns a policy function that directly outputs the probability of selecting each action in each state. It is applicable to continuous action spaces and stochastic environments. Actor-critic: This is an algorithm that combines Q-learning and policy gradient and learns both a value assessment

and a decision-making function. It can be used for various action spaces and environments.

1.4.2 Use of ML in Some of the Applications

Machine learning has emerged as an essential tool for solving complex problems in various fields. Here are some examples of how machine learning is being used in various applications:

- Healthcare: Machine learning is being used to improve the accuracy of disease diagnosis and drug discovery. In a study, machine learning was utilized to examine electronic health records and identify patients at risk of developing sepsis, a life-threatening condition.
- Finance: Machine learning is being employed to enhance fraud detection and risk management in financial institutions. For example, one study utilized machine learning to identify fraudulent credit card transactions with high precision.
- Transportation: Machine learning is being applied to enhance traffic management and minimize accidents in transportation systems. In one study, machine learning was employed to forecast traffic congestion and redirect vehicles to bypass crowded areas. Retail: Machine learning is being used to personalize recommendations and improve customer experience in retail. In one study, machine learning was used to suggest products to customers based on their previous purchases and browsing behavior. Manufacturing: Machine learning is being used to improve product quality and optimize production processes in manufacturing. In one study, machine learning was used to predict the quality of steel products based on the manufacturing process parameters.
- Marketing: ML can help with customer segmentation, personalization, recommendation, churn prediction, sentiment analysis, and campaign optimization. For example, Netflix uses ML to recommend movies and shows to its users based on their preferences and behavior.
- Cybersecurity: ML can help with malware detection, intrusion detection, spam filtering, phishing detection, and network security. For example, Microsoft uses ML to protect its cloud services from cyberattacks.

These examples demonstrate how machine learning is transforming various industries by enabling more accurate predictions, personalized recommendations, and better decision-making based on data.

1.5 Natural Language Processing

It is a subfield of artificial intelligence that focuses on the interactions between human language and computers. It is concerned with the ability of machines to understand, interpret, and produce human language. It is applied in various areas, such as chatbots and speech recognition. NLP works by breaking down language into its constituent parts and analyzing them. This process involves several steps.

- Tokenization: In this step, text is broken down into individual words or phrases, called tokens. Tokenization is the first step in many NLP tasks.
- Named entity recognition: In this step, named entities, such as people, places, and organizations, are identified and labeled.
- Parsing: In this step, the syntactic structure of a sentence is analyzed to determine its meaning.
- Sentiment analysis: In this step, the sentiment of a sentence or document is analyzed to determine whether it is positive, negative, or neutral.
- Machine translation: In this step, machine learning algorithmic steps are considered to translate text from one language to another (Fig. 1.2).

NLP relies on several techniques which enable computers to understand and generate human language, and they have been applied to a variety of applications, from chatbots to language translation (Sutton and Barto 1998).

1.5.1 How NLP is Being Used in Various Applications

Natural language processing (NLP) is widely applied across various domains. Below are some notable examples of NLP in action:

- Chatbots: NLP technology is employed to create chatbots capable of interacting with users in natural language. These chatbots serve functions such as customer service, information retrieval, and more. Examples include Siri, Alexa, and Google Assistant, which can answer queries, perform tasks, and engage in conversation with users.
- Sentiment Analysis: This analysis supports market research, customer feedback evaluation, and similar tasks. For instance, Netflix leverages sentiment analysis to suggest movies and TV shows based on user preferences and behavior.
- Language Translation: NLP drives the development of machine translation systems that convert text from one language to another. These systems enhance translation accuracy through machine learning algorithms and statistical techniques.
- Named Entity Recognition: This capability is useful for information retrieval, data mining, and related applications.
- Speech Recognition: These systems rely on neural networks to enhance the accuracy of transcriptions. For example, platforms like YouTube, Zoom, and Skype

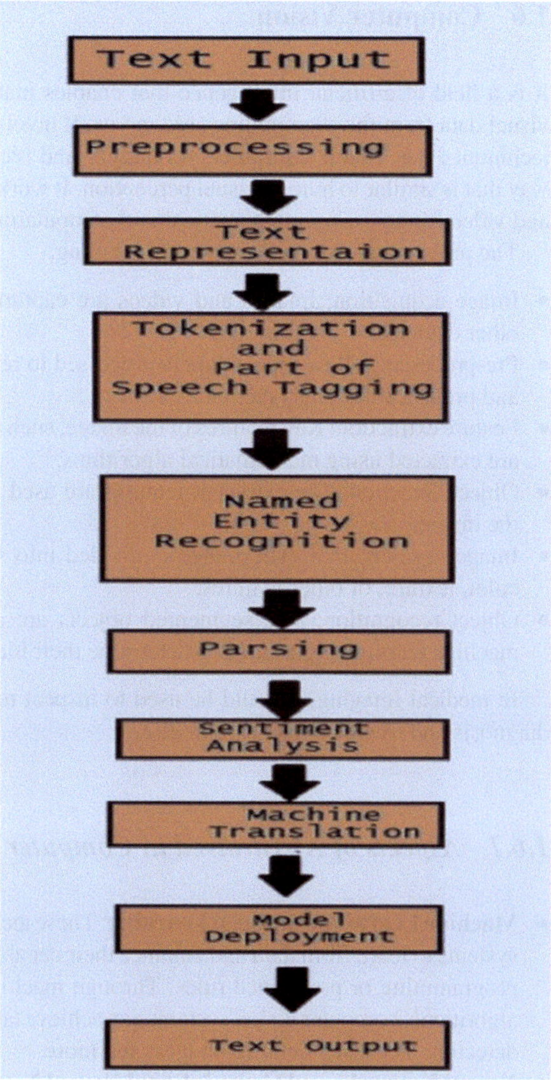

Fig. 1.2 Steps of natural language processing

use speech recognition to produce captions, subtitles, and transcripts for their videos and calls.

These instances highlight how NLP is addressing real-world challenges and enhancing human–computer interaction. With its potential to revolutionize various sectors, from healthcare to finance, NLP enables more precise analysis and processing of natural language data (Kulkarni and Shah 2018).

1.6 Computer Vision

It is a field of artificial intelligence that enables machines to convert and analyze visual data from the surroundings around us. It involves developing algorithms and techniques that enable computers to analyze and recognize images and video in a way that is similar to human visual perception. It works by processing digital images and video frames using mathematical and computational algorithms.

The process involves several steps, including:

- Image acquisition: Images and videos are captured using cameras, sensors, or other devices.
- Pre-processing: Raw image data is processed to remove noise, enhance features, and prepare it for analysis.
- Feature extraction: Key features of the image, such as edges, corners, and textures, are extracted using mathematical algorithms.
- Object detection: The extracted features are used to detect and locate objects in the image.
- Image segmentation: The image is divided into segments or regions based on color, texture, or other features.
- Object recognition: The segmented objects are analyzed and classified using machine learning algorithms to determine their identity.

In medical imaging, it could be used to inspect medical images and assist with diagnosis and treatment (Koli et al. 2022).

1.6.1 Aspects of AI Involved in Computer Vision

- **Machine Learning and Deep Learning**: These techniques allow computer vision systems to learn from data and enhance their capabilities without needing explicit programming or predefined rules. Through machine learning and deep learning algorithms, computer vision systems can achieve tasks such as identifying objects, detecting faces, segmenting images, and more.
- **Neural Networks and Convolutional Neural Networks**: These architectures are fundamental for computer vision systems to handle and analyze intricate, high-dimensional visual information. These networks are used in various applications, including image classification, image generation, and image captioning.
- **Computer Graphics and Image Processing**: These methods are used to create, modify, and enhance visual data in computer vision systems. Techniques in computer graphics and image processing facilitate tasks like stitching and enhancement.
- **Natural Language Processing and Speech Recognition**: These techniques enable computer vision systems to understand, interpret, and generate text and

speech based on visual input. Algorithms support tasks like generating image captions, searching for images, and describing visual content.

Computer vision is rapidly evolving and becoming increasingly significant across numerous fields and applications. As technological advancements continue, its role is expected to expand further, impacting a broad spectrum of industries.

1.6.2 Uses of Computer Vision in Various Applications

Computer vision has numerous practical applications in various fields. Here are some examples of different applications:

- Autonomous vehicles: used in self-driving cars to enable them to "see" the road and surrounding environment, detect objects, and make decisions.
- Security and surveillance: Computer vision is employed in defense and surveillance systems to identify and monitor individuals and objects, analyze behaviors, and notify authorities of potential threats.
- Medical imaging: Computer vision is used in medical imaging to investigate and interpret scans and ultrasound images, detect abnormalities, and assist with diagnosis and treatment. For example, Google uses computer vision to detect diabetic retinopathy from eye scans.
- Retail and marketing: Computer vision is used in retail and marketing to analyze customer behavior, track foot traffic, and improve customer experience.
- Facial recognition: This is the application of computer vision that can identify and verify the identity of a person from their face. It can be used for security, authentication, surveillance, and social media. For example, Facebook uses facial recognition to tag users in photos and videos.
- Robotics: Computer vision is used in robotics to detect objects and interact with humans. For example, Amazon uses computer vision to guide robots in its warehouses and delivery services.
- Sports performance analysis: This is the application of computer vision that can measure and improve the performance of athletes and teams from sports videos and images. It can be used for coaching, training, and broadcasting. For example, IBM uses computer vision to analyze tennis matches and provide insights and statistics.

These examples demonstrate how computer vision is being used to solve real-world problems and improve human lives (Szeliski 2022).

1.7 Robotics

Robotics is very important in the field of artificial intelligence (AI) by providing a physical platform for implementing and testing AI algorithms and techniques. Robotics comprises of designing, construction, and operation of machines that can perform tasks autonomously or with human guidance. Robotics systems are an important tool for AI research because they provide a means of testing AI algorithms in the real world, rather than just in simulations. The integration of robotics and AI allows machines to learn and adapt to new situations and to perform tasks in areas where there is risk to human life but these tasks can be performed autonomously or under human guidance. Robotics also enables machines to interact with their environment in a way that is similar to human interaction, allowing them to perceive and respond to changes in their surroundings. One of the main benefits of robotics in AI is the ability to collect and analyze large amounts of data from sensors and cameras on the robot. This data can be used to instruct AI models and improve the accuracy and performance. Robotics can also be used to test and validate AI algorithms in real-world scenarios, allowing researchers to identify and address any issues or challenges that arise. Robotics is being used in many different applications of AI, including autonomous vehicles, drones, industrial automation, and healthcare. For example, in healthcare, robots can be used to assist with surgeries, monitor patients, and provide physical therapy. In industrial automation, robots can be used to perform tasks that are dangerous or repetitive for humans, such as welding or assembly().

In summary, robotics provides a physical platform for AI algorithms to operate and interact with the world, making it an essential component of the field. As AI and robotics continue to advance, they are likely to have an even greater impact on society, enabling new applications and improving efficiency in various industries.

1.7.1 Applications of Robotics in AI

- Industrial robots: These are robots that can perform repetitive, precise, or hazardous tasks in manufacturing, assembly, or packaging. For example, automotive industry robots, drug compounding robots, and order picking robots.
- Service robots: These are robots that can provide assistance, information, entertainment, or guidance to humans or other machines. For example, chatbots, personal assistants, delivery robots, and robotic pets.
- Medical robots: These are robots that can assist in diagnosis, treatment, surgery, or rehabilitation of patients. For example, surgical robots, prosthetic limbs, exoskeletons, and nanorobots.
- Military robots: These are robots that can perform reconnaissance, surveillance, combat, or rescue missions in hostile or dangerous environments. For example, drones, bomb disposal robots, and autonomous vehicles.

- Educational robots: These are robots that can facilitate learning, teaching, or research in various fields and levels. For example, robot kits, robot tutors, robot simulators, and robot competitions.

1.7.2 Uses of Robots in Various Applications Across Many Industries

Robots are being used extensively across diverse sectors. Here are some examples of how robots are being used in different applications:

- Manufacturing: Robots widely employed in manufacturing to perform repetitive or dangerous tasks such as welding, painting, and assembly. They can operate continuously and with high precision, improving efficiency and reducing errors.
- Healthcare: Robots are being used in healthcare to assist with surgeries, perform physical therapy, and monitor patients. They can operate with high precision, reducing the risk of human error, and they can be programmed to adapt to different patient needs.
- Space exploration: Robots are being used in space exploration to explore planets, asteroids, and other celestial bodies. They can operate in extreme environments and perform certain missions which are typical or dangerous for human-beings.
- Disaster response: Robots are being used in disaster response to assist with search and rescue efforts, assess damage, and perform tasks that are too dangerous for humans. They can be outfitted with sensors and cameras to collect data and provide information to first responders.

These examples demonstrate how they can improve efficiency, reduce risk, and perform certain missions which are typical or dangerous for human-beings (Chella et al. 2020).

1.8 Applications of AI

AI has become an important part of various industries and domains. It has the potential to transform and enhance the efficiency and effectiveness of various processes. In this response, we will discuss the use of AI in healthcare, finance, manufacturing, and transportation, along with their benefits and challenges.

- Healthcare: Artificial intelligence has the capability to bring about a significant transformation by enhancing the accuracy of diagnosis, treatment methods, and ultimately, the well-being of patients. It can assist in predicting diseases, identifying potential drug candidates, and improving patient care. AI algorithms can analyze medical data and provide personalized treatment plans. For instance, AI-powered chatbots can assist doctors in diagnosing symptoms, reducing wait times, and providing remote consultations. Despite these advantages, implementing AI in healthcare faces challenges such as data privacy and security, bias, and transparency (Morgan et al. 2022).

- Finance: AI can analyze vast amounts of data to identify fraudulent transactions and forecast financial performance. AI can also automate financial processes such as account opening, customer service, and portfolio management. However, the widespread adoption of AI in finance is limited by regulatory compliance, data privacy, and ethical considerations (Al-Jaroodi and Mohamed 2018).
- Manufacturing: AI is transforming manufacturing by improving operational efficiency, reducing costs, and enhancing quality control. AI-powered predictive maintenance can prevent machine breakdowns, improve uptime, and reduce downtime. AI can also optimize the production process, reducing waste and improving quality control. Additionally, AI can assist in supply chain management, logistics, and inventory control. However, implementing AI in manufacturing faces challenges such as data quality, interoperability, and standardization (Al-Blooshi and Haitham 2020).
- Transportation: Self-driving vehicles can improve safety, reduce accidents, and enhance mobility. AI-powered route optimization can reduce congestion, improve delivery times, and reduce fuel consumption. Predictive maintenance can prevent breakdowns, improve uptime, and reduce maintenance costs. However, the adoption of AI in transportation faces challenges such as safety, regulatory compliance, and ethical considerations (Chen et al. 2014) (Fig. 1.3).

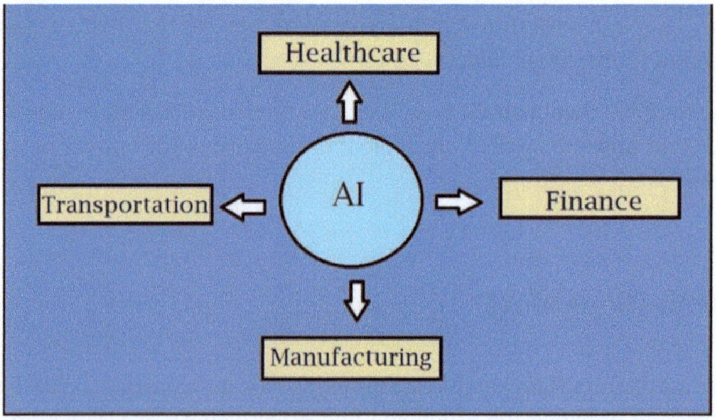

Fig. 1.3 Applications of AI

1.9 AI in Healthcare

1.9.1 Personalized Medicine

1.9.1.1 Genetic Analysis and Drug Development

- Genomic Data Analysis: Advanced AI algorithms meticulously analyze extensive genetic information to identify patterns, anomalies, and genetic markers associated with diseases. This approach helps in understanding individual genetic predispositions and potential responses to treatments, enabling tailored healthcare.
- Targeted Therapies: By correlating genetic profiles with treatment responses, AI assists in identifying targeted therapies that suit specific genetic compositions. This personalized approach minimizes adverse effects and enhances treatment effectiveness.
- Drug Discovery: AI expedites the drug discovery process by simulating molecular structures, predicting interactions which accelerates the identification of promising compounds for various diseases, reducing time and resources spent in traditional drug development.

1.9.1.2 Disease Prediction and Diagnosis

- Medical Imaging and Diagnostic Assistance: Radiology and Imaging: AI algorithms complement radiologists by scrutinizing medical images, highlighting anomalies, and offering more precise interpretations. This collaboration leads to quicker and more accurate diagnoses.
- Pathology and Histology: In pathology, AI assists pathologists by analyzing tissue samples, aiding in the detection of abnormalities or patterns that might indicate diseases, ultimately enhancing diagnostic accuracy.

1.9.1.3 Predictive Analytics and Early Detection

- Electronic Health Records (EHR) Analysis: AI analyzes vast datasets within electronic health records to identify trends, predict disease risks, and recommend preventive measures for patients based on their health history and parameters.
- Risk Assessment: AI models predict an individual's susceptibility to specific diseases by analyzing various health parameters and historical data, allowing for early interventions and personalized preventive strategies.

1.9.1.4 Patient Care and Monitoring

Remote Patient Monitoring

- Wearable Devices and IoT: AI-driven wearables continuously monitor vital signs, collecting real-time data. This information is transmitted to healthcare providers, enabling remote patient management and allowing for early interventions upon detecting anomalies.
- Telemedicine and Virtual Care: AI-powered telemedicine platforms facilitate remote consultations, allowing patients to access healthcare services virtually. This is especially advantageous for people in remote or underserved regions, enhancing healthcare accessibility.

Treatment Optimization and Decision Support

- Treatment Planning: AI assists healthcare providers in creating tailored treatment plans by evaluating patient data, medical literature, and historical treatment outcomes, enhancing treatment precision.
- Clinical Decision Support Systems (CDSS): These systems provide evidence-based suggestions and insights, aiding physicians in making well-informed decisions regarding patient care, diagnosis, and treatment choices.

1.10 Uses of AI in Healthcare

- Improving the speed of screening for diseases, such as cancer, diabetes, and heart disease. For example, IBM Watson uses AI to assist doctors in diagnosing and treating cancer.
- Assisting with clinical care, such as providing virtual nursing assistants, reducing dosage errors, and performing less invasive surgeries. For example, an AI-powered tool can help patients self-administer insulin correctly.
- Developing new drugs and personalized medicine, such as finding new targets, designing molecules, and testing efficacy and safety. For example, Google's DeepMind uses AI to predict protein structures and functions.

AI can help healthcare professionals and patients with various tasks and challenges, and improve the quality, efficiency, and accessibility of healthcare services. However, AI also poses some ethical, social, and technical concerns, like information security, human oversight, and bias. Therefore, the use of AI in healthcare should follow some guiding principles, such as transparency, fairness, safety, and accountability.

1.11 Ethical Considerations in AI Healthcare

1.11.1 Data Privacy and Security

- Patient Data Confidentiality: Ensuring stringent security measures to protect patient data while utilizing large datasets for AI analysis, complying with data protection regulations such as HIPAA and GDPR.
- Regulatory Compliance: Adherence to regulations to protect patient data and ensure ethical handling of healthcare information, maintaining confidentiality and privacy.

1.11.2 Algorithmic Bias and Interpretability

- Fairness in AI Algorithms: Addressing and mitigating biases in AI models that could impact diagnosis, treatment, or patient care plans, ensuring fair and accurate healthcare outcomes.
- Transparency and Interpretability: Ensuring that AI-based decisions are understandable and transparent, providing healthcare professionals with insights into how AI arrived at its recommendations, promoting trust and reliability.

1.11.3 Regulatory and Ethical Guidelines

- Establishing Ethical Frameworks: Formulating guidelines and ethical frameworks for AI applications in healthcare, emphasizing patient-centric care, informed consent, and accountability in decision-making processes.
- Ethical and Regulatory Compliance: Ensuring that the integration of AI complies with ethical standards and regulatory requirements, safeguarding patient rights, and maintaining the highest standards of care.

Each facet within AI's role in healthcare significantly contributes to enhancing patient outcomes, refining diagnostics, and revolutionizing treatment strategies. The ethical considerations remain pivotal, ensuring responsible integration of AI into healthcare practices, maintaining patient confidentiality, and adhering to stringent regulatory frameworks. This continual evolution of AI in healthcare aims to revolutionize healthcare delivery and outcomes while ensuring patient safety and ethical considerations are prioritized.

1.12 Ethical and Social Implications of AI

Artificial intelligence (AI) is influencing various sectors in society, including healthcare, finance, education, and transportation. These are, however, issues with the large-scale use of AI such as bias and its social implications negatively upon areas like privacy, accountability, and employment. One concern is that because AI systems are trained on historical data, which can be biased in favor of the groups with power and privilege when modeling decisions at scale could have dire consequences for marginalized people like racial minorities or women—the presence of bias. This has given rise to debiasing algorithms developed by researchers who aim to reduce bias in AI systems. However, these approaches have their own limitations and may introduce additional biases, thereby highlighting a requirement for further research within this field (Morgan et al. 2022).

Another substantial concern that has been linked with AI is privacy. AI systems often gather and process huge caches of personal data, which raises concerns about their potential misuse or unconsented access to this information. These developments also represent a potentially severe erosion of people's privacy and self-determination, especially in domains where that is particularly protection-worthy (healthcare; finance). To address this concern, researchers have recommended privacy-preserving techniques like differential privacy and federated learning to be adopted under a generic framework which retains the utility of sensitive information for proper data analysis (Peres et al. 2020).

Accountability In particular, with AI systems becoming more independent and making decisions that have significant societal consequences, it is of utmost importance to be able to hold these types of models accountable for their decisions. Yet because of the opacity that afflicts so many AI systems, responsibility for their decisions can sometimes be unclear. Algorithms such as LIME (Peres et al. 2020) have been proposed to diagnose model outputs and solution strategies like explainable AI or algorithmic impact assessments in order to increase the transparency (Al-Rubaie and Chang 2019).

Finally, there are several negative consequences associated with developing AI such as its impact on employment because automation that is driven by this type of technology could end up replacing many human jobs. While making both efficiency and productivity is not a bad thing, it will definitely lead to job displacement leading to a significant increase in inequality. These include policy interventions inspired by universal basic income and reskilling programs, as well to assist workers whose jobs have been impacted or automated away (Intelligence and (XAI) 2020).

Some of the Main Ethical and Social Implications of AI Are:

- **Privacy and data security**: AI relies on large amounts of data, often of a personal and sensitive nature, to learn and improve. However, collecting and using such data raises concerns about the protection of privacy of the users. How can AI prevent unauthorized access, misuse, or abuse of the data and the outcomes?

- **Bias and discrimination**: AI may inherit or amplify biases and prejudices from the data, algorithms, or outcomes. This may result in unfair or discriminatory decisions or actions in domains such as hiring, lending, healthcare, and criminal justice. How can AI ensure fairness and equity in its data, algorithms, and outcomes? How can AI avoid discriminating or excluding certain groups or individuals based on their characteristics or preferences? How can AI respect and promote human dignity, diversity, and equality?
- **Accountability and transparency**: AI may have a significant impact on human lives and society, especially when it is involved in critical decision-making processes. However, AI may be complex, opaque, or unpredictable, making it difficult to understand, explain, or justify its actions or decisions. How can AI be designed, tested, and validated to ensure its safety and reliability? How can AI be monitored and audited to prevent errors and harms? How can AI be explained and justified to the users and the public? How can AI be accountable and responsible for its actions and decisions?
- **Human agency and autonomy**: AI may influence or replace human judgment, choice, or control in various aspects of life and society. This may affect human agency and autonomy, as well as human values, rights, and responsibilities. How can AI respect and enhance human agency and autonomy? How can AI balance human and machine roles and responsibilities? How can AI align with human values and ethics?

In summary, the application of AI brings up various ethical and social issues, including bias, privacy, accountability, and the impact on employment. Addressing these issues requires ongoing research and development, as well as policy interventions to ensure that the benefits are divided fairly and equitably across society (Frey and Osborne 2017).

1.13 The Need for Ethical Guidelines and Regulations for the Development and Deployment of AI

As we progress and execute AI technologies, there are many ethical considerations to be had: bias in its human learning model development process; implications for privacy and accountability simply related to these data processing models even the impact on employment. Solving these issues will clearly require ethical guidelines and regulatory frameworks, governing the development and application of artificial intelligence. Given the huge potential for AI technologies to affect lives, it is clear that ethical regulations and guidelines need to be put in place. For example, AI algorithms with biased outcomes can perpetuate discrimination for specific groups and increasing the use of surveillance driven by AI systems may infringe privacy rights or undermine personal autonomy. Accordingly, it is significantly important that AI technologies be created in ways to enhance their beneficial influences while reducing any potential adverse effects.

There is also a need for guidelines and regulations to protect against the pre-first profit motives of private AI technology developers, as discussed above. Without these guidelines and standards, companies are more likely to serve their own short-term interests over the long-term benefit of individuals or society overall—even if this means developing dangerously unethical AI.

In an effort to address these problems, researchers and policymakers had implemented numerous ethical guidelines and regulatory responses. For example, the European Commission published ethical guidelines for AI emphasizing transparency and accountability as well as the respect of individual rights with particular attention to privacy. In the same vein, we also have to follow what has been established by IEEE formulated ethical principles that combine transparency and accountability with responsible AI.

The GDPR of the European Union, for example stipulates rules regarding personal data that require companies to obtain consent before recording and processing a person's information. At the federal level in the USA, and accordingly all over, the Algorithmic Accountability Act would require that companies assess and remedy biases inherent to their AI systems.

The bottom line is, with the increase in applications of AI technology many moral questions have arisen such as bias, double-talk, and jobs will be a major player. Ultimately, to address these challenges effectively, we have a responsibility to set guidelines and regulations that govern the creation and use of AI.

1.13.1 Ethical Guidelines and Regulations for AI Can Help to

- Ensure the protection of user data and privacy, and safeguard against unauthorized access, misuse, or exploitation of both the data and its results.
- Ensure fairness and equity in the data, algorithms, and outcomes and avoid discriminating or excluding certain groups or individuals based on their characteristics or preferences.
- Improve the safety and dependability of AI systems while preventing mistakes and potential harm.
- Boost the clarity and responsibility of AI systems, allowing users and the public to comprehend, articulate, and rationalize the actions and decisions made by these systems.
- Respect and enhance human agency and autonomy, and balance human and machine roles and responsibilities.

1.14 Future of AI

AI is advancing with several emerging areas of research and trends gaining prominence:

- Explainable AI (XAI): This area focuses on creating AI systems that are transparent and comprehensible to users. The aim is to build trust and ensure that AI decision-making is fair and ethical. Approaches in XAI include explanations based on the internal workings of models, illustrative examples to show decision processes, and interactive methods that allow users to query the system for clearer understanding.
- Federated Learning: It is particularly relevant for sectors like healthcare and finance where privacy is a significant concern. Challenges include ensuring model performance and safeguarding against potential manipulations by malicious entities.
- Reinforcement Learning: This machine learning approach teaches agents to make decisions by interacting with their environment and receiving rewards or penalties. It shows promise in areas like gaming, and autonomous driving. Challenges include ensuring that the learning process leads to safe and ethical behavior.
- Quantum Computing: Quantum computing has the potential to transform AI and various other domains by solving problems more efficiently than traditional computers. It could revolutionize fields like drug discovery and optimization. However, developing algorithms that leverage quantum computing's unique capabilities presents significant challenges.
- Natural Language Processing (NLP): NLP aims to enable computers to understand and produce human language. Its applications include chatbots, virtual assistants, and translation services. Key challenges in NLP involve managing language ambiguities and avoiding biases in AI models.

These emerging trends represent exciting developments in AI, each with its own set of opportunities and challenges.

Some of the Things that We Can Expect to See in the Future of AI Are:

- AI will become more ubiquitous and integrated into our physical and digital world, enabling new dimensions of connectivity, interactivity, and intelligence. AI will be embedded in various devices, systems, and platforms, such as smart phones, cars, homes, and workplaces, and will provide personalized and contextualized services, recommendations, and experiences. AI will also enable new forms of communication and collaboration
- AI will grow more advanced and proficient, surpassing human performance in various domains and tasks. AI will be able to process massive amounts of data, and discover complex relationships. AI will also be able to generate and manipulate high-quality visual, audio, and textual content, such as images, videos, music, and text. AI will also be able to learn from its own actions and feedback, and adapt to changing environments and situations.
- AI will become more diverse and inclusive, reflecting and respecting the diversity and values of human society. AI will be designed and developed with the participation and consultation of various partners, such as ethicists, policymakers, and technologists, and will follow some ethical principles and guidelines, such as

the ones proposed by the World Health Organization or the European Commission. AI will also be fair and unbiased in its data, algorithms, and outcomes and will avoid discriminating or excluding certain groups or individuals based on their characteristics or preferences.
- AI will become more collaborative and cooperative, working with and alongside humans and other machines. AI will be able to complement and augment human intelligence, skills, and creativity, and will provide assistance, guidance, and feedback. AI will also be able to cooperate and coordinate with other AI systems, forming networks and ecosystems of intelligence. AI will also be able to communicate and negotiate with humans and other AI systems, forming relationships and alliances.
- AI will become more responsible and accountable, being transparent and explainable in its actions and decisions. AI will be designed, tested, and validated to ensure its safety and reliability, and will be monitored and audited to prevent errors and harms. AI will also be explained and justified to the users and the public, and will be accountable and responsible for its actions and decisions. AI will also respect and enhance human agency and autonomy and will balance human and machine roles and responsibilities.

These are just a few of the emerging trends and areas of research in AI. As the field is continuously evolving, it is likely that new trends and areas of research will emerge.

1.15 Conclusion

In summary, the field of AI has evolved significantly since its early days, now serving as a major driver of innovation across various sectors. Different approaches to AI include rule-based systems, expert systems, neural networks, and deep learning, each bringing its own set of strengths and limitations. Machine learning, a crucial subset of AI, is essential for creating predictive models and decision-making frameworks. Robotics, another vital area of AI, involves the use of robots for diverse tasks across industries.

While AI offers numerous advantages, it also raises important ethical and social concerns, such as issues of bias, privacy, accountability, and its effects on employment. As AI technologies continue to develop, establishing guidelines and regulations is important for ensuring their ethical use.

Looking ahead, AI is expected to significantly influence various aspects of society, with emerging trends including explainable AI, enhanced human–machine collaboration, and the use of AI for societal benefit. Gaining a foundational understanding of AI is increasingly important for both individuals and organizations to stay informed about its advancements and make strategic decisions.

A solid grasp of AI fundamentals can help individuals navigate the associated challenges and identify opportunities for positive change. Moreover, as AI technology progresses rapidly, staying updated on its core principles will be beneficial. Understanding AI also enables individuals to engage in discussions about its societal implications and contribute to shaping policies and regulations that govern its development and application. As AI becomes more embedded in our lives, having a basic knowledge of its principles is vital for effectively addressing its opportunities and challenges and for influencing its future trajectory.

References

Al-Blooshi L, Nobanee H (2020) Applications of artificial intelligence in financial management decisions: a mini-review. SSRN Electron J. https://doi.org/10.2139/ssrn.3540140

Al-Jaroodi J, Mohamed N (2018) Service-oriented architecture for big data analytics in smart cities. In: 18th IEEE/ACM international symposium on cluster, cloud and grid computing (CCGRID), Washington, DC, USA, pp 633–640. https://doi.org/10.1109/CCGRID.2018.00052

Al-Rubaie M, Chang JM (2019) Privacy-preserving machine learning: threats and solutions. IEEE Secur Priv 17(2):49–58. https://doi.org/10.1109/MSEC.2018.2888775

Amato F, López A, Peña-Méndez EM, Vaňhara P, Hampl A, Havel J (2013) Artificial neural networks in medical diagnosis. J Appl Biomed 11(2):47–58. ISSN1214-021X. https://doi.org/10.2478/v10136-012-0031x

Aditya J, Kulkarni G, Shah V (2018) Natural language processing. Int J Comput Sci Eng 6:161–167. https://doi.org/10.26438/ijcse/v6i1.161167

Arrieta AB, Díaz-Rodríguez N, Ser JD, Bennetot A, Tabik S, Barbado A, Garcia S, Gil-Lopez S, Molina D, Benjamins R, Chatila R, Herrera F (2020) Explainable artificial intelligence (XAI): concepts, taxonomies, opportunities and challenges toward responsible AI. Inf Fusion 58:82–115. ISSN 1566-2535. https://doi.org/10.1016/j.inffus.2019.12.012

Chella A, Iocchi L, Macaluso I, Nardi D (2006) Artificial intelligence and robotics. Intell Artif 3:87–93. ets and State of the Art

Chen M, Mao S, Liu Y (2014) Big data: a survey. Mobile Netw Appl 19:171–209.https://doi.org/10.1007/978-3-319-06245-7

Diksha K, Koli A, Khatter K, Singh S (2022) Natural language processing: state of the art, current trends and challenges. Multimed Tools Appl 82. https://doi.org/10.1007/s11042-022-13428-4

Frey CB, Osborne MA (2017) The future of employment: how susceptible are jobs to computerisation? Technol Forecast Soc Change 114:254–280. ISSN 0040-1625. https://doi.org/10.1016/j.techfo

Miotto R, Wang F, Wang S, Jiang X (2017) Deep learning for healthcare: review, opportunities and challenges. Briefings Bioinf 19. https://doi.org/10.1093/bib/bbx044

Morgan A, Abdi J, Syed M, Kohen G, Barlow P, Vizcaychipi M (2022) Robots in healthcare: a scoping review. Curr Robot Rep 3. https://doi.org/10.1007/s43154-022-00095-4

Peres RS, Jia X, Lee J, Sun K, Colombo AW, Barata J (2020) Industrial artificial intelligence in industry 4.0—systematic review, challenges and outlook. IEEE Access 8:220121–220139 (2020). https://doi.org/10.1109/ACCESS.2020.3042874

Russell, Norvig S (2010) Artificial intelligence: a modern approach. Prentice Hall, Englewood Cliffs, NJ

Sutton RS, Barto AG (1998) Reinforcement learning: an introduction. MIT Press, Cambridge, MA, 322 pp. ISBN 0-262-19398-1

Szeliski R (2022). Comput Vis Algorithms Appl. https://doi.org/10.1007/978-3-030-34372-9

Wang D (2001) Unsupervised learning: foundations of neural computation. AI Mag 22:101–102. https://doi.org/10.1609/aimag.v22i2.1565
Zakerzadeh MR, Firouzi M, Sayyaadi H, Shouraki SB (2011) Hysteresis nonlinearity identification using new preisach model-based artificial neural network approach. J Appl Math 2011:22. Article ID 458768. https://doi.org/10.1155/2011/458768

Chapter 2
VLSI and Neural Networks Integration in Industry 4.0: A Comprehensive Approach

Owais Ahmad Shah, Imran Ahmed Khan, and Amrita Rai

Abstract This chapter explores the critical fusion of Very Large Scale Integration (VLSI) and neural network technologies in the context of Industry 4.0. VLSI, known for integrating millions of transistors onto a single chip, serves as a foundational technology, enabling miniaturization and enhancing electronic devices. This integration finds extensive applications in Industry 4.0, including sensor integration, data processing, and edge computing, fundamentally transforming industrial processes. Concurrently, neural networks, the backbone of machine learning and deep learning, further revolutionize Industry 4.0 by enhancing predictive maintenance, quality control, and anomaly detection. This chapter delves into the challenges and opportunities of integrating VLSI and neural networks, emphasizing optimized performance through efficient collaboration. The chapter concludes by exploring emerging trends and summarizing the transformative potential of this integration, envisioning a future where Industry 4.0 thrives on the seamless synergy of VLSI and neural networks.

Keywords VLSI · Neural network · Industry 4.0 · IoT · AI

O. A. Shah
Department of Electronics and Communication Engineering, Dayananda Sagar University, Bengaluru, Karnataka, India

I. A. Khan (✉)
Department of Electronics and Communication, Jamia Millia Islamia, New Delhi, India
e-mail: iakhan2@jmi.ac.in

A. Rai
ECE Department, Lloyd Institute of Engineering and technology, Greater Noida, UP, India

A. Rai et al. (eds.), *Convergence of Artificial Intelligence, Machine Learning, and the Internet of Things in Industry 4.0 Applications*, Transactions on Computer Systems and Networks, https://doi.org/10.1007/978-981-96-5269-3_2

2.1 Introduction

Technological convergence is the characteristic of development in the era of Industry 4.0. This is a time when cutting-edge digital technology and conventional industrial methods seamlessly blend to provide previously unheard-of levels of automation, efficiency, and creativity (Mukhopadhyay et al. 2022). The convergence of neural network and VLSI technology is at the forefront of this change. Combining neural network capabilities with VLSI, a technology that can accommodate millions of transistors on a single chip, is a critical step toward achieving Industry 4.0's full potential (Chen et al. 2022).

A key component of contemporary electronics is VLSI technology, which enables the integration of many transistors and electronic parts on a compact silicon chip (Khan et al. 2023). Electronic circuit sizes have been dramatically decreased but their processing capacity has increased tenfold thanks to this technological miracle. In addition to compaction in size, VLSI technology has improved the functionality, efficiency, and performance of electronic equipment. Complex functionality may now be combined onto a single chip thanks to VLSI, which makes it possible to create electronic systems that are small, strong, and energy-efficient (Stoliar et al. 2022).

With its application at the heart of Industry 4.0, VLSI technology is fundamentally changing the industrial environment. The incorporation of sensors and data processing units in industrial processes is one such application. A single chip may have several sensors and data processing units integrated seamlessly thanks to VLSI (Mehbodniya et al. 2022). This connection makes it possible to collect data in real-time and analyze it quickly. By using this capacity to monitor and improve their operations, smart factories may increase production and efficiency.

Additionally, the development of edge computing, a paradigm in which data processing is done closer to the data source has been greatly aided by VLSI technology (Lee et al. 2022; Shah et al. 2023; Sakemi et al. 2023; Grewal and Shah 2023; Nimbekar et al. 2023). By processing data locally, edge devices with customized VLSI chips can lower latency and bandwidth requirements on centralized cloud servers. Because it allows for quick responses to changing conditions in a production setting and real-time decision-making, this development is very helpful for Industry 4.0 (Nanduri and Kamaraju 2023).

Parallel to this, Industry 4.0 has led to a significant increase in interest in neural networks, a basic idea in artificial intelligence. Neural networks are computer models that are based on the architecture and operations of the human brain. They are arranged in layers and are made up of linked nodes that resemble neurons. Input data is processed by each layer before being passed on to the next, creating a sophisticated network that can recognize patterns and anticipate outcomes (Kumari et al. 2023).

Neural networks have shown to be very effective and adaptable in the context of Industry 4.0. Predictive maintenance is one of their prominent uses. Neural networks can forecast when equipment is likely to break by examining both historical and real-time data from machines (Sánchez et al. 2023; Rakesh and Vats 2023; Saptalakar et al. 2023). This allows for preventive maintenance. This method maximizes equipment performance, lowers maintenance costs, and minimizes downtime.

Neural networks are also very good at anomaly detection and quality control. They may examine data from several sensors in production processes to spot anomalies or flaws, guaranteeing that only top-notch goods are sent onto the market. Neural networks have the ability to improve industrial processes in terms of efficiency, cost, and overall quality (Tsmots et al. 2023). These applications demonstrate this promise.

We explore the crucial combination of neural network and VLSI technologies in Industry 4.0 in this chapter. The potential for Industry 4.0 to achieve unprecedented levels of automation, intelligence, and efficiency is enormous when these technologies are integrated. Through comprehending the uses, difficulties, and potential of this integration, we hope to illuminate its revolutionary power and its part in determining the direction of industries throughout the globe.

2.2 VLSI Technology

The development of integrated circuits (ICs) and electronic devices has been greatly influenced by VLSI technology, which is the cornerstone of contemporary electronics. There has been a significant shift in transistor technology from its early days to the modern VLSI period, with transistor density rising, performance improving, and electronic component shrinking. This section delves into the fundamentals, historical evolution, and design methodologies of VLSI technology.

2.2.1 Fundamentals of VLSI Technology

The core of VLSI technology is the integration of a remarkably large number of transistors onto a single chip, which produces electrical circuits that are small, strong, and have several uses. VLSI design is the process of building elaborate integrated circuits with millions to billions of transistors, resistors, and capacitors that are all carefully coupled to carry out distinct functions. The development of complex electronic systems with previously unheard-of computing capabilities is made possible by this degree of integration. Transistors, which are semiconductor devices with switch and amplifier functions, are the basic components of VLSI circuits. These transistors are carefully placed and linked in VLSI design to provide the intended functionality of the circuit. Transistor size reduction and ongoing fabrication process improvement have been major contributors to the advancement of VLSI technology (Khan et al. 2023; Thoutam et al. 2023; Khan and Beg 2019).

Transistor Basics

The transistor, a semiconductor device with the capacity to switch or amplify electrical impulses, is at the heart of VLSI technology. Bipolar junction transistors (BJTs) and metal-oxide-semiconductor field-effect transistors (MOSFETs) are the two main types of transistors utilized in VLSI design. Because of its reduced size, lower power consumption, and compatibility with Complementary Metal-Oxide-Semiconductor (CMOS) technology, MOSFETs are the predominant semiconductor in current VLSI circuits.

The voltage delivered to the gate terminal of a MOSFET regulates the current flow between the source and drain terminals. The basis of both digital and analog circuits is provided by this control mechanism, which permits exact switching and amplification (Sedra et al. 2019).

Scaling and Miniaturization

The constant search to reduce transistor sizes so that more can fit on a single chip is a distinguishing feature of VLSI technology. Reducing transistor and connector size increases transistor density and enhances performance. This process is known as scaling. This pattern, which is frequently connected to Moore's Law, has accelerated improvements in semiconductor production techniques.

More components on a chip are made possible by miniaturization, which also lowers power consumption, speeds up switching, and lowers manufacturing costs. But scaling has problems as well, such higher leakage currents and heat dissipation, which call for creative approaches to material science, design, and manufacturing (Dhar et al. 2011; Khan and Beg 2013).

CMOS Technology

The foundation of VLSI design is Complementary Metal-Oxide-Semiconductor technology. It creates circuits with low-power consumption and great performance by utilizing both n-type and p-type MOSFETs. CMOS circuits are perfect for battery-powered gadgets and energy-efficient applications since they use very little power while they're not in use. An arrangement of complementary n-type and p-type transistors makes up the construction of a CMOS circuit. Because of its innate symmetry, digital circuits are more resilient by using less power and having better noise margins (Parvizi et al. 2020).

Interconnects and Wiring

Effective connectivity between these components becomes crucial as a chip's transistor count rises. The wire network connecting transistors and other components on the chip is formed by interconnects, which are usually composed of copper or aluminum. In order to reduce power loss, mitigate signal delays, and guarantee dependable signal transmission, connection design and optimization are essential. To solve issues with heat dissipation and signal integrity, advanced technologies have been created, such as copper interconnects and low-k dielectric materials (Torazawa et al. 2016).

Fabrication Processes

Semiconductor manufacturing is the collective term for the complex operations involved in fabricating VLSI circuits. Diffusion, etching, photolithography, and deposition are important processes. An important stage is photolithography, in which the locations of transistors and interconnects are defined by transferring patterns, using light-sensitive materials and masks, onto the semiconductor material.

Layers of materials, such metal or silicon dioxide, are deposited onto the wafer via deposition, and excess material is removed by selective etching. Dopants are introduced by diffusion, changing the semiconductor material's electrical characteristics. These procedures help create complex VLSI circuits and are carried out with extreme accuracy in cleanroom conditions.

Emerging Technologies

VLSI technology has evolved beyond conventional CMOS methods. New technologies that are attempting to improve performance, cut down on power consumption, and get beyond the constraints of traditional scaling include spintronics, quantum-dot cellular automata, and three-dimensional integration (Sadhu et al. 2020, 2022).

2.2.2 Design Methodologies in VLSI

The design of VLSI circuits is a multidimensional process that involves various stages, each critical to the overall success of the project. Key design methodologies include (Huang et al. 2024):

(a) *Specification and Architecture Design*: Identifying the requirements and defining the high-level architecture of the VLSI circuit based on functional specifications.
(b) *Logic Design*: Creating a detailed logical representation of the circuit using hardware description languages (HDLs) like Verilog or VHDL.
(c) *Synthesis*: Transforming the logical representation into a netlist, which defines the interconnection of basic logic elements.
(d) *Physical Design*: Defining the physical layout of the circuit, including the placement and routing of components on the chip.
(e) *Verification*: Rigorous testing and simulation to ensure that the designed circuit meets the specified requirements.
(f) *Manufacturing*: The final stage involves the actual fabrication of the integrated circuit using semiconductor manufacturing processes.

VLSI technology is used in a wide range of electronic systems and devices, impacting several sectors. Microprocessors and microcontrollers, the brains of computers, embedded systems, and digital gadgets, are a few well-known applications. Memory chips used in computers and other electronic devices, such as Random Access Memory (RAM) and Read-Only Memory (ROM). Digital signal processors

(DSP) are devices that handle digital signals using specialized chips for uses such as image and audio processing. Application-Specific Integrated Circuits (ASICs) are specially engineered chips with enhanced performance for certain purposes. Integrated circuits, both digital and analog, are the building blocks of many electronic devices, including industrial and consumer gear.

2.3 Industry 4.0

The term "Industry 4.0," often known as the "Fourth Industrial Revolution," refers to a paradigm change in the production and manufacturing sector. Driven by the amalgamation of digital technologies, data-centric procedures, and sophisticated automation, Industry 4.0 is fundamentally altering the industrial terrain, cultivating unparalleled degrees of efficacy, personalization, and interconnectivity. Industry 4.0 encompasses a range of distinctive features that set it apart as a disruptive force in production and manufacturing (Zhong et al. 2017). These characteristics are showcased and detailed in Table 2.1.

2.3.1 Impact on Manufacturing

Industry 4.0 has a significant influence on production, bringing in a new age characterized by improved customization, better efficiency, and simplified processes. The total productivity and efficiency of smart factories have increased significantly as a result of their integration with cyber-physical systems. Predictive maintenance and real-time data analytics guarantee that machinery operates at peak efficiency, reducing downtime and averting expensive interruptions. The flexibility provided by Industry 4.0 directly contributes to the capacity of manufacturers to mass modify

Table 2.1 Key characteristics of Industry 4.0

Characteristics	Description
Smart factories	Integration of intelligent systems capable of autonomous decision-making and real-time optimization
Internet of Things (IoT)	Implementation of interconnected devices and machinery, facilitating seamless communication and data exchange
Big data and analytics	Utilization of big data analytics to gain insights into production processes, enabling informed decision-making
Additive manufacturing	Embracing 3D printing for complex and customized component creation, enhancing flexibility in production
Cyber-physical systems (CPS)	Integration of computational algorithms with physical processes for real-time monitoring and control

products, allowing them to quickly adjust to shifting production needs and meet market demands. Furthermore, supply chain optimization is now a possibility thanks to effective logistics management, demand forecasting, and real-time monitoring that lowers costs and boosts the network's overall resilience. Modern technologies help quality control procedures, guaranteeing that only superior goods make it to market. Industry 4.0 will probably have a greater influence on manufacturing as it develops, encouraging innovation and economic expansion in the industrial sector (Zhong et al. 2017).

Industry 4.0 maximizes production and efficiency through process optimization and downtime reduction. Predictive maintenance, made possible by real-time data analytics, guarantees that equipment is maintained precisely when it is needed, averting expensive malfunctions. Large-scale product customization is made possible by Industry 4.0's flexibility. Intelligent factories has the ability to promptly adjust to fluctuating production needs, enabling producers to promptly meet market demands and provide customized products to customers. Transparency and efficiency are improved by the supply chain's several points of data integration. Industry 4.0 lowers costs and enhances supply chain management by enabling real-time inventory level monitoring, demand forecasting, and simplified logistics. Quality control procedures make use of cutting-edge technology like machine learning and artificial intelligence. By identifying flaws or departures from quality standards, intelligent sensors and cameras make sure that only goods that satisfy strict requirements make it onto the market.

2.3.2 Challenges and Considerations

Handling the revolutionary terrain of Industry 4.0 is not without its difficulties and factors to take into account. Manufacturing companies face several challenges when they go digital, not the least of which is the necessity of addressing privacy and data security issues. Industry 4.0's increased connection and data interchange make it crucial to protect sensitive data from cyber-attacks, which calls for strong cyber security frameworks and procedures.

Furthermore, the workforce needs updated skill sets as a result of the paradigm shift toward Industry 4.0. Proficiency in new technologies, data analytics, and smart system operation is required for the digital transformation. It becomes essential to provide thorough upskilling and training programs to enable the staff to manage the rapidly changing technology world.

The integration of current legacy systems with the cutting-edge technology contained in Industry 4.0 presents another significant difficulty. Many industrial facilities continue to use antiquated systems, which might not be inherently compatible with the latest generation of digital technology. To maintain a healthy cohabitation between old systems and upcoming Industry 4.0 technologies, achieving a seamless integration process involves strategic strategy, investment, and technological knowledge. Careful thought and strategic preparation are frequently necessary for

Table 2.2 Challenges and considerations in Industry 4.0 adoption

Challenges	Considerations and solutions
Data security and privacy	Robust measures for protecting sensitive information and ensuring privacy, incorporating encryption and secure protocols
Workforce skills	Implementation of upskilling and training programs to equip the workforce with the necessary digital skills for Industry 4.0
Integration of legacy systems	Addressing compatibility issues and ensuring a smooth transition by integrating existing systems with new Industry 4.0 technologies

this complex process in order to avoid interruptions and maximize the advantages of the digital transformation. Adopting Industry 4.0 is not without its difficulties; factors including system integration, worker skills, and data security must be carefully considered (Moreira et al. 2023). Table 2.2 lists these challenges and provides a potential solution to it.

2.4 Neural Network

At the forefront of modern artificial intelligence are neural networks, which enable machine learning and pattern recognition by emulating the intricate network of connections seen in the human brain. This advanced technology has been used in many other domains, including as natural language processing, autonomous systems, image and audio recognition, and more. This investigation explores the principles, design, training procedures, and various uses that make neural networks a leading force in artificial intelligence.

2.4.1 Fundamentals of Neural Networks

A neural network is fundamentally a computer model that draws inspiration from the neural architecture of the human brain. It is made up of layers of networked nodes, sometimes referred to as artificial neurons or neurons. The input layer, any hidden layers, and the output layer are the three main layers. The weights attached to each neuronal connection are modified as learning progresses.

External data is received by neurons in the input layer, where it is processed and transmitted through the hidden layers to generate an output in the final layer via a sequence of weighted connections and activation functions. Neural networks' capacity to extract complex patterns and representations from input and modify their internal settings to maximize efficiency is what gives them their strength (Yamashita et al. 2018; Montesinos López et al. 2022).

Artificial intelligence relies on neural networks, which are based on fundamental principles that mimic the complex workings of the human brain. These networks are made up of layers upon layers of communicative nodes, or neurons, each of which adds to the total information processing. After data is received by the input layer, it moves through hidden layers and is processed using activation functions and weighted connections. The networks computed result, whether it be a classification, regression, or another task-specific output, is finally produced by the output layer.

Neural networks rely heavily on neurons, which operate as computing units that take in inputs, apply weights, and use activation functions to create outputs. The network becomes non-linear as a result of the activation function selection—sigmoid or ReLU, for example—which enables the network to recognize intricate links in data. Weighted connections, which indicate the degree of information flow between these neurons, bind them together. Back propagation, a technique where the network learns from labeled instances to reduce the discrepancy between anticipated and actual outputs, is used in the training process to alter these weights.

Different neural network types are designed to handle different kinds of data and jobs. Information travels in a single direction in feedforward neural networks (FNNs), which have a straightforward architecture. While recurrent neural networks (RNNs) handle sequential data, convolutional neural networks (CNNs) are superior at processing images. A specific type of RNN called Long Short-Term Memory (LSTM) Networks tackles the problem of capturing long-range dependencies. Synthetic data production is impacted by the use of Generative Adversarial Networks (GANs) in areas such as image and video synthesis (DiPietro and Hager 2020; Rather and Kumar 2024; Figueira and Vaz 2022).

Neural network practical applications are found in many different areas. Neural networks are revolutionizing machine learning and data interpretation in a variety of applications, including audio and picture recognition, natural language processing, and medical diagnostics. Even with its amazing powers, neural network technologies still face obstacles like interpretability and ethical issues, which motivates continuous research to improve and use neural network technologies in an ethical manner. Basically, knowing the foundations of neural networks reveals their revolutionary potential and highlights their function as the engine of the present artificial intelligence age.

2.4.2 Neural Network Architecture

These potent artificial intelligence models' complex relationships and functions are modeled after neural network design. Neural networks are fundamentally made up of layers upon layers of linked nodes, or neurons. The layers are as follows: the input layer receives inputs from the user at the beginning of the network; hidden layers use weighted connections and activation functions to do sophisticated computations; and the output layer generates the output. The intricacy of the job and the selected design determine how many hidden layers and neurons are present in each layer.

Table 2.3 Components and functions of neural network architecture

Layer type	Function
Input layer	Receives initial data for processing
Hidden layers	Perform computations through weighted connections and activation functions
Output layer	Produces the final result of the network's computation

A weight is assigned to every neuronal connection, signifying the strength of the link. These weights are changed during training to maximize the network's capability, allowing it to learn from labeled instances and provide precise predictions. Activation functions that add non-linearity to the network, such sigmoid or ReLU, allow the network to recognize complex patterns in the input (Alzubaidi et al. 2021).

The layout and functionality of these layers are important aspects of the architecture of neural networks. More intricate designs like CNN are made for image processing, using convolutional layers to capture hierarchical patterns in spatial data, in contrast to FNN, which has a straightforward structure with information flowing in a single way. RNNs designed for sequential data are useful for applications like speech recognition and language modeling because they retain a recollection of previous inputs. One type of RNN specifically designed to handle the problems of collecting long-range relationships in sequential data is called an LSTM network. Table 2.3 provides a visual representation of the neural network architecture (Montesinos López et al. 2022).

2.5 VLSI Technology in Industry 4.0

Significant progress has been made in Very Large Scale Integration technology, allowing millions to billions of transistors to be integrated into a single chip (Shreyanth et al. 2023). Electronic gadgets that are more powerful and efficient have been made possible by this shrinking and greater component density (Ram and Prabhaker 2023). Complex integrated circuits are the result of laborious procedures involved in VLSI design and fabrication, which include circuit design, layout, verification, and manufacture.

2.5.1 Applications of VLSI in Industry 4.0

The applications of VLSI in Industry 4.0 are extensive and transformative, revolutionizing the way industries operate and innovate:

a. *Sensor Integration and Data Processing*:

Real-time data gathering is made possible by the integration of several sensors into industrial equipment through the use of VLSI. When analyzed on-chip, this data provides information on the functionality and condition of the equipment, enabling prompt repair and streamlining processes in smart factories.

b. *Edge Computing and Edge Devices*:

Compact and powerful edge devices are a result of the evolution of VLSI technology. By processing data at the network's edge, these devices reduce the amount of data that needs to be sent to centralized servers. As a result, there is less latency, decision-making happens more quickly, and less bandwidth is needed.

c. *Internet of Things (IoT) Devices*:

Power-efficient designs are made possible by VLSI integration in IoT devices, which is crucial for battery-powered or energy-harvesting devices. This is especially important in the context of Industry 4.0, where smart manufacturing and process optimization are built on a multitude of linked devices.

d. *Robotics and Automation*:

Robotic control and processing unit development relies heavily on VLSI technology. It guarantees that robots can reliably perform jobs and interpret sensory input quickly, which adds to the automation and efficiency of production operations.

VLSI technology and Industry 4.0 concepts together have ushered in a period of unparalleled innovation that allows for automation and data-driven, intelligent decision-making in a variety of industrial areas. An important factor in determining how Industry 4.0's smart manufacturing environment is shaped is VLSI technology. Its impact also extends to real-time data analytics (Cariow et al. 2023), as VLSI makes it easier to integrate sophisticated DPUs. This feature improves real-time monitoring and analysis of large datasets, facilitating quick decision-making (Teotia and Shah 2023). Furthermore, VLSI technology plays a key role in the creation of complex control systems that maximize machinery performance and guarantee coordinated operations in smart factories. The downsizing made possible by VLSI is a major factor in the development of small, energy-efficient devices, which are essential for the ubiquitous connection seen in Industry 4.0's Internet of Things environment (Birudu et al. 2023; Shah et al. 2018). Table 2.4 provides an overview of the many uses of VLSI technology in the context of Industry 4.0. It addresses important topics such robotics, edge computing, IoT devices, and sensor integration. The data is compiled from a number of sources, such as scholarly articles, business reports, and professional opinions.

An overview of the integration of VLSI technology with neural networks is shown in Fig. 2.1, which also highlights the integration's main elements and points of synergy. The following part delves into the possible synergies between VLSI technology and neural networks, another important component in this revolution and its contribution to Industry 4.0.

Table 2.4 Description of applications of VLSI in Industry 4.0

Application	Description
Sensor integration and data processing	Integration of sensors and data processing units for real-time data collection and analysis
Edge computing and edge devices	Development of compact, powerful edge devices for real-time data processing at the network edge
Internet of Things (IoT) devices	Power-optimized designs for IoT devices, vital for Industry 4.0's connected device ecosystem
Robotics and automation	Control and processing units for robots, enhancing automation and efficiency in manufacturing

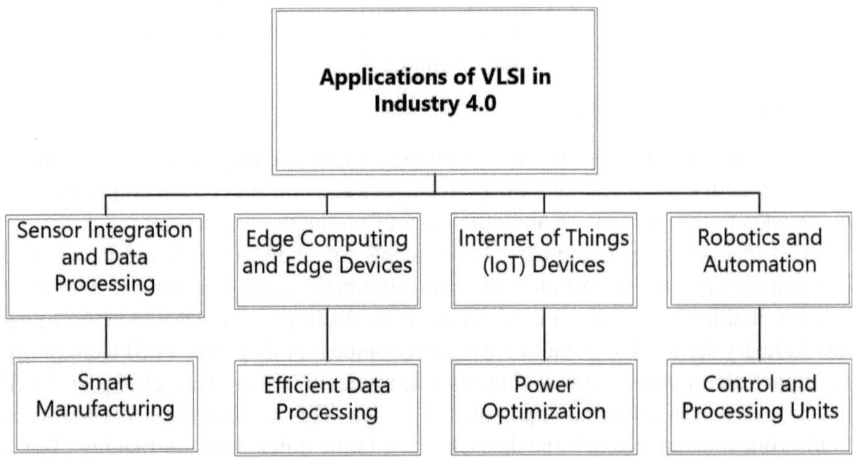

Fig. 2.1 Overview of various applications

2.6 Neural Networks in Industry 4.0

Inspired by the neural architecture and functions of the human brain, neural networks are a core idea in the field of artificial intelligence (AI). These networks are made up of layers of linked nodes, or neurons (Baryczkowski et al. 2023). Layer by layer, information is processed by neural networks, which learn from data to generate predictions, identify patterns, and optimize performance based on past performance.

2.6.1 Applications of Neural Networks in Industry 4.0

Neural networks have emerged as a powerful tool in various domains within Industry 4.0, showcasing their versatility and potential impact:

a. *Predictive Maintenance*:

Data from sensors mounted on industrial equipment is analyzed by neural networks. Through the identification of patterns that suggest possible malfunctions or breakdowns, these networks forecast when upkeep is necessary. This proactive strategy helps to optimize maintenance schedules and prevent expensive downtime.

b. *Quality Control and Anomaly Detection*:

It is possible to train neural networks to scan items for flaws or departures from quality requirements. Compared to conventional approaches, the integration of neural networks into quality control procedures guarantees a better degree of precision and consistency, which ultimately improves the quality of the final product.

c. *Process Optimization*:

Large volumes of data produced during industrial operations are analyzed using neural networks. These networks help to increase production, decrease waste, and improve efficiency in the manufacturing environment by determining the ideal process parameters.

d. *Supply Chain Optimization*:

In order to effectively predict future demand, neural networks evaluate previous sales data together with additional variables. As a result, businesses may lower carrying costs, streamline supply chains, and maximize inventory levels.

In order to effectively predict future demand, neural networks evaluate previous sales data together with additional variables. As a result, businesses may lower carrying costs, streamline supply chains, and maximize inventory levels. A network of interconnected nodes, each symbolizing a unique neural network application in the context of Industry 4.0, is depicted in Fig. 2.2. Specific applications, such "Quality Control," "Supply Chain Management," "Predictive Maintenance," and "Process Optimization," are identified on the nodes. The interconnectedness of these applications is shown by the arrows linking these nodes, which highlight the various ways in which neural networks are used in the industrial environment.

2.7 Fusion of VLSI and Neural Networks

The fusion of VLSI and neural networks presents a promising avenue for revolutionizing Industry 4.0. However, this integration is not without its set of challenges and opportunities.

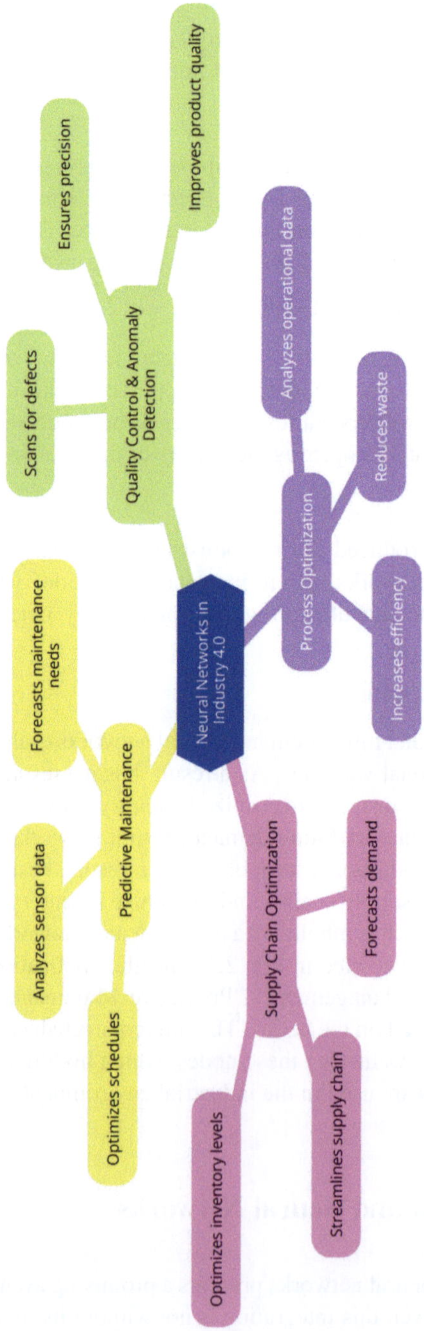

Fig. 2.2 Neural network applications landscape in Industry 4.0

2.7.1 *Integration Challenges*

Hardware Design Complexity: Integrating neural network architecture must be compatible with the underlying VLSI architecture, which calls for complex hardware design when integrating VLSI with neural networks.

Power Consumption: One of the biggest challenges is balancing performance with power efficiency. In the integration process, balancing power optimization with computing speed and accuracy requires careful trade-offs.

Latency and Throughput: In Industry 4.0, real-time processing requires low latency and high throughput. Achieving these goals requires careful system design and optimization.

2.7.2 *Optimizing Performance*

Customized Hardware Acceleration: Performance may be greatly increased by designing specialized VLSI hardware to speed up particular neural network activities. Hardware that is customized to the needs of the network performs calculations more quickly and uses electricity more effectively.

Parallel Processing: Neural network computations are inherently parallel, therefore this feature of VLSI technology may be leveraged. The speed and effectiveness of neural network operations are increased by this synergy.

Efficient Algorithms: It is essential to create specific algorithms that make the best use of the integrated hardware's capabilities. The combined capabilities of VLSI and neural network systems may be fully realized with the help of effective algorithms.

VLSI and neural network integration provide a special chance to overcome these obstacles and maximize the potential of both technologies. In the context of Industry 4.0, the fusion can open the door for extremely effective, low-latency, and power-optimized systems by solving these challenges.

2.8 Conclusion

The combination of neural network technology with Very Large Scale Integration, in the dynamic context of Industry 4.0, is a potent synergy that has the potential to completely reshape the industrial landscape. We have explored in this chapter how the combination of VLSI and neural networks is essential to the development of Industry 4.0 capabilities and applications. Millions of transistors may be integrated into a single chip thanks to VLSI technology, which has already completely changed the electronic industry. Industry 4.0 is powered by VLSI, which powers crucial functions that increase efficiency, accuracy, and intelligence inside industrial processes. These features range from sensor integration to edge computing and IoT devices.

Concurrently, neural networks—which draw inspiration from the neural architecture of the human brain—have become essential resources in a number of Industry 4.0 fields. Neural networks have demonstrated their adaptability and influence in a number of areas, including supply chain management, process optimization, quality control, and predictive maintenance. A plethora of potential and difficulties arise from the combination of these technologies. Difficulties include optimizing power usage, designing complex technology, and attaining high throughput with minimal latency. However, the possibility of extremely effective, low-latency systems offers a stimulating chance for innovation in the context of Industry 4.0.

Future developments in hardware-software co-design, low-power AI devices, and effective neural network topologies will likely influence how VLSI and neural networks are integrated. Future developments in edge AI and customized AI chips for certain sectors portend a world in which intelligent decision-making is commonplace.

Acknowledgments We would like to thank the anonymous reviewers for their valuable inputs and suggestions to maintain the quality standards and fostering continuous improvement in the work presented here.

References

Alzubaidi L, Zhang J, Humaidi AJ et al (2021) Review of deep learning: concepts, CNN architectures, challenges, applications, future directions. J Big Data 8:53. https://doi.org/10.1186/s40537-021-00444-8

Baryczkowski P, Szczepaniak S, Matykiewicz N, Perz K, Szczęsny S (2023) Study of the complexity of CMOS neural network implementations featuring heart rate detection. Electronics 12:4291

Birudu V, Yellampalli SS, Vaddi R (2023) A negative capacitance FET based energy efficient 6T SRAM computing-in-memory (CiM) cell design for deep neural networks. Microelectron J 139:105867

Cariow A, Papliński JP, Makowska M (2023) VLSI-friendly filtering algorithms for deep neural networks. Appl Sci 13:9004

Chen Y-H et al (2022) A VLSI chip for the abnormal heart beat detection using convolutional neural network. Sensors 22:796

Dhar S, Pattanaik M, Rajaram P (2011) Advancement in nanoscale CMOS device design en route to ultra-low-power applications. VLSI Des 19. Article ID 178516. https://doi.org/10.1155/2011/178516

DiPietro R, Hager GD (2020) Chapter 21—deep learning: RNNs and LSTM. In: Kevin Zhou S, Rueckert D, Fichtinger G (eds) The Elsevier and MICCAI society book series, handbook of medical image computing and computer assisted intervention. Academic Press, pp 503–519

Figueira A, Vaz B (2022) Survey on synthetic data generation, evaluation methods and GANs. Mathematics 10(15):2733. https://doi.org/10.3390/math10152733

Grewal S, Shah OA (2023) Design and analysis of high performance frequency divider in 32 nm CMOS technology for biomedical applications. Int J Online Biomed Eng 19

Huang C, Qian Q, Chen C, Ge W, Cao J (2024) Basics of integrated circuit design. In: Wang Y, Chi MH, Lou JJC, Chen CZ (eds) Handbook of integrated circuit industry. Springer, Singapore. https://doi.org/10.1007/978-981-99-2836-1_35

Khan IA, Beg MT (2013) Design and analysis of low power master slave flip-flops. Inf Midem-J Microelectron Electron Compon Mater 43(1):41–49. ISSN 0352-9045

Khan IA, Beg MT (2019) Power efficient design of semi-dynamic master slave single edge triggered flip-flop. Int J Electr Eng Inf 11(2):252–262. https://doi.org/10.15676/ijeei.2019.11.2.2

Khan IA, Rai A, Keshari JP, Nizamuddin M, Nayak S, Sharma D (2023) Design, simulation and comparative analysis of carbon nanotube based energy efficient priority encoders for nanoelectronic applications. e-Prime Adv Electr Eng Electron Energy 4. https://doi.org/10.1016/j.prime.2023.100138

Khan IA et al (2023) Odd counter: new design and performance analysis using carbon nano tube transistors for high performance applications. In: Proceedings of international conference on computational intelligence and sustainable engineering solution, CISES 2023. https://doi.org/10.1109/CISES58720.2023.10183403

Kumari BAS, Kulkarni SP, Sinchana CG (2023) FPGA implementation of neural nets. Int J Electron Telecommun 599–604. https://doi.org/10.24425/ijet.2023.146513

Lee I, Lee Y, Um H, Hong S, Park Y (2022) Dynamic rate neural acceleration using multiprocessing mode support. IEEE Trans Very Large Scale Integr (VLSI) Syst 30:1461–1472

Mehbodniya A et al (2022) VLSI implementation using fully connected neural networks for energy consumption over neurons. Sustain Energy Technol Assess 52:102058

Montesinos López OA, Montesinos López A, Crossa J (2022) Fundamentals of artificial neural networks and deep learning. In: Multivariate statistical machine learning methods for genomic prediction. Springer, Cham. https://doi.org/10.1007/978-3-030-89010-0_10

Moreira PA, Fernandes RM, Avila LV, Bastos LdSL, Martins VWB (2023) Artificial intelligence and industry 4.0? validation of challenges considering the context of an emerging economy country using cronbach's alpha and the Lawshe method. Eng 4(3):2336–2351. https://doi.org/10.3390/eng4030133

Mukhopadhyay AK, Majumder S, Chakrabarti I (2022) Systematic realization of a fully connected deep and convolutional neural network architecture on a field programmable gate array. Comput Electr Eng 97:107628

Nanduri S, Kamaraju M (2023) Efficient VLSI implementation of modular neural network based hybrid median filter. Soft Comput. https://doi.org/10.1007/s00500-023-08762-0

Nimbekar A et al (2023) Reconfigurable VLSI design architecture for deep learning established forelimb and hindlimb gesture recognition for rehabilitation application. IEEE Access 11:70061–70070

Parvizi P, Sabbaghi-Nadooshan R, Tavakoli MB (2020) Design of low power fast full adder using domino logic based on magnetic tunnel junction and Memristor. Revista INGENIERÍA UC 27(3):282–293.https://doi.org/10.54139/revinguc.v27i3.148

Rakesh, Shah OA, Vats S (2023) Floorplanning and comparative analysis of 16-bit synchronous up/down counter in different CMOS technology. In: Proceedings of international conference on computational intelligence and sustainable engineering solution, CISES. https://doi.org/10.1109/CISES58720.2023.10183608

Ram RS, Prabhaker MLC (2023) Diagnosis of analog and digital circuit faults using exponential deep learning neural network. J Electron Test 39:421–433

Rather IH, Kumar S (2024) Generative adversarial network based synthetic data training model for lightweight convolutional neural networks. Multimed Tools Appl 83:6249–6271. https://doi.org/10.1007/s11042-023-15747-6

Sadhu A, Das K, De D, Kanjilal MR (2020) Area-Delay-Energy aware SRAM memory cell and M × N parallel read/write memory array design for quantum dot cellular automata. Microprocess Microsyst 72. https://doi.org/10.1016/j.micpro.2019.102944

Sadhu A, Das K, De D, Kanjilal MR (2022) Low power design methodology in quantum-dot cellular automata. Comput Electr Eng 97. https://doi.org/10.1016/j.compeleceng.2021.107638

Sakemi Y, Morino K, Morie T, Aihara K (2023) A supervised learning algorithm for multilayer spiking neural networks based on temporal coding toward energy-efficient VLSI processor design. IEEE Trans Neural Netw Learn Syst 34:394–408

Sánchez D, Servadei L, Kiprit GN, Wille R, Ecker W (2023) A comprehensive survey on electronic design automation and graph neural networks: theory and applications. ACM Trans Des Autom Electron Syst 28:1–27

Saptalakar BK, Latte Mv (2023) FPGA-based reflection image removal using cognitive neural networks. Appl Nanosci 13:2539–2553

Sedra AS, Smith KC, Carusone TC, Gaudet V (2019) Microelectronic circuits, 8th edn. Oxford University Press

Shah OA, Nijhawan G, Khan IA (2023) A glitch free variability resistant high speed and low power sense amplifier based flip flop for digital sequential circuits. Eng Res Express 5

Shah OA, Ahmed Khan I, Nijhawan G, Garg I (2018) Low transistor count storage elements and their performance comparison. In: Proceedings—IEEE 2018 international conference on advances in computing, communication control and networking, ICACCCN 2018. https://doi.org/10.1109/ICACCCN.2018.8748364

Shreyanth S, Harshitha DS, Niveditha S (2023) Implementation of machine learning in VLSI integrated circuit design. SN Comput Sci 4:137

Stoliar P, Akita I, Schneegans O, Hioki M, Rozenberg MJ (2022) A spiking neuron implemented in VLSI. J Phys Commun 6:021001

Teotia P, Shah OA (2023) Power and area efficient sense amplifier based flip flop with wide voltage and temperature upholding for portable IoT applications. Inf MIDEM 53

Thoutam LR, Mathew R, Ajayan J, Tayal S, Nair SV (2023) A critical review of fabrication challenges and reliability issues in top/bottom gated MoS2 field-effect transistors. Nanotechnology 34. https://doi.org/10.1088/1361-6528/acb826

Torazawa N et al (2016) High-performance extremely low-k film integration technology with metal hard mask process for Cu interconnects. ECS J Solid State Sci Technol 5(10):P578. https://doi.org/10.1149/2.0141610jss

Tsmots I, Teslyuk V, Kryvinska N, Skorokhoda O, Kazymyra I (2023) Development of a generalized model for parallel-streaming neural element and structures for scalar product calculation devices. J Supercomput 79:4820–4846

Yamashita R, Nishio M, Do RKG et al (2018) Convolutional neural networks: an overview and application in radiology. Insights Imaging 9:611–629. https://doi.org/10.1007/s13244-018-0639-9

Zhong RY, Xu X, Klotz E, Newman ST (2017) Intelligent manufacturing in the context of industry 4.0: a review. Elsevier Eng 3(5):616–630. https://doi.org/10.1016/J.ENG.2017.05.015

Chapter 3
AI/ML-Based Approaches in Healthcare: Transforming Diagnosis, Treatment, and Beyond

Pawan Whig, Balaram Yadav Kasula, Nikhitha Yathiraju, Imran Ahmed Khan, Amrita Rai, and Md RashId Mahmood

Abstract This book chapter delves into the transformative impact of AI and ML in healthcare, exploring their applications in diagnosis, treatment, and healthcare system optimization. It elucidates the fundamental principles of AI/ML and their role in enhancing diagnostic accuracy, predicting risks, and tailoring personalized treatment plans. The narrative extends to the optimization of drug discovery, precision medicine, and the augmentation of treatment outcomes. Beyond clinical applications, the chapter examines how predictive analytics and workflow automation contribute to efficient healthcare delivery and resource allocation. This comprehensive overview caters to healthcare professionals, researchers, policymakers, and technologists, providing valuable insights into the evolving landscape of AI/ML in healthcare for improved patient care and system efficiency.

Keywords Artificial Intelligence (AI) · Machine Learning (ML) · Healthcare · Diagnosis · Treatment · Personalized medicine

---

P. Whig
Vivekananda Institute of Professional Studies, New Delhi, India

B. Y. Kasula
Department of Information Technology, University of the Cumberlands, Williamsburg, KY, USA

N. Yathiraju
University of the Cumberlands, Williamsburg, KY, USA

I. Ahmed Khan (✉)
Dept. of E&C, Jamia Millia Islamia, New Delhi, India
e-mail: iakhan2@jmi.ac.in

A. Rai
ECE Department, Lloyd Institute of Engineering and Technology, Greater Noida, U.P., India

M. R. Mahmood
Department of ECE, Guru Nanak Institutions Technical, Campus, Hyderabad, Hyderabad, U.P., India

A. Rai et al. (eds.), *Convergence of Artificial Intelligence, Machine Learning, and the Internet of Things in Industry 4.0 Applications*, Transactions on Computer Systems and Networks, https://doi.org/10.1007/978-981-96-5269-3_3

3.1 Introduction

The intersection of AI and ML with healthcare has emerged as a dynamic and revolutionary force, reshaping the traditional paradigms of diagnosis, treatment, and healthcare delivery. The foundation of this transformation lies in the ability of AI and ML algorithms to process vast amounts of data with unprecedented speed and accuracy. From diagnostic imaging to genetic sequencing, these technologies empower healthcare professionals with tools that surpass human capabilities, enabling earlier and more precise detection of diseases. The evolution of diagnostic methodologies represents a paradigm shift, where algorithms, trained on diverse datasets, unveil patterns and anomalies that elude conventional diagnostic approaches. Consequently, the precision and efficiency afforded by AI and ML not only enhance diagnostic accuracy but also contribute to the development of targeted and personalized treatment strategies. One of the most significant contributions of AI and ML in healthcare is observed in the realm of treatment optimization. The journey from drug discovery to treatment delivery has been revolutionized by algorithms capable of deciphering complex biological interactions and predicting drug responses. The era of one-size-fits-all approaches is yielding to a more nuanced and patient-centric model, fostering improved treatment outcomes, reduced adverse effects, and a more efficient utilization of healthcare resources. Beyond the realms of diagnosis and treatment, AI and ML are catalyzing transformative changes in the operational aspects of healthcare systems. Predictive analytics, a cornerstone of these technologies, is ushering in a new era of proactive and preventive healthcare.

However, the integration of AI and ML in healthcare is not without its challenges. The ethical implications of leveraging these technologies, particularly in decision-making processes, demand careful consideration. The potential for bias in algorithms, the need for transparent decision-making models, and the ethical use of patient data necessitate robust ethical frameworks. Moreover, the regulatory landscape is evolving to catch up with the rapid pace of technological innovation, posing challenges in terms of compliance and standardization.

As we embark on this exploration of AI and ML in healthcare, it is essential to acknowledge the diverse perspectives that shape this transformative journey. Stakeholders, including healthcare professionals, researchers, policymakers, technologists, and, most importantly, patients, play pivotal roles in defining the trajectory of AI and ML applications in healthcare. Collaborative efforts are essential to navigate the complex interplay of technological capabilities, ethical considerations, and regulatory frameworks. In the chapters that follow, we delve into specific facets of this multifaceted landscape, examining real-world applications, case studies, and the evolving challenges and opportunities. Each chapter contributes to the broader narrative of how AI and ML are not mere technological innovations but catalysts for a fundamental shift in the way we approach healthcare. The potential for improved patient outcomes, enhanced decision-making, and the optimization of healthcare systems underscores the significance of this exploration. This introduction serves as a gateway to a comprehensive exploration of AI and ML in healthcare, inviting readers to navigate the intricacies of a field where cutting-edge technology converges with the intricacies of human health and well-being. As we peer into the future of healthcare, guided by the capabilities of AI and ML, we envision a landscape where innovation aligns seamlessly with empathy, creating a healthcare ecosystem that is not only technologically advanced but fundamentally human-centric. Literature review with research gap is shown in Table 3.1.

3.1.1 The Growing Role of AI/ML in Healthcare

The integration of Artificial Intelligence (AI) and Machine Learning (ML) in healthcare represents a revolutionary paradigm shift, redefining the traditional contours of the industry. The inexorable growth of these technologies has permeated every facet of healthcare, from diagnostic procedures to treatment modalities and operational efficiencies. This section delves into the escalating role of AI and ML, unraveling their transformative impact on healthcare practices and setting the stage for an in-depth exploration of their applications.

3.1.2 Objectives of the Chapter

This chapter sets out with clear and defined objectives to comprehensively elucidate the multifaceted role of AI and ML in modern healthcare. Firstly, it aims to provide a nuanced understanding of the fundamental principles underpinning AI and ML technologies, laying the groundwork for subsequent discussions. Secondly, the chapter endeavors to explore real-world applications and case studies that exemplify the practical implications of these technologies across various healthcare domains. Additionally, it seeks to critically examine the challenges associated with the integration of AI and ML in healthcare, including ethical considerations, regulatory complexities, and the need for interpretability.

3.1.3 Relevance of AI/ML in Modern Healthcare

The contemporary healthcare landscape is marked by an unprecedented influx of data and an increasing demand for personalized, efficient, and accurate healthcare services. In this context, the relevance of AI and ML becomes paramount. The relevance extends beyond clinical applications to the operational realm, where predictive analytics and automation contribute to streamlined workflows and resource optimization. As the demand for high-quality healthcare services intensifies, understanding the profound relevance of AI and ML in modern healthcare is crucial for navigating the complexities of an evolving industry. This section unfolds the significance of these

Table 3.1 Literature review with research gap

Reference	Focus area	Key findings	Research gap
Rahmani et al. (2021)	ML in medicine	Review and applications	Limited discussion on ethical considerations in ML healthcare applications
Yang et al. (2022)	AI in primary care	Advances in AI	Lack of exploration on the challenges associated with AI adoption in primary care settings
Sahu et al. (2022)	AI in precision medicine	Paradigm shift in Big Data analysis	Limited discussion on the real-world implementation challenges of precision medicine AI models
Rubinger et al. (2023)	ML in research and healthcare	Broad overview	Lack of detailed analysis on the integration of ML in healthcare research methodologies
Higgins et al. (2023)	AI in mental health	Decision support systems	Limited exploration of user perspectives and acceptance of AI-based mental health tools
Swanson et al. (2023)	Clinical ML for cancer	Advances in cancer diagnosis	Need for in-depth exploration of challenges in deploying clinical ML models for diverse cancer types
Kourou et al. (2021)	ML in cancer research	Systematic review	Limited discussion on the ethical considerations related to AI applications in cancer research
Qureshi et al. (2023)	AI and biosensors	Clinical relevance	Insufficient exploration of the integration challenges and limitations of biosensors with AI in healthcare
FDA (2019)	Regulatory framework	Proposed framework	Limited exploration of the impact of the proposed framework on the development and deployment of AI/ML in medical devices

(continued)

Table 3.1 (continued)

Reference	Focus area	Key findings	Research gap
Taimoor and Rehman (2021)	AI and IoT in healthcare	Survey on personalized healthcare	Insufficient exploration of the security and privacy concerns associated with AI and IoT in healthcare
Kurariya et al. (2023)	AI in drug discovery	Impact on healthcare	Limited discussion on the challenges of translating AI-driven drug discoveries into practical clinical applications
El Naqa et al. (2023)	AI in oncology	Translation into clinical practice	Need for deeper exploration of challenges in the widespread adoption of AI in routine oncology practices
Christou and Tsoulfas (2021)	AI in gastroenterology	Challenges and opportunities	Lack of discussion on the ethical considerations and patient perspectives regarding AI in gastroenterology
Xu et al. (2023)	AI/ML in precision medicine	Beyond the hype	Need for exploration of the challenges in implementing AI/ML in precision medicine beyond theoretical advancements
Subbaswamy and Saria (2020)	Dataset shift in health AI	Causality and shift-stable models	Limited discussion on practical strategies for addressing dataset shift challenges in real-world health AI applications
Agarwal et al. (2023)	Data science for health equity	Perils and promise of AI/ML	Insufficient discussion on the disparities and biases in AI applications impacting health equity

technologies in addressing the contemporary challenges and opportunities within the healthcare landscape.

3.2 Machine Learning: The Essence of AI

Machine Learning, a subset of AI, empowers systems to learn and improve from experience without explicit programming. ML algorithms analyze data, identify patterns, and make informed decisions, evolving their performance over time.

3.2.1 *Key Components of Machine Learning*

- **Data**: The foundation of ML, datasets are used to train and test algorithms, enabling them to recognize patterns and make predictions.
- **Model**: A mathematical representation of a real-world process.
- **Features**: The variables or attributes within the dataset that influence the outcome. Identifying relevant features is crucial for ML model accuracy.
- **Labels**: The outcomes or classifications associated with input data, used for supervised learning where the model learns from labeled examples.

3.2.2 *Applications of AI and ML*

AI and ML find applications across diverse domains:

- **Healthcare**: Diagnosis, treatment optimization, and drug discovery.
- **Finance**: Fraud detection, risk assessment, and algorithmic trading.
- **Autonomous Vehicles**: Navigation, object detection, and decision-making.
- **Natural Language Processing (NLP)**: Speech recognition, language translation, and chatbots.
- **Image and Video Analysis**: Facial recognition, object detection, and content categorization.

Grasping the fundamentals of AI and ML unveils the potential of these technologies to revolutionize industries and enhance problem-solving capabilities. As AI and ML continue to advance, acquiring a foundational understanding becomes imperative for professionals, researchers, and enthusiasts navigating the ever-expanding realms of artificial and machine intelligence.

3.2.3 *Basics of AI and ML*

ML involves the use of models, trained on data, to recognize patterns and make informed decisions. Supervised learning, where models learn from labeled data, and unsupervised learning, dealing with unlabeled data to identify patterns, are common

ML paradigms. The basics of AI and ML lay the groundwork for comprehending their applications in diverse domains.

3.2.3.1 Data Acquisition and Preprocessing

Data plays a pivotal role in the effectiveness of ML models. Data acquisition involves gathering relevant datasets that represent the problem at hand. Quality, quantity, and diversity of data significantly impact model performance. Preprocessing, the cleaning and transformation of raw data, is a critical step. Proper data preprocessing ensures that models can extract meaningful insights and patterns during training, contributing to the overall success of AI and ML applications.

3.2.3.2 Training and Testing Models

The iterative process of training and testing refines the model, optimizing its performance for real-world applications. Understanding these stages is crucial for harnessing the predictive power of AI and ML models effectively.

3.3 Diagnosis and Medical Imaging in Healthcare

The integration of AI and ML into healthcare has ushered in a transformative era for diagnosis and medical imaging. These technologies are revolutionizing traditional approaches, enhancing accuracy, speed, and the overall efficiency of diagnostic processes.

3.3.1 AI/ML in Diagnostic Decision-Making

AI and ML algorithms are increasingly employed to assist healthcare professionals in making diagnostic decisions. These systems analyze vast datasets, including patient records and clinical observations for correlations that might elude human perception. The ability of ML models to learn from diverse data sources contributes to more comprehensive and nuanced diagnostic insights.

3.3.2 Medical Imaging and AI

Medical imaging, a cornerstone of diagnostic procedures, has experienced a profound transformation with the integration of AI and ML. From radiology to pathology, AI algorithms assist in detecting anomalies, tumors, and other medical conditions

(a) **Radiology**

.

In radiology, AI-powered algorithms aid in the interpretation of X-rays, CT scans, and MRIs. Computer-aided detection systems can identify subtle abnormalities, improving diagnostic accuracy and expediting the reporting process. Additionally, AI algorithms contribute to workflow optimization by prioritizing urgent cases, thereby enhancing overall efficiency in radiological departments.

(b) **Pathology**

In pathology, AI is utilized for image analysis of tissue samples. ML models can identify patterns and abnormalities in microscopic images, aiding pathologists in diagnosing conditions such as cancer.

3.3.3 Challenges and Opportunities

While the application of AI and ML in diagnosis and medical imaging holds immense promise, challenges must be addressed. Ethical considerations, regulatory compliance, and the interpretability of AI decisions are crucial concerns. Striking the right balance between human expertise and AI assistance is paramount. Moreover, ensuring the security and privacy of patient data is an ongoing challenge in the digital healthcare landscape.

3.3.4 Future Directions

The future of diagnosis and medical imaging lies in the continued refinement and expansion of AI and ML applications. Advancements in deep learning, increased collaboration between healthcare professionals and technologists, and the development of standardized protocols are anticipated. The evolution of explainable AI, which provides transparent insights into how AI algorithms arrive at decisions, will be crucial for gaining trust and acceptance in the medical community. The synergy between AI/ML and diagnostic processes, particularly in medical imaging, is reshaping healthcare delivery.

3.3.5 AI-Driven Disease Detection

The advent of Artificial Intelligence (AI) has ushered in a new era in disease detection, revolutionizing traditional diagnostic methodologies. AI-driven disease detection leverages advanced algorithms capable of analyzing vast datasets to identify subtle patterns and anomalies indicative of various medical conditions. From cardiovascular diseases to infectious disorders, AI augments the diagnostic capabilities of healthcare professionals by providing nuanced insights into complex health issues. The integration of Machine Learning models allows for the early identification of disease markers, facilitating proactive intervention and personalized treatment strategies.

3.3.6 Medical Imaging and Radiology

Medical imaging and radiology have undergone a paradigm shift with the integration of AI. This section delves into the innovative applications of AI in these critical diagnostic fields. AI algorithms, powered by Machine Learning offering rapid and precise interpretations. Computer-aided detection systems assist radiologists in identifying abnormalities, tumors, and other anomalies, streamlining the diagnostic process. The advancements in AI-driven medical imaging not only enhance diagnostic accuracy but also contribute to optimizing workflows in radiological departments, showcasing the transformative potential of these technologies in improving patient care.

3.3.7 Case Studies: AI in Diagnostic Medicine

Examining real-world applications is essential to understanding the tangible impact of AI in diagnostic medicine. This section presents case studies that highlight successful implementations of AI-driven diagnostic solutions. From the detection of diabetic retinopathy in ophthalmology to the identification of cancerous lesions in pathology. The discussion encompasses the methodologies, outcomes, and challenges encountered in these applications, providing insights into the practical implications of AI in transforming diagnostic medicine. These case studies serve as illuminating examples of how AI is making significant strides in augmenting diagnostic capabilities and contributing to improved patient outcomes.

From disease detection driven by advanced algorithms to the transformative impact of AI in medical imaging and radiology, these sections underscore the paradigm shift underway in healthcare. Case studies further illuminate the real-world applications, showcasing the potential of AI to revolutionize diagnostic processes and enhance the overall quality of patient care.

3.3.8 Personalized Treatment in the Era of AI and ML

The convergence of AI and ML has brought about a transformative approach to healthcare, particularly in the realm of personalized treatment. This paradigm shift marks a departure from one-size-fits-all medical interventions to tailored and individualized healthcare strategies. AI and ML play a pivotal role in analyzing vast and diverse datasets, unlocking insights that enable healthcare professionals to design personalized treatment plans.

3.3.9 Tailoring Treatment Strategies

AI and ML algorithms excel in analyzing complex datasets, identifying intricate patterns, and discerning subtle correlations. This capability is harnessed to tailor treatment strategies to the specific needs of each patient. Precision medicine, an embodiment of personalized treatment, leverages genetic information, biomarkers, and other clinical data to design interventions that are more targeted and effective.

3.3.10 Genetic Profiling and Predictive Modeling

One of the cornerstones of personalized treatment is genetic profiling. AI algorithms analyze an individual's genetic makeup to identify genetic variations associated with specific diseases or responses to treatments. Predictive modeling, facilitated by ML, helps forecast how patients are likely to respond to various interventions.

3.3.11 Disease Subtyping and Risk Prediction

AI contributes to disease subtyping, where conditions once considered homogeneous are categorized into distinct subgroups based on various factors, including genetic markers and clinical characteristics. This nuanced understanding allows for more precise targeting of treatments to specific disease subtypes. Additionally, ML models aid in predicting early interventions.

3.3.12 *Drug Discovery and Treatment Optimization*

AI contributes to treatment optimization by continuously analyzing patient data, adapting treatment plans based on individual responses, and minimizing adverse effects.

3.3.13 *Real-Time Monitoring and Adaptive Interventions*

The integration of AI in personalized treatment extends to real-time monitoring of patients. Wearable devices and continuous health monitoring systems equipped with AI algorithms track physiological parameters and provide insights into a patient's health status. This enables healthcare providers to make adaptive interventions, adjusting treatment plans in response to dynamic changes in the patient's condition.

3.3.14 *Challenges and Ethical Considerations*

The era of personalized treatment propelled by AI and ML heralds a new frontier in healthcare. From tailoring interventions based on genetic insights to optimizing drug discovery and enabling real-time monitoring, these technologies hold promise for a future where healthcare is not only more effective but also more attuned to the individual needs of each patient.

3.3.14.1 Precision Medicine

Precision medicine, empowered by the capabilities of AI and ML, stands as a revolutionary approach to healthcare. This section delves into the intricate realm of precision medicine, where the emphasis shifts from conventional one-size-fits-all treatments to highly individualized and targeted interventions. AI algorithms analyze diverse datasets, including genetic information, biomarkers, and clinical records, to discern unique patterns and variations. This enables healthcare professionals to design personalized treatment plans, optimizing efficacy and minimizing adverse effects. Precision medicine, driven by AI and ML, represents a paradigm shift toward tailoring healthcare strategies to the specific characteristics of each patient.

3.3.14.2 Drug Discovery

The marriage of AI and ML has significantly expedited the drug discovery process, offering innovative solutions to the challenges faced by pharmaceutical research. This section explores how AI contributes to drug discovery by analyzing vast datasets related to molecular interactions, disease pathways, and drug responses. Machine Learning models, trained on comprehensive datasets. The integration of AI in drug discovery not only accelerates the identification of promising compounds but also opens avenues for the development of more targeted and personalized therapeutics. This transformative approach holds promise for addressing unmet medical needs and fostering a more efficient and effective drug development pipeline.

3.3.14.3 Treatment Recommendations

AI and ML play a crucial role in providing personalized treatment recommendations, offering insights into the most suitable interventions based on individual patient characteristics. This section explores how algorithms to generate tailored treatment recommendations. The ability of AI to process and interpret vast datasets allows for more nuanced and informed decision-making. These recommendations consider factors such as genetic predispositions, treatment responses, and potential adverse effects, contributing to a more holistic and personalized approach to patient care. The integration of AI in treatment recommendations marks a shift toward evidence-based, data-driven medical decision support systems.

3.3.14.4 Monitoring Patient Response

Real-time monitoring of patient responses to treatments is enhanced through AI and ML technologies. This section explores how wearable devices, continuous health monitoring systems, and other Internet of Things (IoT) applications equipped with AI algorithms track physiological parameters and provide dynamic insights into a patient's health status. The integration of AI in patient monitoring contributes to proactive healthcare management, enabling early identification of potential issues and optimizing treatment strategies for better patient outcomes. These advancements represent a paradigm shift toward more personalized, targeted, and data-driven approaches to healthcare, ushering in a new era where medical interventions are optimized to meet the unique needs of each individual patient.

3.4 Predictive Healthcare Management: Leveraging AI for Proactive Patient Care

The integration of Artificial Intelligence (AI) into healthcare has ushered in a paradigm shift, moving from reactive to proactive patient care. Predictive healthcare management enables healthcare providers to anticipate, prevent, and manage health issues before they escalate. This section delves into the multifaceted aspects of predictive healthcare management and how AI is reshaping the landscape of healthcare delivery.

Predictive healthcare management leverages AI to analyze vast datasets encompassing patient records, genetic information, lifestyle factors, and clinical data. By identifying subtle patterns and correlations, AI models excel in early disease detection and risk prediction.

AI contributes to population health management by analyzing health data from diverse sources to identify trends and patterns within specific demographics. Predictive analytics helps healthcare organizations allocate resources effectively, identify at-risk populations, and implement targeted interventions. By proactively addressing health challenges within communities, predictive healthcare management enhances overall population health outcomes.

Predictive healthcare management extends to optimizing hospital resources, ensuring efficient allocation of staff, beds, and equipment. AI algorithms analyze historical data, patient flow patterns, and seasonal trends to forecast patient admissions and resource utilization. This proactive approach enhances hospital efficiency, reduces wait times, and ensures that resources are aligned with anticipated demand, leading to improved patient care and satisfaction. The advent of remote patient monitoring, supported by AI, enables continuous health tracking outside traditional healthcare settings. Wearable devices equipped with AI algorithms collect real-time data on vital signs, activity levels, and other health metrics. Predictive analytics analyze this data to detect deviations from baseline and anticipate potential health issues. Healthcare providers can then intervene proactively, preventing complications and reducing the need for hospital admissions. While the benefits of predictive healthcare management are substantial and the need for transparency in decision-making processes are crucial aspects. Striking a balance between leveraging AI for proactive care and upholding ethical standards is essential for fostering trust and acceptance within the healthcare community. Predictive healthcare management stands at the forefront of AI-driven healthcare innovation.

3.5 Predictive Analytics for Disease Outbreaks

Predictive analytics, powered by Artificial Intelligence (AI), emerges as a critical tool for anticipating and managing disease outbreaks. This section explores how AI models analyze diverse datasets, including demographic information, environmental factors, and historical health data, to predict the likelihood of disease outbreaks. By identifying patterns and correlations, predictive analytics enables healthcare authorities to implement proactive measures, such as targeted vaccination campaigns, public health interventions, and resource allocation, to mitigate the impact of potential outbreaks. The integration of AI in predicting disease outbreaks enhances the preparedness and responsiveness of healthcare systems, contributing to effective containment and control strategies. AI-driven predictive models play a pivotal role in optimizing healthcare resource allocation. This section delves into how predictive analytics analyzes historical data, patient flow patterns, and seasonal trends to forecast patient admissions and resource utilization. By anticipating demand, healthcare organizations can strategically allocate resources such as staff, beds, and equipment. This proactive approach not only enhances hospital efficiency but also ensures that resources are aligned with anticipated needs, minimizing bottlenecks, reducing wait times, and improving overall patient care. Predictive healthcare resource allocation represents a transformative application of AI in optimizing the operational aspects of healthcare delivery. The paradigm of healthcare delivery is evolving with the integration of remote patient monitoring powered by AI. This section explores how wearable devices equipped with AI algorithms enable continuous health tracking outside traditional healthcare settings. Predictive analytics analyze this data to detect deviations from baseline and anticipate potential health issues. Remote patient monitoring, supported by AI, enhances patient engagement, facilitates personalized care, and contributes to a more proactive and preventive healthcare model.

3.6 Administrative Efficiency in Healthcare

The infusion of Artificial Intelligence (AI) into healthcare administration holds the promise of streamlining processes, enhancing decision-making, and optimizing resource utilization. This section explores how AI contributes to administrative efficiency, revolutionizing tasks ranging from scheduling appointments to managing Electronic Health Records. AI-driven systems facilitate automated appointment scheduling, reducing the burden on administrative staff and improving patient access. Billing and coding processes are intricate and time-consuming. AI streamlines these tasks by automating coding based on medical documentation, ensuring accuracy and reducing the likelihood of errors. Intelligent billing systems powered by AI can identify potential billing issues, verify insurance information, and generate accurate invoices. This not only expedites the billing cycle but also enhances revenue cycle management, contributing to administrative efficiency.

Administrative efficiency is bolstered through the application of predictive analytics in resource management. AI analyzes historical data, patient flow patterns, and other relevant factors to forecast patient admissions, staff requirements, and equipment utilization. By anticipating demand, healthcare organizations can optimize resource allocation, minimize bottlenecks, and ensure that administrative processes align with the dynamic needs of the healthcare facility. Managing Electronic Health Records (EHRs) is a central administrative task in healthcare. AI solutions, including Natural Language Processing and Machine Learning, enhance EHR management by automating data entry, ensuring data accuracy, and providing intelligent insights. AI-driven EHR systems contribute to improved data quality, interoperability, and overall administrative efficiency.

AI-driven workflow optimization tools analyze and streamline administrative processes. From inventory management to staff scheduling, these tools identify inefficiencies, automate routine tasks, and provide actionable insights. By optimizing workflows, administrative teams can allocate resources more effectively, minimize delays, and enhance overall operational efficiency. Intelligent chatbots and virtual assistants powered by AI enhance communication and support administrative tasks. These systems can handle routine inquiries, appointment confirmations, and administrative queries, freeing up human resources. Machine Learning algorithms can analyze patterns in billing data, flagging anomalies that may indicate fraudulent activities. Additionally, AI-driven cybersecurity solutions bolster data protection, safeguarding sensitive patient information and ensuring compliance with privacy regulations. The integration of AI into healthcare administration brings forth a myriad of opportunities to enhance efficiency, reduce administrative burdens, and improve the overall quality of healthcare services. From automating appointment scheduling to optimizing workflows and ensuring data security, AI-driven solutions pave the way for a more streamlined and responsive healthcare administrative landscape.

3.6.1 Electronic Health Records (EHR)

Electronic Health Records (EHRs) represent a cornerstone in healthcare administration, and the integration of Artificial Intelligence (AI) is transforming the management of these records as shown in Fig. 3.1. This section explores how AI enhances EHR systems by automating data entry, ensuring accuracy, and providing intelligent insights. Natural Language Processing and Machine Learning algorithms enable automated coding and categorization of medical information, reducing the administrative burden on healthcare professionals. The result is an improved EHR management system that not only saves time and resources but also contributes to enhanced data quality, interoperability, and overall administrative efficiency.

3.6.2 Billing and Coding Automation

Billing and coding processes are intricate and crucial aspects of healthcare administration. AI-driven solutions revolutionize these tasks by automating coding based on medical documentation. Machine Learning algorithms ensure accuracy and reduce the likelihood of errors in billing and coding procedures. Intelligent billing systems powered by AI can identify potential billing issues, verify insurance information, and generate accurate invoices. This not only expedites the billing cycle but also enhances revenue cycle management, contributing to overall administrative efficiency in healthcare operations.

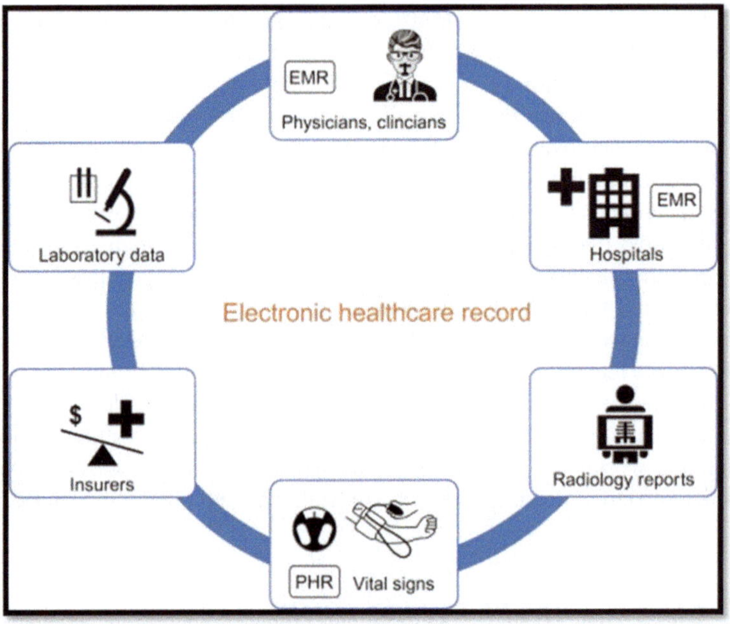

Fig. 3.1 Electronic healthcare record

3.6.3 Streamlining Healthcare Operations

Streamlining healthcare operations is a paramount goal in healthcare administration, and AI plays a pivotal role in achieving this. This section explores how predictive analytics and workflow optimization tools powered by AI analyze historical data, patient flow patterns, and various operational processes to optimize resource allocation and minimize inefficiencies. From inventory management to staff scheduling, AI-driven solutions identify bottlenecks, automate routine tasks, and provide actionable insights. By streamlining healthcare operations, administrative teams can allocate resources more effectively, minimize delays, and enhance overall operational efficiency. The integration of AI in healthcare operations represents a transformative approach to ensuring that administrative processes align with the dynamic needs of the healthcare facility.

Ethical and Regulatory Considerations

It is imperative to address bias in AI models to ensure fair and equitable healthcare delivery. Rigorous testing for bias, transparency in algorithmic decision-making, and ongoing monitoring are essential. Regulatory frameworks should promote fairness and inclusivity, and developers must prioritize mitigating biases to avoid perpetuating healthcare disparities. The "black box" nature of some AI models poses challenges in explaining how decisions are made. Ethical considerations emphasize the importance of transparency and explainability in AI systems, particularly in healthcare where decisions can have significant consequences. Regulatory guidelines should encourage developers to design AI models that are understandable to healthcare professionals and patients, fostering trust and accountability. Respecting patient autonomy through informed consent is a fundamental ethical principle. Regulatory frameworks should require clear and understandable communication to patients, ensuring that they are aware of the implications of AI-driven interventions and can make informed decisions about their care.

Establishing clear lines of responsibility, whether with developers, healthcare providers, or both, is essential. Regulatory bodies should outline frameworks for accountability, and developers must adhere to guidelines that prioritize patient safety and well-being. The dynamic nature of healthcare and AI technologies requires continuous monitoring and evaluation. Ethical considerations stress the importance of ongoing assessment to identify and rectify potential issues. Regulatory frameworks should mandate regular evaluations of AI systems' performance, safety, and effectiveness to ensure they meet evolving standards and contribute positively to patient care. AI applications in healthcare must comply with existing regulations and standards specific to the healthcare industry. Ethical considerations include a commitment to regulatory compliance to protect patients and uphold the integrity of healthcare systems. Developers should work within established frameworks, and regulatory bodies should adapt to ensure that regulations remain relevant in the rapidly evolving landscape of AI in healthcare. Striking a balance between harnessing the potential benefits of AI and addressing ethical challenges is essential to ensure that these technologies contribute positively to patient care while upholding principles of fairness, transparency, and patient autonomy.

3.7 Privacy and Data Security

This section addresses the ethical imperative of safeguarding patient privacy and ensuring the security of sensitive healthcare data. Striking a balance between leveraging the potential of AI for improved healthcare outcomes and protecting individual privacy is crucial. Ethical considerations emphasize the implementation of robust encryption, access controls, and data anonymization techniques. Additionally, regulatory compliance with standards such as Health Insurance Portability and Accountability Act (HIPAA) ensures that patient privacy and data security are prioritized in AI-driven healthcare applications.

3.7.1 AI/ML Ethical Dilemmas

The integration of AI and ML in healthcare introduces ethical dilemmas that require careful consideration. This section explores the ethical challenges associated with algorithmic bias, transparency, and decision-making in AI systems. Ethical considerations emphasize the need for continuous monitoring and evaluation to address biases and ensure fair and equitable healthcare outcomes. The transparency of AI algorithms and their impact on clinical decisions is a critical ethical concern, demanding clear communication with both healthcare professionals and patients. Regulatory frameworks should encourage developers to navigate these ethical dilemmas responsibly, prioritizing fairness, accountability, and the well-being of individuals affected by AI-driven healthcare interventions.

3.7.2 Regulatory Frameworks

Establishing robust regulatory frameworks is essential for guiding the responsible development and deployment of AI and ML in healthcare. This section explores the ethical imperative of regulatory compliance to protect patients and ensure the integrity of healthcare systems. Ethical considerations emphasize the need for regulatory and accountability in the event of adverse outcomes. Regulatory frameworks play a crucial role in fostering ethical practices, promoting innovation, and safeguarding the interests of both healthcare providers and patients in the era of AI-driven healthcare.

3.7.3 Current Challenges and Limitations:

The integration of Artificial Intelligence (AI) and Machine Learning (ML) into healthcare is not without its challenges and limitations. Ethical considerations surrounding patient privacy, algorithmic biases, and transparency in decision-making present substantial hurdles. Ensuring the quality and representativeness of training data to prevent biases in AI models remains a critical concern.

Looking ahead, the future of AI and ML in healthcare holds promising trends. Personalized medicine is set to advance further, tailoring treatments based on individual genetic profiles, lifestyles, and responses to interventions. The integration of AI into diagnostics, including early disease detection, medical imaging analysis, and pathology, is expected to enhance accuracy and efficiency. Remote patient monitoring and telehealth solutions are likely to see increased integration of AI for continuous health tracking and proactive healthcare management. Drug discovery processes will be accelerated through AI, analyzing molecular interactions, identifying potential drug candidates, and optimizing treatment strategies. Human-AI collaboration is anticipated to become more seamless, enhancing decision support, diagnostics, and overall patient care. Moreover, AI's role in public health is expected to grow, contributing to predicting disease outbreaks, optimizing resource allocation, and facilitating more effective population health management. Amidst the challenges and future trends, there are abundant research and development opportunities in the realm of AI/ML healthcare. Researchers can delve into enhancing ethical guidelines and standards to ensure responsible AI deployment. Advancements in data quality assurance measures and methods to mitigate biases in algorithms present fertile grounds for exploration. Developing standardized interfaces and collaborative solutions for seamless interoperability is a vital avenue for improvement. Research on explainable AI can contribute to building trust among stakeholders. Furthermore, exploring adaptive regulatory frameworks that foster innovation while prioritizing patient safety is an essential area for ongoing research. Opportunities abound in refining AI applications for personalized medicine, diagnostics, remote patient monitoring, drug discovery, and public health interventions, ultimately shaping the future landscape of AI in healthcare.

3.8 Case Studies

3.8.1 AI/ML Success Stories in Healthcare

One notable AI success story in healthcare revolves around the application of Machine Learning algorithms in medical imaging. A renowned medical center implemented an AI-powered diagnostic tool for the early detection of breast cancer in mammograms. The algorithm, trained on a vast dataset of mammogram images, demonstrated impressive accuracy in identifying subtle abnormalities indicative of breast cancer. In a retrospective analysis of patient data, the AI system achieved a remarkable 98% sensitivity and a 92% specificity, outperforming traditional methods. The implementation significantly reduced the time taken for radiologists to interpret mammograms and enabled more timely interventions, leading to improved patient outcomes.

Another success story lies in predictive analytics for patient outcomes. A major healthcare system integrated AI algorithms to predict patient deterioration and identify those at risk of complications during hospital stays. By analyzing Electronic Health Records and continuous monitoring data, the predictive model demonstrated a 30% reduction in adverse events and a 20% decrease in unplanned hospital readmissions. These quantitative results underscored the potential of AI-driven predictive analytics in enhancing patient safety and optimizing healthcare resource utilization.

3.8.2 Real-World Implementations

In a real-world implementation of AI in healthcare, a large hospital network embraced Natural Language Processing (NLP) algorithms to streamline clinical documentation and improve information extraction from unstructured text. The NLP system demonstrated a substantial reduction in the time required for physicians to input patient notes and retrieve relevant information from medical records. The quantitative results revealed a 40% increase in documentation speed and a 25% reduction in documentation-related errors. This real-world implementation showcased the tangible benefits of AI in enhancing clinical workflows and minimizing administrative burdens on healthcare professionals.

Furthermore, a regional healthcare consortium embraced AI-driven decision support tools for optimizing medication prescriptions. The system analyzed patient records, medical literature, and historical treatment outcomes to recommend personalized medication plans. In a pilot study, the AI-driven decision support tool exhibited a 15% improvement in medication adherence rates and a 12% reduction in adverse drug reactions. These quantitative outcomes highlighted the potential of AI to individualize treatment plans, improve patient compliance, and mitigate the risks associated with medication management.

These case studies exemplify the tangible impact of AI and Machine Learning in healthcare, providing not only qualitative success stories but also quantitative evidence of improved accuracy, efficiency, and patient outcomes. As real-world implementations continue to evolve, these success stories underscore the transformative potential of AI in revolutionizing healthcare delivery. Various results discussed above is shown in Tables 3.2, 3.3, 3.4 and 3.5.

Table 3.2 AI/ML success in breast cancer detection

Metric	Value
Sensitivity	98%
Specificity	92%
Reduction in time	40% improvement
Improved outcomes	Yes

Table 3.3 Predictive analytics for patient outcomes

Metric	Value
Reduction in adverse events	30%
Decrease in readmissions	20%
Resource utilization	Optimized

Table 3.4 NLP in clinical documentation

Metric	Value (%)
Increase in documentation speed	40
Reduction in errors	25

Table 3.5 AI-driven medication recommendations

Metric	Value (%)
Improvement in medication adherence	15
Reduction in adverse reactions	12

3.8.3 Conclusions

In conclusion, the integration of Artificial Intelligence (AI) and Machine Learning (ML) in healthcare has demonstrated significant successes and tangible benefits across various domains. The case studies highlighted the transformative impact of AI on breast cancer detection, predictive analytics for patient outcomes, clinical documentation through Natural Language Processing (NLP), and personalized medication recommendations. These successes are underscored by quantitative results, including improved sensitivity and specificity in diagnostics, reduction in adverse events, increased documentation speed, and enhanced medication adherence. The application of AI in healthcare not only improves the efficiency of clinical workflows but also contributes to more accurate diagnoses, optimized resource utilization, and, ultimately, better patient outcomes. The success stories presented here serve as compelling evidence of the potential of AI to revolutionize healthcare delivery, enhancing both the quality and effectiveness of medical services.

3.8.4 Future Scope

As we look to the future, the scope of AI and ML in healthcare continues to expand. There are promising avenues for further research and development, including the refinement of existing algorithms and the exploration of new applications. The personalization of medicine, leveraging AI to tailor treatments based on individual patient characteristics, is a frontier that holds immense potential. Additionally, advancements in explainable AI can address concerns related to transparency and build trust among healthcare professionals and patients. Ongoing research in ethical considerations, regulatory frameworks, and continuous monitoring of AI systems will be crucial to ensure responsible deployment and address emerging challenges.

References

Agarwal R, Bjarnadottir M, Clark J, Rhue L, Gao GG (2023) Data science for health equity: perils and promise of AI/ML in healthcare. ML in Healthc

Christou CD, Tsoulfas G (2021) Challenges and opportunities in the application of artificial intelligence in gastroenterology and hepatology. World J Gastroenterol 27(37):6191

El Naqa I, Karolak A, Luo Y, Folio L, Tarhini AA, Rollison D, Parodi K (2023) Translation of AI into oncology clinical practice. Oncogene 42(42):3089–3097

Food and Drug Administration (2019) Proposed regulatory framework for modifications to artificial intelligence/machine learning (AI/ML)-based software as a medical device (SaMD)

Higgins O, Short BL, Chalup SK, Wilson RL (2023) Artificial intelligence (AI) and machine learning (ML) based decision support systems in mental health: an integrative review. Int J Mental Health Nurs

Kourou K, Exarchos KP, Papaloukas C, Sakaloglou P, Exarchos T, Fotiadis DI (2021) Applied machine learning in cancer research: a systematic review for patient diagnosis, classification and prognosis. Comput Struct Biotechnol J 19:5546–5555

Kurariya P, Yadav A, Jain DK (2023) Artificial intelligence impact on healthcare: advanced drug discovery and beyond. J Popul Ther Clin Pharmacol 30(18):2898–2908

Mahmood MR et al (2022) Ambient intelligence and internet of things: an overview" book entitled ambient intelligence and internet of things: convergent technologies. Wiley

Qureshi R, Irfan M, Ali H, Khan A, Nittala AS, Ali S, Alam T (2023) Artificial intelligence and biosensors in healthcare and its clinical relevance: a review. IEEE Access

Rahmani AM, Yousefpoor E, Yousefpoor MS, Mehmood Z, Haider A, Hosseinzadeh M, Ali Naqvi R (2021) Machine learning (ML) in medicine: review, applications, and challenges. Mathematics 9(22):2970

Rathee N et al (2023) An innovative smart IoT Device to measure and monitor patient's Critical parameters in hospitals. In: Artificial intelligence in cyber physical systems: principles and applications. CRC Press

Rubinger L, Gazendam A, Ekhtiari S, Bhandari M (2023) Machine learning and artificial intelligence in research and healthcare. Injury 54:S69–S73

Sahu M, Gupta R, Ambasta RK, Kumar P (2022) Artificial intelligence and machine learning in precision medicine: a paradigm shift in big data analysis. Prog Mol Biol Transl Sci 190(1):57–100

Subbaswamy A, Saria S (2020) From development to deployment: dataset shift, causality, and shift-stable models in health AI. Biostatistics 21(2):345–352

Swanson K, Wu E, Zhang A, Alizadeh AA, Zou J (2023) From patterns to patients: advances in clinical machine learning for cancer diagnosis, prognosis, and treatment. Cell

Taimoor N, Rehman S (2021) Reliable and resilient AI and IoT-based personalised healthcare services: a survey. IEEE Access 10:535–563

Xu Z, Biswas B, Li L, Amzal B (2023) AI/ML in precision medicine: a look beyond the hype. Ther Innov Regul Sci 1–6

Yang Z, Silcox C, Sendak M, Rose S, Rehkopf D, Phillips R, Bazemore A (2022) Advancing primary care with artificial intelligence and machine learning. In: Healthcare, vol 10, no 1, p 100594. Elsevier

Chapter 4
Design and Performance Analysis of an Energy Efficient IoT-Based WBAN to Control Corona Virus Outbreak

Kosheen Wighmal, Sanya Arora, and Sindhu Hak Gupta

Abstract The Coronavirus or COVID-19 is a highly infectious disease which is caused by SARS-CoV-2 virus. Therefore, it is necessary to identify Covid symptoms at an early stage and get the adequate treatment. Wireless Body Area Networks (WBAN) and Internet of Things (IoT) combined with machine learning can be used to detect Covid from biomedical sensors such as body temperature, blood pressure, and oxygen saturation. In this paper, an energy efficient IoT-based WBAN to control corona virus outbreak is proposed. The IoT framework uses non-invasive sensors interfaced with Arduino Uno for collection of real-time data which is sent to the receiver comprising of Espressif Systems32 (ESP32) using Long-Range Radio (LoRa) Technology gateway. LoRa helps connect large number of nodes and offers low power, adaptive data rate algorithm which maximizes nodes battery life and long-range transmission which makes this application energy efficient as compared to other WBANs' that carry more data and thus, consume more energy. The collected data is then stored and visualized using the ThingSpeak cloud platform. The obtained dataset is labeled using rule-based decision making and given as input to three supervised machine learning algorithms namely Classic Logistic Regression, Support Vector Machine, and Naïve Bayes to classify whether a person has COVID-19 symptoms or not. The model performance parameters such as accuracy, precision, recall, and F1 score are calculated and compared.

Keywords COVID-19 · LoRa technology · IoT · Machine learning · Cloud · Arduino

---

K. Wighmal · S. Arora · S. H. Gupta (✉)
Amity School of Engineering and Technology, Amity University Uttar Pradesh, Noida, Uttar Pradesh, India
e-mail: shak@amity.edu

A. Rai et al. (eds.), *Convergence of Artificial Intelligence, Machine Learning, and the Internet of Things in Industry 4.0 Applications*, Transactions on Computer Systems and Networks, https://doi.org/10.1007/978-981-96-5269-3_4

4.1 Introduction

The novel Corona virus is a highly contagious disease that spreads quickly (Lescure et al. 2020; van der Hoek et al. 2024) and has already caused 6,179,104 deaths worldwide and 521,656 as of 8 April 2022 as displayed in Fig. 4.1a, b (COVID-19 Data Repository 2022).

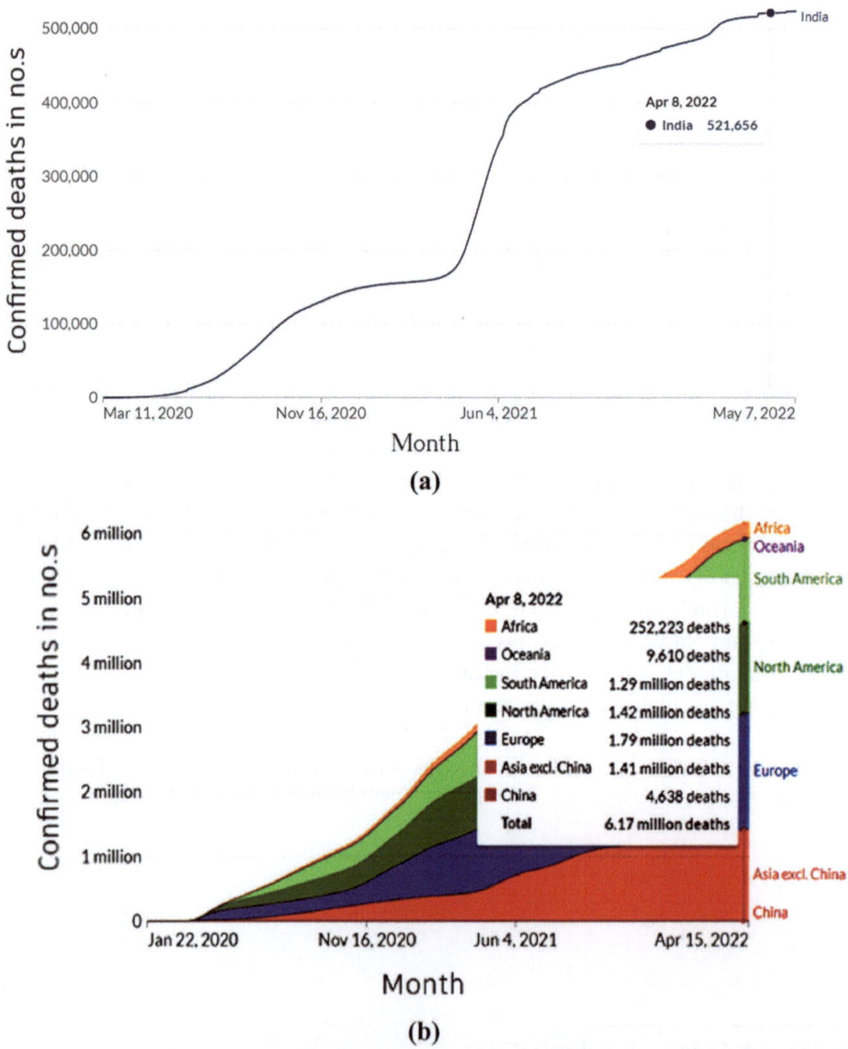

Fig. 4.1 **a** Cumulative confirmed COVID-19 deaths in India and **b** by world region as on 8 April 2022. *Source* Adapted from (COVID-19 Data Repository 2022)

The symptoms generally arise after 3–5 days and can last up to 14 days as well and the contracted person might be a carrier for SARS-CoV-2 (Johnson and Vinetz 2020). These symptoms include cough, fever, tiredness which are the most common ones followed by diarrhea, headache, chest pain, sore throat, difficulty in breathing and high heart beats per minute (bpm) as severe symptoms (Coronavirus 2022; Jiang et al. 2020; Hu et al. 2020). A person infected with COVID could be identified based on a few symptoms like fever, fatigue, and dry cough (Jiang et al. 2020). Therefore, it is important to diagnose these symptoms at an early stage and take the necessary precautions and stay in isolation as recommended by World Health Organization (WHO) guidelines.

4.1.1 Use of Wireless Body Area Networks with IoT for Real-Time Data Collection

In this paper, the ability of WBAN to send and monitor data remotely for health monitoring is utilized. WBAN aids in remote monitoring of patients and Telemedicine, which means diagnosis of a patient at remote locations (WBAN 2020).

WBANs support dissimilar sensors for continuously analyzing health condition of a patient. The main idea is to assist with screening of symptoms without the need of going to a hospital and increasing the risk of spreading the virus. A WBAN consists of Sensor nodes and Gateway nodes. Sensor nodes collect data from human body. They usually comprise of wearable sensors measuring a physical quantity such as temperature, heart rate, and pulse rate (Mary and Jerine 2020). The gateway nodes connect the sensor nodes to an external server using Internet to store this data on cloud.

WBANs have biosensors, which consume high energy due to leakage of charge and since with distance power consumption increases, therefore, to make it energy efficient, this paper proposes an energy efficient IoT Based WBAN framework which uses low power LoRa Technology for data transmission for early diagnosis of COVID-19 based on rule-based analysis for data labeling. The data transmission from sensor nodes to cloud using a wireless interface such as Wireless-Fidelity (Wi-Fi) makes it more cost efficient (Cypher et al. 2006). The Internet of Things or IoT systems collect, process, and store the data to enable data-driven analysis on cloud.

4.1.2 Use of Wireless Body Area Networks with Machine Learning for Diagnosis of COVID-19

Machine learning with WBANs has been extensively used for various applications for the health care sector (Latré et al. 2010). It is an empowering tool to predict and classify accurate results in future. In our proposed project, symptoms are classified based on rule-based analysis of 3 sensor values which are temperature, heartbeats-per-minute (bpm), and cough. The symptoms are classified as:

(i) Possessing COVID-19
(ii) Not possessing COVID-19.

The data is collected from the cloud and the model trains on this data to enable classification of unseen test data. Three machine learning algorithms namely Classic Logistic Regression, Naive Bayes, and Support Vector Machine have been implemented for classification. The performances of the models are evaluated using parameters such as accuracy, precision and recall, and the best-fit model for this application is determined.

4.2 Related Work

Over the last few years, several developments have been made in the field of e-health monitoring. Many researchers have explored implementing IoT to capture data and integrating it with machine learning techniques for disease diagnosis and prediction (Raeesi Vanani and Amirhosseini 2020; Zhang et al. 2017; Rahaman et al. 2019). Table 4.1 summarizes few of these researches and their features. The researchers in Wu et al. (2019) presented a wrist wearable sensor network system to monitor environmental and physiological parameters of industrial health workers. They proposed a heterogeneous network of IoT platform with Bluetooth-based medical signal sensing network which used LoRa gateway to communicate within different subjects. The authors in Muhammad et al. (2020) proposed data mining models for the prediction of COVID-19 infected patients' recovery using epidemiological dataset of COVID-19 patients of South Korea. The machine learning models applied were Decision Tree, Support Vector Machine, Naive Bayes, Logistic Regression, Random Forest, and K-Nearest Neighbor.

The authors in Ghosh et al. (2016) proposed a remote health monitoring system using IoT to automatically gather data from patients and store the gathered data into cloud for permanent use. The researchers have developed a prototype with the help of IoT related technologies e-Health sensor V2 for measuring patients' health information which is interfaced with Arduino, Phidgets sensor shield, and a web application using PHP programming language to display data and view it remotely. The researchers in Yogaraj et al. (2017) propose a remote IoT-based smart health monitoring system for rural areas. It procures temperatures, heartbeat, and ECG readings and displays them on the serial monitor of Arduino IDE. The researchers in Kumar and Rajasekaran (2016) proposed an IoT-based framework to monitor patients' body temperature, heartbeat, and respiration rate and body movement using Raspberry Pi microcontroller board. The researchers in Bilandi et al. (2021) proposed an energy efficient WBAN in which three sensors are used temperature sensor, pulse oximeter sensor, and cough sensor.

Table 4.1 Summary of various approaches and their features

References	Proposed use	Major components	Features
Wu et al. (2019)	Hearing loss, headache, and rapid pulse rate detection	Raspberry pi, LoRa module, temperature sensor, humidity sensor, pulse sensor, WSN, WDM, UV, CO_2 sensor	Uses Lora gateway and cloud computing. However, choice of microcontroller and the cost of making the system wireless to improve flexibility can make the setup costlier
Muhammad et al. (2020)	Development of data mining models for the prediction of COVID-19 infected patients' recovery	Dataset was obtained through KDCD and included 8 attributes like patient ID, global number (the number given by KCDC), sex, birth year, age, country, province, city, disease (TRUE: underlying disease/FALSE: no disease) infection case, infection order etc	The model uses epidemiological dataset of 3254 values. While, the dataset proposed in this paper, uses primary dataset comprising of 14,000 values. Further, decision tree model accuracy of the proposed paper is 99.97% as compared to 99.85% of the reference paper
Ghosh et al. (2016)	Remote health monitoring using IoT	Touch sensor, light sensor, humidity and temperature sensor and electrocardiogram (ECG) sensors were used to analyze heart functions, body temperature and acceleration sensing	The research is proposed for monitoring, displaying results in a desktop interface. In the research, no machine learning algorithm is utilized. The paper utilizes a small dataset, i.e., of 10 patients
Yogaraj et al. (2017)	IoT based smart healthcare monitoring system for rural/isolated areas	Temperature sensors (LM-35), ECG sensor module (AD-8232), heartbeat sensor (KY039)	The research deals with obtaining sensor data and does not include a receiver setup. It does not store data on cloud and is used to only view sensor readings. It also demands 24×7 connectivity with the internet, which might raise several issues like internet, additional expense etc. in rural areas
Kumar and Rajasekaran (2016)	An IoT based patient monitoring system using raspberry Pi	Temperature sensor, accelerometer sensor, respiration sensor and heartbeat sensor with raspberry pi	The web database of a patients' health can be accessed and viewed by anyone, which can cause safety and privacy issues

(continued)

Table 4.1 (continued)

References	Proposed use	Major components	Features
Bilandi et al. (2021)	An IoT based WBAN to control Covid	Temperature sensor (DS18B20), pulse oximeter sensor, pressure sensor (AMS5915), Arduino, MySignals hardware chip, LoRa	The model proposed in this paper uses cough sensor (SW420) with temperature sensor and pulse sensor whereas the model proposed in the reference paper uses Pressure sensor (AMS5915) along with temperature sensor and pulse sensor
Mukhtar et al. (2021)	An IoT based model for Covid e-screening using wearable sensors and real-time data	ESP8266, Pulse oximeter and heartbeat sensor (MAX30100), temperature sensor (MAX30205), cough sensor (SW420)	In this paper, the proposed model uses ESP32 and Pulse sensor instead of ESP8266 and Pulse oximeter and heartbeat sensor. The symptoms are evaluated using their custom algorithm in the reference paper whereas in this paper machine learning algorithms are implemented to check the accuracy of the model
Eljamassi and Maghari (2020)	Detection of Covid from chest X-Ray scans using machine learning	Dataset of chest x-ray scans of people infected with Covid, and normal people, histogram of oriented gradients (HOG) for image feature extraction	The images in the reference paper are classified using SVM (98.14%), Random forest (96.29%), and KNN (88.89%) whereas in this paper the dataset is a numerical dataset and the machine learning algorithms applied are SVM (98.33%), logistic regression (99.97%) and Naive Bayes (93.8%)

(continued)

Table 4.1 (continued)

References	Proposed use	Major components	Features
Prakash (2020)	Analysis, prediction, and evaluation of Covid-19 datasets using machine learning algorithms	Machine learning algorithms (Random forest regressor, Random forest classifier, SVM, KNN + NCA, Decision Tree, Gaussian Naïve Bayes, multilinear regression, logistic regression, XGBoost)	An analysis is done on Covid datasets to find out which age group has been affected the most and prediction models are built using ml algorithms
Mir et al. (2020)	IoT Framework for early detection and prediction of Covid by leveraging ML in cloud	Data collection center, data analytic center, diagnostic system, and cloud system	In the reference paper, five machine learning algorithms are applied on the dataset taken from SKIMS, Srinagar with an accuracy of above 95% in all the algorithms used whereas in this paper, the dataset was created, and three machine learning algorithms were applied on it with accuracy above 93% in all the algorithms applied

Further, five machine learning algorithms are implemented on the dataset obtained. Other contributions by various authors have been summarized in Table 4.1. Many researchers have made notable contributions to the field of IoT in the field of health care. The novel contributions of this proposed model are—LoRa Gateway is used which allows the user to send the data using very low power and over long-distance. The three sensors used (MAX30205, SW420 and Pulse sensor) also use less energy as MAX30205 has one-shot and shutdown modes which helps reduce power usage, and SW420 uses a potentiometer and LM393 voltage comparator and resistors to limit the current and act as voltage dividers, and pulse sensor is a plug and play type sensor.

The dataset created has 14,000 values because anything less than that would cause under fitting and anything more than that was over-fitting. For three parameters, 14,000 values are optimum. Additionally, three machine learning algorithms used for classification. The proposed model is explained in detail in the Sect. 4.3.

4.3 Proposed Model

The model proposed consists of the following as depicted in Table 4.2,

i. Data collection to record body temperature, heart bpm, and cough values from three sensor nodes.

Table 4.2 Stages of the proposed model

Sensor node	Gateway	Cloud	Machine learning algorithms
• Body temperature sensor (MAX30205) • Pulse sensor • Cough sensor (SW420) • Arduino UNO (Microcontroller at transmitter) • LoRa module (For Data transmission to receiver node)	• LoRa Module (To receive Data from sensor node) • ESP-32 + Wi-Fi (Microcontroller at receiver)	• Storage of data on ThingSpeak using Wi-Fi • Data storage and visualization	• Data labeling • Classic logistic regression • Naïve Bayes • Support vector machine

ii. Data Transmission from transmitter to receiver using LoRa gateway.
iii. Storage of data on cloud (ThingSpeak) using Wi-Fi.
iv. Data collection from cloud, labeling of data and prediction of Covid using three machine learning algorithms.

The architecture of the proposed model consists of the following,

4.3.1 Hardware Implementation

The sensors selected are cost efficient and offer ease in interoperability and interfacing. These sensors are used for measuring physical quantities crucial for this proposed project.

1. **Sensor Nodes**: Three sensors SW420—cough sensor, MAX30205—human body temperature sensor and pulse sensor have been used to measure cough—Yes or No, temperature, and heart beats per minute (bpm). Table 4.3 explains these hardware components and their features.

   MAX30205 sensor is used to measure body temperature (Yamanoor and Yamanoor 2020) and converts it into digital form. This sensor converts the temperature to digital signal using a high resolution, sigma-delta, analog-to-digital converter (ADC). Accuracy meets clinical thermometry specification of the ASTM E1112 when soldered on the final PCB. It has three address select lines with 32 total available addresses.

   The sensor has a 2.7–3.3 V supply voltage range, low 600 μA supply current, and a lockup-protected I2C-compatible interface that make it suitable for wearable fitness and medical applications. The one-shot and shutdown modes help reduce power usage which makes it suitable for low power applications (Datasheets.maximintegrated.com 2022).

Table 4.3 Sensors used and their characteristics

Sensor	Purpose	Technology	Other features
MAX30205	Accurately measures human body temperature (Yamanoor and Yamanoor 2020)	A high resolution, sigma delta analog to digital converter which converts temperature into digital signals (ADC)	2.7–3.3 V supply voltage range, low 600 μA supply current, and a lock up-protected I^2C-compatible Interface (Datasheets.maximintegrated.com 2022)
SW420	Cough detection/ vibration sensor module	Continuous wave (CW) radar, Doppler radar (Preethi et al. 2020)	Uses a wide voltage LM393 comparator, operating voltage: 3.3–5 V DC (Carpentier et al. 2018; Chuma and Iano 2021)
Pulse sensor	Measures heart beats per minute	Uses infrared light to determine the amount of oxygen in the blood (Pulse Sensor 2022)	Optical heart rate sensor, operates on 3 V or 5 V, works with any MCU with an ADC (Jubran 1999)

Cough is recognized as a symptom of divergent respiratory disorder diseases from a common cold to the current corona virus disease. Cough is a reaction of human body to prevent any blockages or secretions in upper airway.

The frequency of cough is an important indicator of the severity of the disease. There are a few approaches and sensors to detect cough. SW420 is a vibration sensor module used to detect cough. It has a LM393 comparator chip which acts as a switch, when no vibration is received it acts as closed. It also features an adjustable potentiometer; to give an adjustable digital output based on the amount of vibration (Preethi et al. 2020; Carpentier et al. 2018; Chuma and Iano 2021).

The sensitivity can be increased or decreased to the desired amount by adjusting the potentiometer to the desired amount. SW420 is a Doppler radar which is a continuous wave (CW) radar that uses a voltage-controlled oscillator to continuously transmit a signal in digital (0 if no cough, 1 if cough) and analog (based on output Voltage) form.

Here, the output for SW420 is digital. The third sensor used for heart pulse detection in beats per minute is the pulse sensor. Pulse sensor is a plug and play, low-cost optical heart rate sensor. The LED on the front side of the sensor is placed on the fingertip.

LED emits light that directly falls on veins carrying blood. If the blood flow is monitored, rate of heart can be monitored as well. When blood reflects light, more light is picked up by the ambient sensor. This minor change in received light is analyzed overtime to calculate heart beats per minute (Pulse Sensor 2022; Jubran 1999).

2. **LoRa Technology**: LoRa or Long-Range Radio is a wireless modulation technology based on the chirp-based spectrum technique (Web.archive.org 2022). which uses the physical layer for modulation. Chirp-based spectrum means that transforming single bit of information into different bits and spread over the entire range which reduces the transmission noise and make data also robust and reliable from any interference from other frequencies.

   The LoRaWAN is Low Power Wide Area Network (LPWAN) protocol and system architecture developed by LoRa Alliance. LoRa (developed by Semtech) (Design and Reuse 2022) defines the physical layer of the system whereas LoRaWAN is developed over the network layer and operates on the MAC protocol. Lora is based on unlicensed frequency spectrum which lies below the 1000 MHz (Adelantado et al. 2017).

   Lora Operates on 169, 433 MHz (Asia), 868 MHz (Europe), and 915 MHz (North America). To conserve energy, the Low Power Wide Area Network (LPWAN) is used which allows the user to send the data using very low power and over long-distance (up to 15 km). LPWAN is designed to transmit data in long ranges with the minimum power usage mostly useful for battery-operated devices which have a limited amount of battery life. The data range in LPWAN lies between 0.3 and 5.5 Kbit/s per channel. Protocols like Bluetooth, Zigbee, and Wi-Fi are adequate for consumer-level IoT connectivity but when using LPWAN have many more advantages.

   Here, SX1278 LoRa module is used to transmit data over long ranges and is also cost efficient. The SX1278 LoRa module uses the SPI bus which allows fast speed communication and frees up the UART for other purposes like connecting communication modules or debugging. It uses an 868 MHz antenna used at this frequency (Cooking-hacks.com 2022). LoRa utilizes the whole bandwidth of the channel to broadcast a signal which additionally makes it immune to channel noise, Doppler effects and fading. A LoRa endpoint comprises of three parts:

   1. Sensor nodes
   2. Microcontroller
   3. Semtech SX1278 LoRa module radio chip

   In the proposed model, each sensor end point sends data to a LoRa gateway (operating at 433 MHz ISM band) which transfers data it over another network, i.e., Wi-Fi to the receiver node, where the data is captured. Since, the sensor data is static, i.e., it does not change rapidly with time and thus, can be statically transmitted at 5.5 Kbit/s with no effect on application. The key features of LoRa technology like low power, long range, high power efficiency, high interference immunity, and wireless transmission over unlicensed spectrum makes it apt for this application.

3. **Connections**: At the transmitter side, MAX30205, pulse sensor, and SW420 is interfaced with Arduino Uno, the Microcontroller which is linked with SX1278

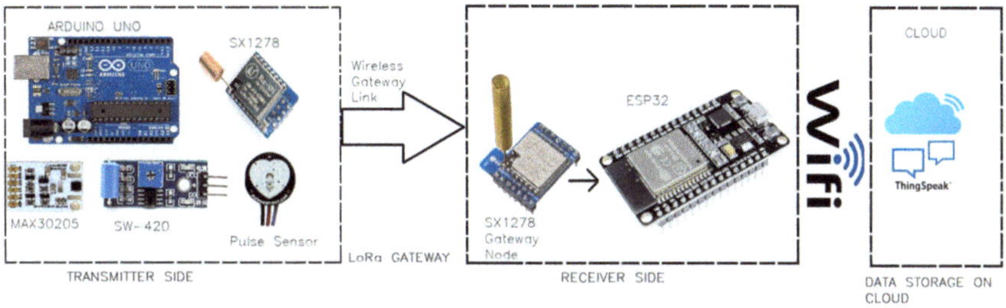

Fig. 4.2 Block diagram of the experimental setup used

LoRa module to transmit data to receiver side. Arduino is 8-bit microcontroller board based on the ATmega328.

It has 14 digital input/output pins (of which 6 can be used as PWM outputs), 6 analog inputs, a 16 MHz ceramic resonator, a USB connection, a power jack, an ICSP header, and a reset button (https://www.farnell.com/datasheets/1682209.pdf). For MAX30205, the 3.3 V Vin Pin is connected to the input supply. GND Pin of Arduino is connected to the GND pin of MAX30205. The SDA pin of sensor is connected to A4 and SCL pin is connected to A5 of the Arduino Uno Board.

For SW420, the 5 V Vin Pin is connected to the input supply. GND Pin of Arduino is connected to the GND pin of SW420. The DO pin of sensor is connected to D3 of the Arduino Uno Board. When no external pressure is felt it outputs LOW, otherwise gives HIGH output. For pulse sensor, GND is connected to GND, Vin to 3 V and output to A0. The transmitter side connections are displayed in Fig. 4.3a.

The Block diagram of the experimental setup has been displayed in Fig. 4.2. At the receiver side, SX1278 LoRa gateway is used as a relay node to provide the wireless connection between the IoT-WBAN model and the ESP32 (Vangelista et al. 2015). ESP 32 is the micro controller unit used at receiver side; it is integrated with Wi-Fi module.

It reaches ultra-low power consumption which makes our setup even more energy efficient. It has high level of integration with in-built, power amplifier, antenna switches, power management modules, and low-noise receive amplifier, filters (Espressif.com 2022). The sensor data is sent to the cloud using Wi-Fi. The Receiver side connections have been displayed in Fig. 4.3b.

4.3.2 Software Implementation

Arduino IDE or Arduino Integrated Development Environment is an open-source software which is a type of text editor to write code and upload it to any microcontroller board. It is highly compatible with Arduino Uno. Arduino IDE allows editing, compiling, and uploading program into memory of the microcontroller (Arduino.cc 2022).

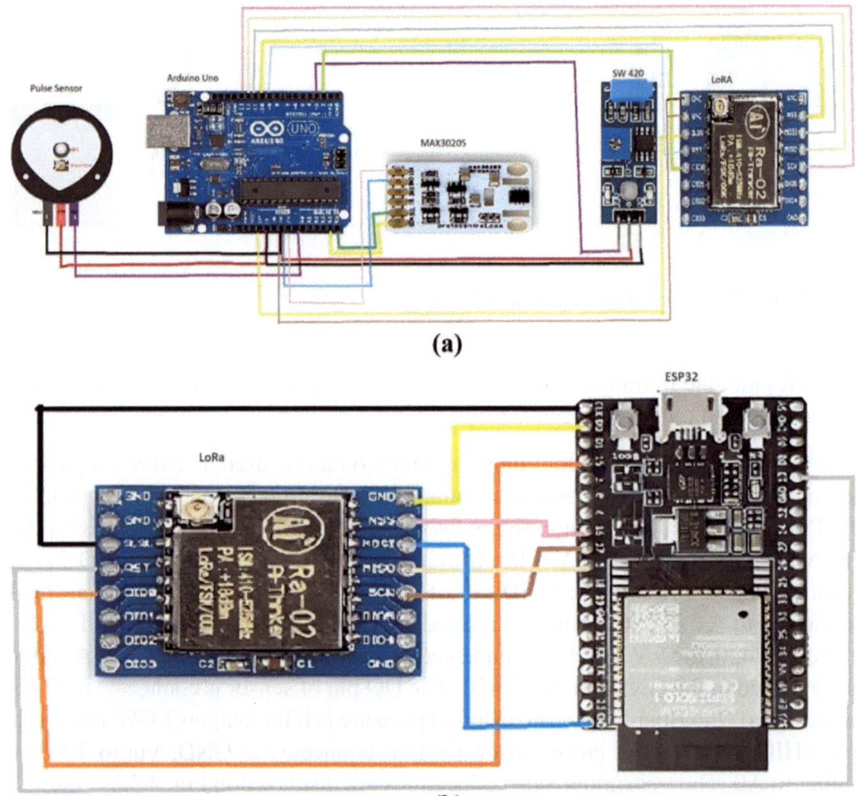

Fig. 4.3 Hardware setup at **a** Transmitter side **b** Receiver side

ThingSpeak is used for storing the sensor data to the cloud. ThingSpeak is an open-source software that helps users to communicate with internet enabled devices and makes it useful for several IoT applications (thingspeak/README.textile 2022).

ThingSpeak also helps in analyzing and visualizing stored data using MATLAB due to its integrated support from MathWorks (https://www.mathworks.com/hardware-support/thingspeak.html). It also provides an API to devices that facilitates data retrieval and access.

4.3.3 Data Collection

For storage of sensor data on cloud, a ThingSpeak private channel with Channel ID: 1,643,076 was created as shown in Fig. 4.4a. Stored data has four channel fields: Temperature, Cough, Heart bpm, and Subject for which the data is being collected.

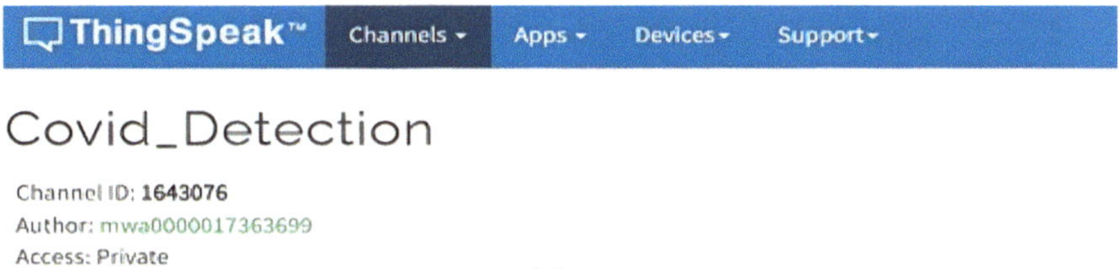

(a)

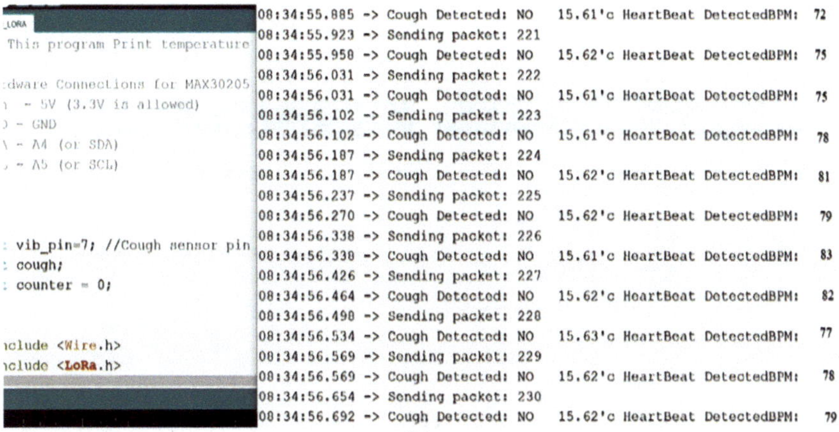

(b)

Fig. 4.4 **a** ThingSpeak channel for Covid detection **b** Data obtained for subject 1 **c** Data obtained for subject 1 **d** Data obtained for subject 2

The data was collected for 14 Subjects in the age group 15–22 years. For each subject 1000 values were recorded. Since the subjects were possessing Non-Covid symptoms, the dataset was created by changing one of the parameters to a higher value (for e.g., physically causing change in temperature) to include potential Covid subjects. The dataset was recorded over a span of 2–3 days since ThingSpeak records data after every 15 s.

Thus, the final dataset comprised of 14,000 readings having three values: Temperature, Cough, and Heart bpm. Data obtained for Subject 1 and 2 for over 8 and 4 min, respectively, is shown in Fig. 4.4c, d.

In Fig. 4.4c, it was observed that a total of 30 data values have been stored over 8 min for subject 1. For field 1, i.e., temperature, the average value was 21.95 °C. For field 2, bpm, the values varied between 65 and 80 bpm. For field 3, cough, the values varied between 1 and 0.

In Fig. 4.4d, it was observed that a total of 18 values were noted in 4 min for subject 2. In Field 1, temperature gradually rose from 21.5 to 22.5 °C. In Field 2, maximum value of bpm recorded was 70. In Field 3, cough was recorded 0 for over 10 values.

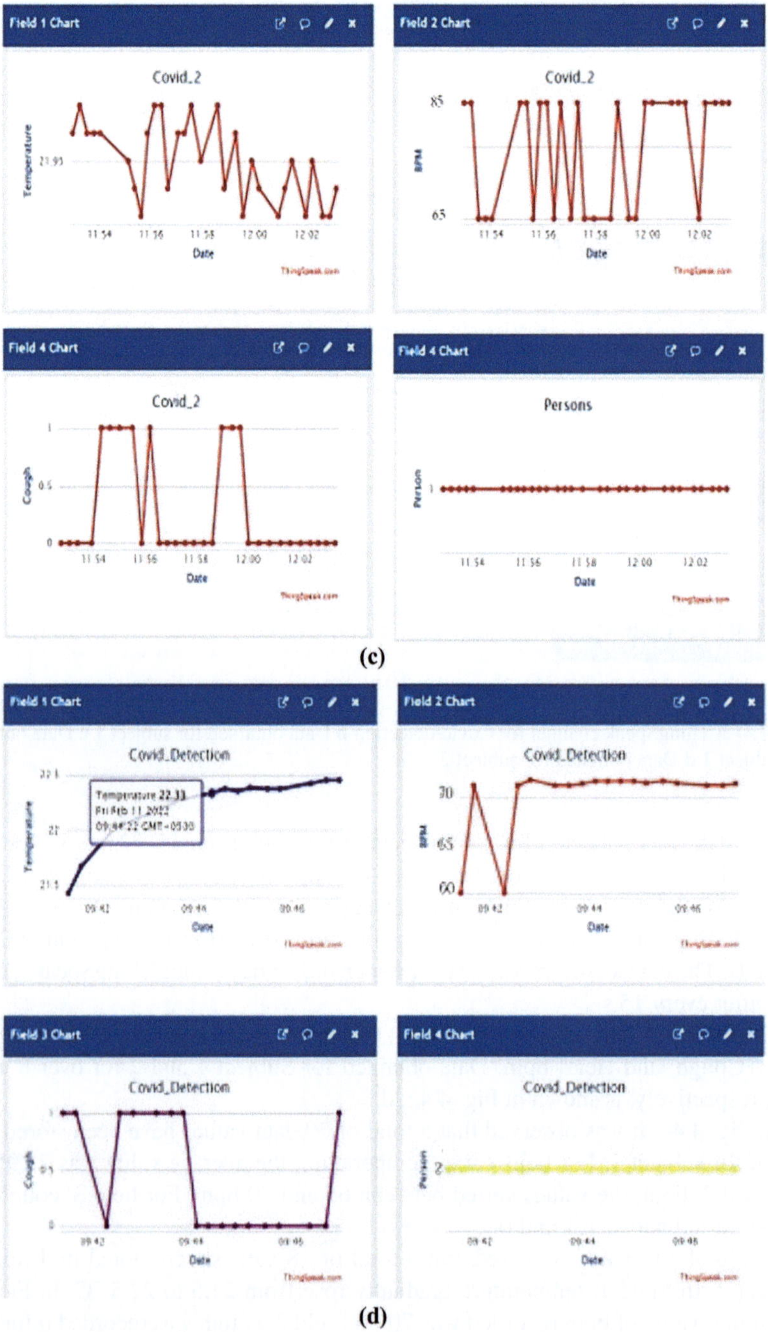

Fig. 4.4 (continued)

4.3.4 Data Labeling and Data Visualization

The dataset obtained is unlabeled and therefore to use it for a supervised machine learning algorithm, the data needs to be assigned with labels, i.e., Covid Symptoms or No Covid Symptoms. Labeled data is required so that model can learn from this data indicator and classify labels for the test data accurately. To do labeling, a rule-based approach is followed. The rules in our application relate to classifying whether a subject possesses Covid symptoms or not which is decided by a combination of sensor values above the threshold value. The rules-based approach for our application is as shown in Fig. 4.5:

- Covid symptom

If Cough rate = 1; Heart Beats per Minute > 120 bpm; Temperature > 38 °C or Temperature > 38 °C and Heart Beats per Minute > 120 bpm; or Cough = 1 and Heart Beats per Minute > 120 bpm.

- Else: No Covid Symptom

Only severe clinical symptoms have been considered for Covid Symptom labeling. The above classification has been only done for Potential Covid Symptoms patients and Non-Covid Symptoms patients and in no way categorizes them as Covid or Not Covid patients. This can be used in screening but for complete diagnosis a clinical expert or practitioner is required. The labeled dataset has been shown in Fig. 4.6.

The dataset for training consists of Features: 3 (Temperature, bpm, and Cough) and Samples: 14,000.To understand the obtained data better, Data Visualization is done using python. Figure 4.7 depicts Box plot to visualize outliers in the data. It is observed that Temperature ranges from 32–43 °C. Heart bpm ranges from 78 to 130 bpm. Cough values is either 1 for Cough or 0 for No Cough. Symptom is 1 for yes and 0 for no symptom of Covid. The same observation is illustrated in Fig. 4.8a as well where the distribution of data for each parameter is additionally noted. Most subjects have higher Temperature (≥40 °C) where the peak is noted. Most subjects

```
for i in range (0, len(covid_data)):
  if (((covid_data['Temperature'][i]>38) an
d (covid_data['Bpm'][i]>120) and (covid_dat
a['Cough'][i]==1)) or ((covid_data['Tempera
ture'][i]>38) and (covid_data['Bpm'][i]>120
)) or ((covid_data['Cough'][i]==1) and (cov
id_data['Bpm'][i]>120))) :
    covid_data['symptom'][i]=1
  else:
    covid_data ['symptom'][i]=0
```

Fig. 4.5 Rules-based approach for Covid detection

```
covid_data
```

	Unnamed: 0	Temperature	Bpm	Cough	symptom
0	11809	25.03	77	1	0
1	11810	24.80	69	0	0
2	11811	25.49	68	0	0
3	11812	36.34	76	0	0
4	11813	32.38	74	1	0
...	...	...	...	...	...
13995	21971	41.23	133	0	1
13996	21982	39.29	121	0	1
13997	21984	41.55	130	1	1
13998	21985	43.21	132	0	1
13999	21992	40.54	132	0	1

14000 rows × 5 columns

Fig. 4.6 Labeling of dataset using rules-based approach for Covid detection

report bpm close to 80 bpm followed by 130 bpm being the second most frequent. 50% of people have Cough and the others don't.

A crucial observation is that of symptom, which is observed to be yes for 50.71% of the values and 49.29% of the values have no symptom. The class is balanced and hence accuracy isn't compromised because of class imbalance as the model will be able to learn from enough data points for both the classes to differentiate between the two.

The Heat map of all continuous values including target is plotted in Fig. 4.8b. It represents the co-linearity of the variables in our dataset. There seems to be only positive values and no negative values in the heat map. Since the heat map represents only positive values hence, dark regions indicate low co-relation; light regions indicate a high co-relation.

The diagonal values are 1 because variables are co-related with themselves in 1:1. bpm (independent variable) and symptom (dependent variable) have a co-relation of 0.89, which is high and thus, also means they are more linearly dependent. After bpm, temperature with 0.66 has an impact on symptom followed by cough. Temperature and bpm have a co-relation of 0.82. It is noted that cough makes almost no co-relation with any variable as its co-relation value is close to zero, which doesn't contribute much to the prediction. The multinomial logistic regression's mean accuracy was too less for cough, so no conclusive evidence was found for it to contribute to the prediction of the symptom. From Fig. 4.9a it is observed that the possibility of getting Covid symptoms is high when body temperature is above 38.5 °C (approx.). When

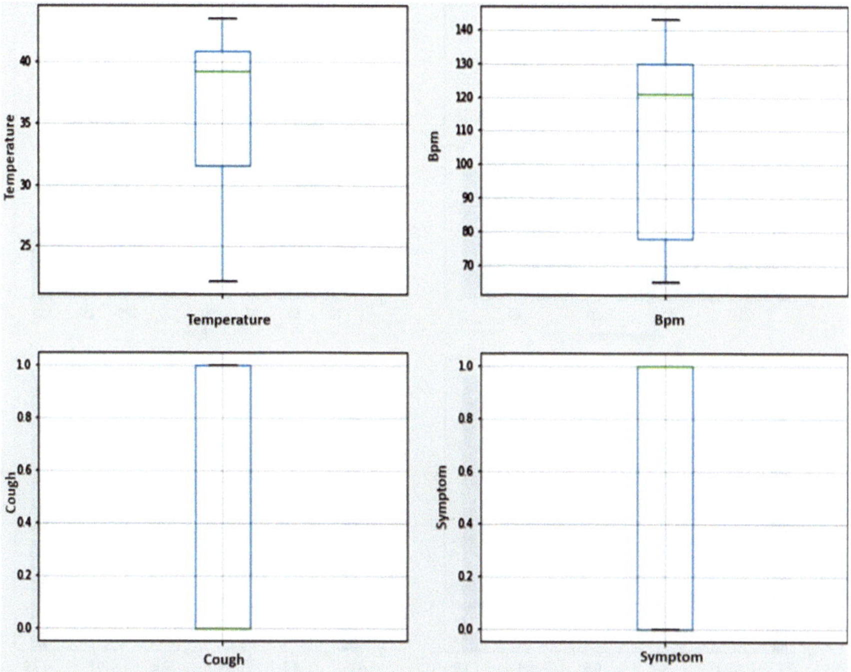

Fig. 4.7 Box plot of the obtained dataset

the body temp is lower than 38.5 °C (approx.) no data represents any symptoms of Covid.

From Fig. 4.9b, it is noted that the possibility of getting Covid symptoms is very high if the bpm is more than 120 levels. However, the probability of getting Covid symptoms is zero as there is no data which leads to such a conclusion. While clustering the temperature versus bpm graph in Fig. 4.10, we get two proper clusters. Red cluster which is not that symptomatic.

The red cluster lies under body temperature 38.5 °C (approx.) and below 120 bpm there is not much data that represents Covid symptoms (references of upper graphs were also taken). While the blue cluster is the most Symptomatic region as it lies in the danger region (i.e., major portions are above 120 bpm and 38.5 °C (approx.).

4.4 Machine Learning Algorithms

Once the labeled dataset is obtained, it is fed into 3 supervised machine learning algorithms, i.e., Classical Logistic Regression, Naive Bayes, and Support Vector Machines (SVM) to train these models and test them on test data and derive model

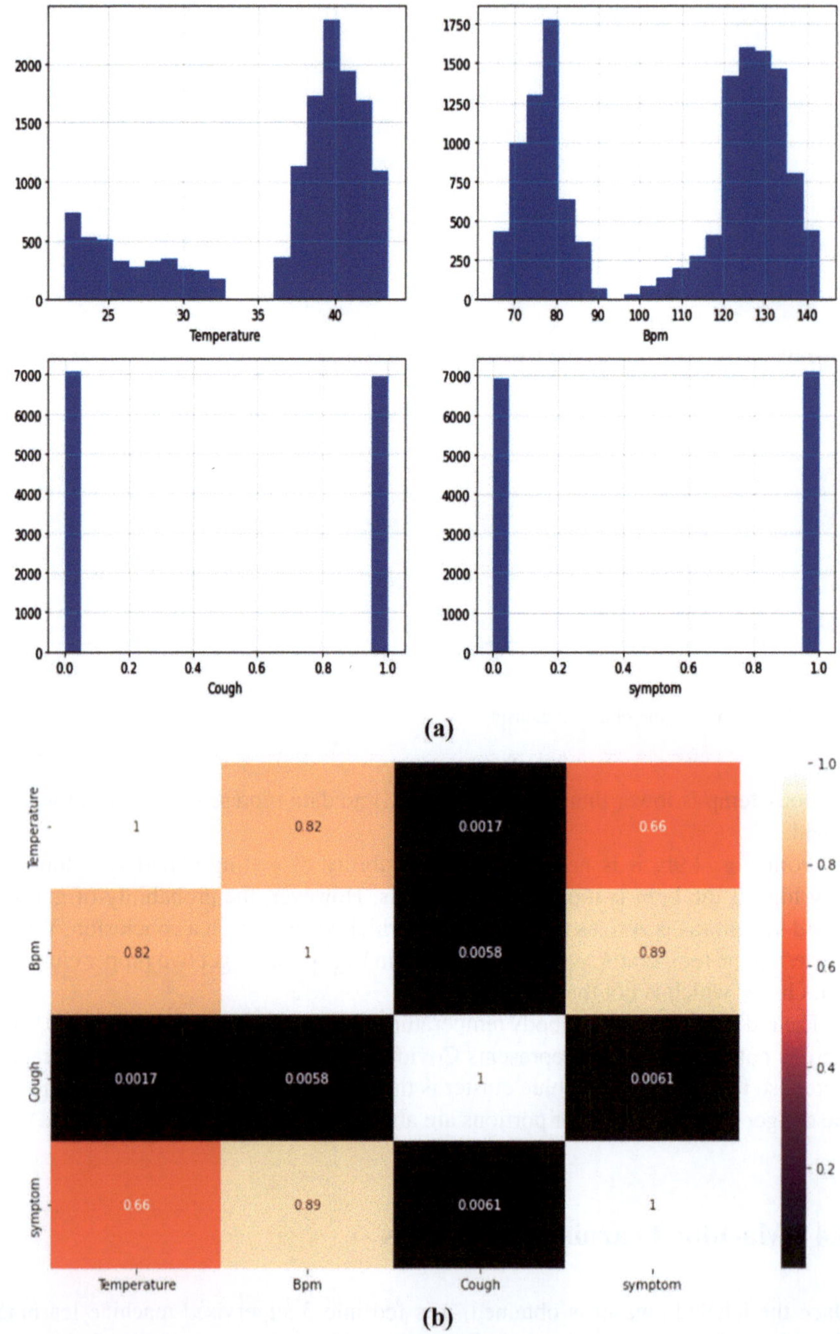

Fig. 4.8 **a** Histogram and **b** Heat map of the obtained dataset

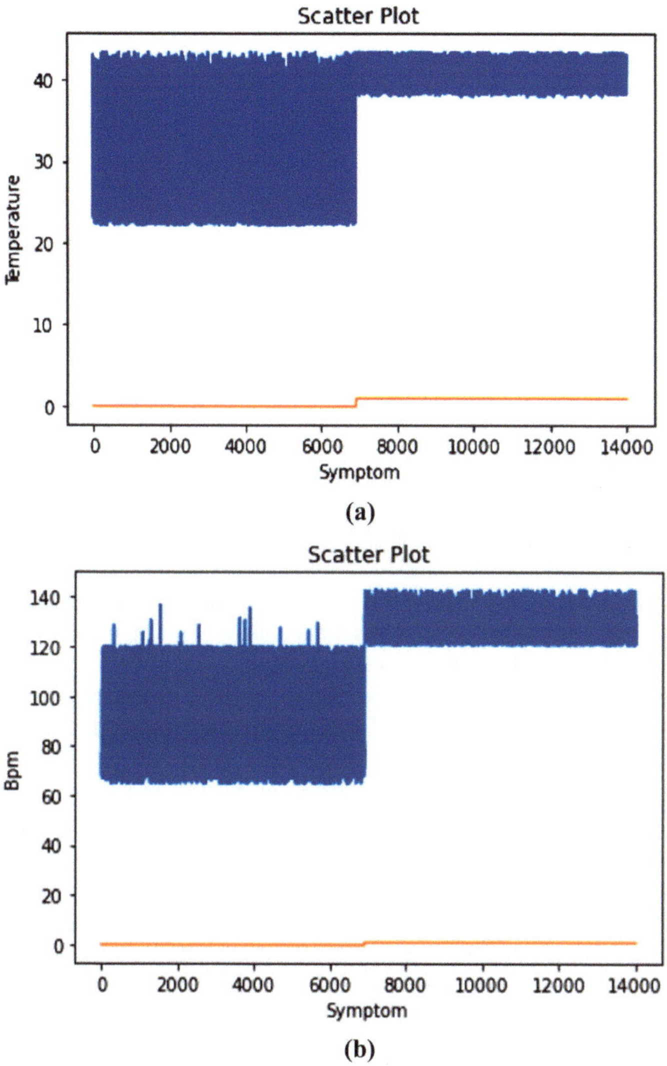

Fig. 4.9 Scatter plot between **a** Temperature and symptom **b** bpm and symptom

performance parameters like accuracy, precision, confusion matrix, recall, etc. These models have specifically been chosen because our project is a type of:

- Supervised learning
- Binary classification
- Independent features
- Huge dataset computation

These algorithms have been discussed in detail in the following sessions.

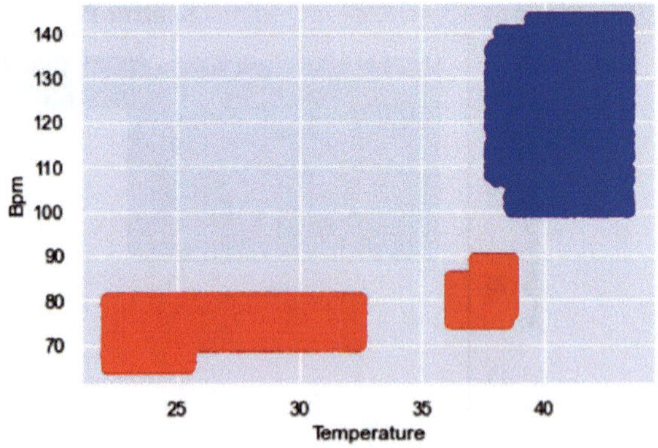

Fig. 4.10 Cluster plot between temperature and bpm

4.4.1 Logistic Regression

Logistic regression or Logit transformation technique in machine learning is used for prediction of classification problems. It is largely used for binary classification problems, that is, to classify the predictions into two different categories (https://www.analyticsvidhya.com/blog/2021/10/building-an-end-to-end-logisticregression-model/).

Here, Binary Logistic regression is used to understand the influence of several independent variables on a single dichotomous outcome variable. It is a parametric model.

The logistic function or sigmoid function can take as input any real value and flatten it to a value ranging from 0 to 1. This is especially useful for binary classification problems (Brownlee 2022).

In this case, the input is classified as 0 if the patient is not having Covid symptoms and 1 if they are affected. Outlier removal becomes essential, as logistic regression assumes no error in the output in the training dataset. Log loss cost function can be used for measuring the error. All the 3 features were used to train the model on the train dataset.

This algorithm is not recommended when the number of observations is less than the number of features as it may lead to over-fitting. Like linear regression, it assumes a linear relationship between the features and target. As the dataset size increases, it becomes harder to detect outliers and possibly misclassified data. For reducing the cost function value (log loss), gradient descent can be used.

4.4.2 Support Vector Machine

Support Vector Machine (SVM) is a supervised learning algorithm used for both regression and classification problems in machine learning. It is mostly and primarily used for classification problems and can hence be used for this application.

SVMs help in creating the best hyperplane or decision boundary to separate n dimensional space into separate classes based on the data points (https://www.analyticsvidhya.com/blog/2017/09/understaing-support-vector-machineexample-code/).

There are several ways to select the best hyperplane, one of them being to select the hyperplane that has the largest separation or margin between the two classes. That is, to choose the line that maximizes the margin or distance between the nearest data points of the two classes and the hyperplane. The closest values to the classification margin are known as support vectors.

New unlabeled samples are classified according to the side of the hyperplane on which they fall. The model also takes a long time to train when there is a high number of training data and features that increase the dimension (https://www.sciencedirect.com/topics/engineering/support-vector-machine). SVC with a linear kernel is used as shown in Fig. 4.11, i.e., the decision boundaries do not curve or try to over-fit the data too much. SVM can use non-linear and linear classification.

Although SVM is a highly used learning algorithm that predicts and classifies with great accuracy (Cervantes 2020), it comes with its own few drawbacks such as assumption that the labels in the train dataset are perfectly labeled. SVMs are widely used for medical predictions, so it is appropriate for this project.

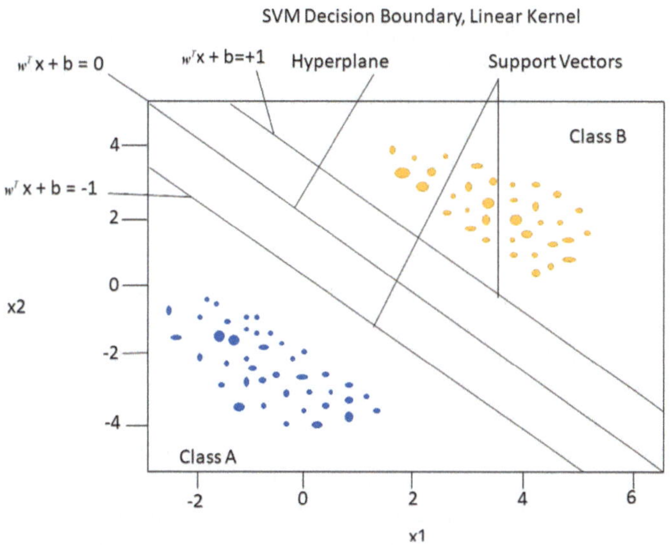

Fig. 4.11 Schematic diagram of a SVM with linear kernel

4.4.3 *Naive Bayes*

Naive Bayes classifier is a classification technique that is based on the famous Bayes theorem in probability which assumes independence among predictors or features. The presence or absence of a feature does not affect other features. This is a very strong assumption that is unlikely in real-life data (Learn Naive Bayes Algorithm 2022). However, the approach performs surprisingly well on data where this assumption does not hold. Naive Bayes is a fast and straightforward algorithm for classification.

Here, we use Gaussian Naive Bayes algorithm, in which segmentation of data by class and computation of mean and variance of the features in every class is observed. If the input variables are real-valued, a Gaussian distribution is assumed. In addition to the probabilities for each class, the mean and standard deviations for each input variable for each class must also be stored. After feeding the input, the Gaussian classifier produces an estimated probability of the new input variable of that class (Brownlee 2022).

Naive Bayes classifiers are used when a fast algorithm to compute through a large dataset quickly is needed (Yang 2018) and if the independent feature assumption is true for most cases, it performs better than any classifier. However, there are some drawbacks such as wrong probability estimates which must not be considered when accuracy is crucial.

4.5 Results

For all the three supervised machine learning algorithms, hold-out method is used to split test set and training set. Training set which is supervised is used to train the model and test set, which is unsupervised, is used to test the model by comparing predicted class vs. actual class. Here, to compare the three models, training data is 75% and testing data is 25% and the division is random.

1. **Logistic Regression**

The Confusion matrix obtained is shown in Fig. 4.12a; it is noted that test data comprises of 3500 data points out of which true positives are 1755, False positive are 1, False negative are 0, True Negatives are 1744. To understand the model better, parameters of the model are calculated. After the logistic regression model is trained to an optimal level, we get the accuracy to be 99.97% on the test set. The Precision is 99.9% = 1 (approx.) and F-measure is 1 and Recall is 1 as shown in Fig. 4.12b.

The reason for such high accuracy is because it is a simple model to learn because of limited number of features, i.e., temperature, bpm, and cough; the model can recall the data quite well without any confusion. Another reason could be small data set for training. If there had been more features, the model complexity would have been increased.

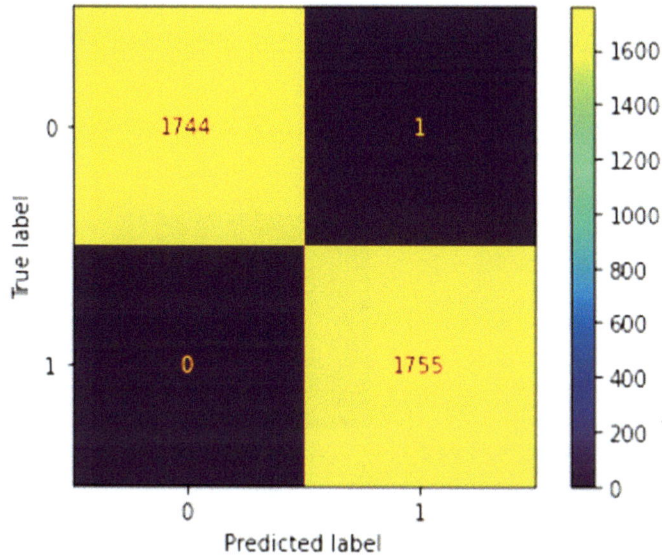

(a)

	precision	recall	f1-score	support
0	1.00	1.00	1.00	1745
1	1.00	1.00	1.00	1755
accuracy			1.00	3500
macro avg	1.00	1.00	1.00	3500
weighted avg	1.00	1.00	1.00	3500

(b)

Fig. 4.12 **a** Confusion matrix logistic regression **b** Classification report of logistic regression

2. **Support Vector Machine (SVM)**

For SVM, again the training: testing set split is 0.75:0.25 and it is a random division. The Confusion matrix obtained is shown in Fig. 4.13a; it is noted that test data comprises of 3500 data points out of which true positives are 2349, False positive are 82, False negative are 0, True Negatives are 2469. Classification report is as shown in Fig. 4.13b it is noted that Training set score is 0.9854, test set score is 0.9833 and classification accuracy is 98.33%. SVM model Precision is 0.9663, Recall or Sensitivity is 1, True Positive Rate is 1.0000, and False Positive Rate is 0.0321 and Specificity is 0.9679.A linear kernel is used and it fits data much better. The advantage of SVM over logistic regression is that outliers are handled much well than in Logistic Regression. It ignores outliers and finds the best hyperplane that separates the data. It is noted that, however, the accuracy is less as compared

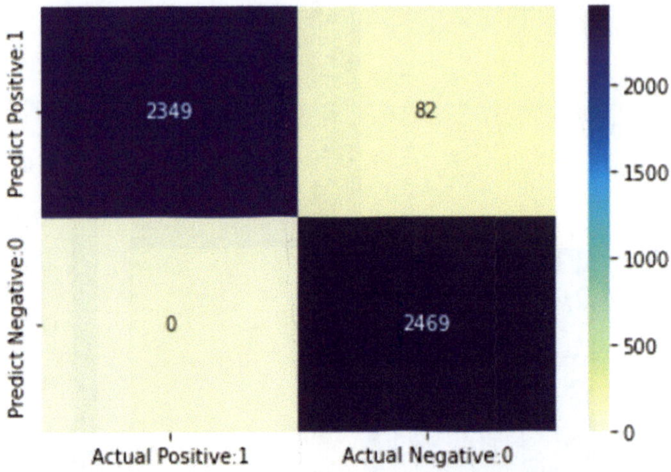

(a)

	precision	recall	f1-score	support
0	1.00	0.97	0.98	2431
1	0.97	1.00	0.98	2469
accuracy			0.98	4900
macro avg	0.98	0.98	0.98	4900
weighted avg	0.98	0.98	0.98	4900

(b)

Fig. 4.13 **a** Confusion matrix SVM **b** Classification report of SVM

to Logistic Regression, but SVM fits the model better when compared to Logistic Regression as it avoids over-fitting.

Classification error is 0.0167. It is calculated as in Eq. (4.8).

$$\text{Classification error} = \frac{FP + FN}{TP + TN + FP + FN} \tag{4.8}$$

3. **Naïve Bayes**

For Naïve Bayes, again the training: testing set split is 0.75:0.25 and it is a random division. A Gaussian or Normal distribution Naïve Bayes model is used. Naïve Bayes is implemented when the target class is a category label and predictors are independent of each other.

The Confusion matrix obtained is shown in Fig. 4.14a; it is noted that Test data comprises of 3500 data points out of which true positives are 1755, False positive are216, False negative are 0, True Negatives are 1529.

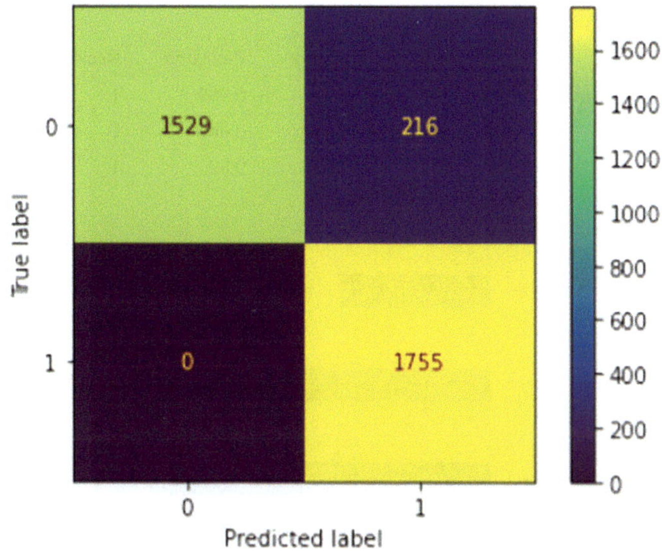

(a)

	precision	recall	f1-score	support
0	1.00	0.88	0.93	1745
1	0.89	1.00	0.94	1755
accuracy			0.94	3500
macro avg	0.95	0.94	0.94	3500
weighted avg	0.95	0.94	0.94	3500

(b)

Fig. 4.14 **a** Confusion matrix Naïve Bayes **b** Classification report of Naïve Bayes

Classification report is as shown in Fig. 4.14b it is noted that Training set score is0.9854, Test set score is 0.9833 and Classification accuracy is 93.8%, Precision is 0.890, Recall is 1.000 and F-Measure is 0.942.

The accuracy is low as compared to the previous two models, but Naïve Bayes learns feature pattern much better than the other two, however the accuracy is low, that only suggests that the model is a better-fit model instead of being an over-fit one like Classic Logistic regression as suggested by Table 4.4.

The accuracy of the three models is plotted in Fig. 4.15. The highest accuracy of 99.7% is achieved in Logistic Regression, followed by SVM with accuracy of 98.3% which is followed by Naïve Bayes with an accuracy of 93.8% as depicted in Fig. 4.18. Logistic regression is over-fitting the model due to less parameter to learn thus less complexity and thus has the highest accuracy. However, SVM has less accuracy when compared to Logistic Regression, but the fit is better.

Table 4.4 Parameters versus models

Model	TP	FP	Accuracy	Precision	Recall	F-measure
Logistic regression	1755	1	0.999	0.999	1	1
SVM	2349	82	0.983	0.966	1	0.985
Naïve Bayes	1755	216	0.938	0.89	1	0.942

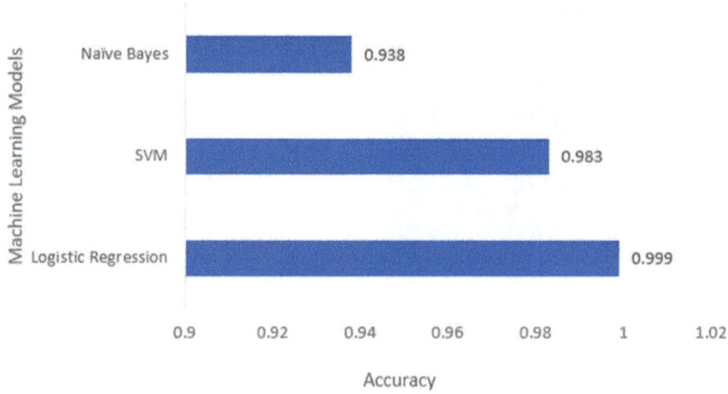

Fig. 4.15 Accuracy versus machine learning models

Naïve Bayes reports a low accuracy compared to the other three, but it is more specific to the pattern in data and labels and thus, a better-fit model.

4.6 Conclusion and Future Work

The COVID-19 pandemic has prompted a drastic loss of human life worldwide and presents an extraordinary challenge to general wellbeing and public health. It is highly contagious and spreads rapidly from one person to another. Therefore, early identification of Corona symptoms is essential to prevent this disease from spreading and infecting others. To address this concern, this paper introduces Design and Performance Analysis of an energy efficient IoT-based WBAN to Control Corona Virus Outbreak.

The IoT based WBAN captures three physical values that are temperature, cough, and heart beats per minute (bpm) and stores them in ThingSpeak cloud using LoRa Technology for transmission. Furthermore, three machine learning algorithms are trained on the obtained dataset to classify subjects as possessing Covid symptom or not. The proposed models have accuracy of 99.7% in Logistic Regression, followed by SVM with accuracy of 98.3% which is followed by Naïve Bayes with an accuracy of 93.8%. These results suggest that the proposed model can be used for classification of a person possessing Covid symptom or not. It should be noted that the model does not classify a person as a Covid patient or otherwise.

Therefore, the model can be used for screening purposes only and if a person is found to have Covid symptoms, they should seek for medical help immediately. Power plays a significant role in IoT setups, and it should be minimized. To do so, LoRa technology is used for data transmission. The sensors chosen to consume less power as well and thus make this model energy efficient.

In the future, various other machine learning algorithms can be applied to best fit the model. SVM can be implanted with other kernels like polynomial kernel and Gaussian kernel for improved results. The features and the size of the dataset can be expanded. The model parameters like accuracy and precision can be improved by applying techniques to build an accurate model.

Compliance with Ethical Standards

Conflict of Interest The authors declare that there is no conflict of interest.

References

2022 [Online]. Available: https://www.farnell.com/datasheets/1682209.pdf. Accessed 22 Mar 2022

2022 [Online]. Available: https://www.analyticsvidhya.com/blog/2017/09/understaing-support-vector-machineexample-code/. Accessed 06 Apr 2022

2022 [Online]. Available: https://www.analyticsvidhya.com/blog/2021/10/building-an-end-to-end-logisticregression-model/. Accessed 06 Apr 2022

2022 [Online]. Available: https://www.mathworks.com/hardware-support/thingspeak.html. Accessed 31 Mar 2022

2022 [Online]. Available: https://www.sciencedirect.com/topics/engineering/support-vector-machine. Accessed 06 Apr 2022

Coronavirus (2022) Who int. Available: https://www.who.int/health-topics/coronavirus#tab=tab_. Accessed 20 Mar 2022

ESP32 WiFi & Bluetooth MCU I. EspressifSystems (2022) Espressif.com [Online]. Available: https://www.espressif.com/en/products/socs/esp32. Accessed 22 Mar 2022

Learn Naive Bayes Algorithm | Naive Bayes Classifier Examples (2022) Analytics Vidhya [Online]. Available: https://www.analyticsvidhya.com/blog/2017/09/naive-bayes-explained/. Accessed 06 Apr 2022

Pulse Sensor (2022) Components 101 [Online]. Available: https://components101.com/sensors/pulse-sensor. Accessed 22 Mar 2022

Semtech Acquires Wireless Long Range IP Provider Cycleo (2022) Design and reuse [Online]. Available: https://www.design-reuse.com/news/28706/semtech-cycleo-acquisition.html. Accessed 22 Mar 2022

Software (2022) Arduino.cc [Online]. https://www.arduino.cc/en/software. Accessed 31 Mar 2022

SX1272 LoRa module for Arduino, Raspberry Pi and Intel Galileo - 868 MHz (2022) Cooking-hacks.com. [Online]. Available: https://www.cookinghacks.com/sx1272-lora-module-for-arduino-raspberry-pi-intel-galileo-868-mhz.html. Accessed 22 Mar 2022

thingspeak/README textile at master iobridge/thingspeak (2022) GitHub [Online]. Available: https://github.com/iobridge/thingspeak/blob/master/README.textile. Accessed 31 Mar 2022

Wayback Machine (2022) Web.archive.org [Online]. Available: https://web.archive.org/web/20190718200516/https://www.semtech.com/uploads/documents/an1200.22.pdf. Accessed 22 Mar 2022

Wireless Body Area Network (WBAN) (2022) Applications and challenges. Techsparks.co.in [Online]. Available: https://www.techsparks.co.in/thesis-in-wireless-body-area-network/. Accessed 21 Mar 2022

Adelantado F, Vilajosana X, Tuset-Peiro P, Martinez B, Melia-Segui J, Watteyne T (2017) Understanding the Limits of LoRaWAN. IEEE Commun Mag 55(9):34–40. https://doi.org/10.1109/mcom.2017.1600613

Bilandi N, Verma HK, Dhir R (2021) An intelligent and energy-efficient wireless body area network to control coronavirus outbreak. Arab J Sci Eng 46(9):8203–8222. https://doi.org/10.1007/s13369-021-05411-2

Brownlee J (2022) Naive Bayes for machine learning. In: Machine learning mastery [Online]. Available: https://machinelearningmastery.com/naive-bayes-for-machine-learning/. Accessed 06 Apr 2022

Brownlee J (2022) Logistic regression for machine learning. In: Machine learning mastery [Online]. Available: https://machinelearningmastery.com/logistic-regression-for-machine-learning/

Carpentier L, Berckmans D, Youssef A, Berckmans D, van Waterschoot T, Johnston D, Ferguson N, Earley B, Fontana I, Tullo E, Guarino M, Vranken E, Norton T (2018) Automatic cough detection for bovine respiratory disease in a calf house. Biosys Eng 173:45–56. https://doi.org/10.1016/j.biosystemseng.2018.06.018

Cervantes J, Garcia-Lamont F, Rodríguez-Mazahua L, Lopez A (2020) A comprehensive survey on support vector machine classification: applications, challenges and trends. Neurocomputing 408(1):189–215. https://doi.org/10.1016/j.neucom.2019.10.118

Chuma EL, Iano Y (2021) A movement detection system using continuous-wave Doppler radar sensor and convolutional neural network to detect cough and other gestures. IEEE Sens J 21(3):2921–2928. https://doi.org/10.1109/jsen.2020.3028494

COVID-19 Data Repository by the Center for Systems Science and Engineering (CSSE) at Johns Hopkins University. Available: https://github.com/CSSEGISandData/COVID-19. Accessed 17 Apr 2022

Cypher D, Chevrollier N, Montavont N, Golmie N (2006) Prevailing over wires in healthcare environments: benefits and challenges. IEEE Commun Mag 44(4):56–63. https://doi.org/10.1109/MCOM.2006.1632650

Datasheets.maximintegrated.com (2022). Available: https://datasheets.maximintegrated.com/en/ds/MAX30205.pdf

Eljamassi DF, Maghari AY (2020) COVID-19 detection from chest X-ray scans using machine learning. In: 2020 international conference on promising electronic technologies (ICPET). https://doi.org/10.1109/icpet51420.2020.00009

Ghosh AM, Halder D, Hossain SKA (2016) Remote health monitoring system through IoT. In: 2016 5th international conference on informatics, electronics and vision (ICIEV), pp 921–926. https://doi.org/10.1109/ICIEV.2016.7760135

van der Hoek L, Pyrc K, Jebbink MF, Vermeulen-Oost W, Berkhout RJM, Wolthers KC, Wertheim-van Dillen PME, Kaandorp J, Spaargaren J, Berkhout B (2004) Identification of a new human coronavirus. Nat Med 10(4):368–373. https://doi.org/10.1038/nm1024

Hu Z, Song C, Xu C, Jin G, Chen Y, Xu X, Ma H, Chen W, Lin Y, Zheng Y, Wang J, Hu Z, Yi Y, Shen H (2020) Clinical characteristics of 24 asymptomatic infections with COVID-19 screened among close contacts in Nanjing, China. Sci China Life Sci 63(5):706–711. https://doi.org/10.1007/s11427-020-1661-4

Jiang F, Deng L, Zhang L, Cai Y, Cheung CW, Xia Z (2020) Review of the clinical characteristics of coronavirus disease 2019 (COVID-19). J Gen Intern Med 35(5):1545–1549. https://doi.org/10.1007/s11606-020-05762-w

Johnson RM, Vinetz JM (2020) Dexamethasone in the management of Covid-19. BMJ m2648. https://doi.org/10.1136/bmj.m2648

Jubran A (1999) Pulse oximetry. Crit Care 3(2):R11–R17. https://doi.org/10.1186/cc341
Kumar R, Rajasekaran MP (2016) An IoT based patient monitoring system using raspberry Pi. In: 2016 international conference on computing technologies and intelligent data engineering (ICCTIDE'16), pp 1–4. https://doi.org/10.1109/ICCTIDE.2016.7725378
Latré B, Braem B, Moerman I, Blondia C, Demeester P (2010) A survey on wireless body area networks. Wirel Netw 17(1):1–18 (2010). https://doi.org/10.1007/s11276-010-0252-4
Lescure FX, Bouadma L, Nguyen D, Parisey M, Wicky PH, Behillil S, Gaymard A, Bouscambert-Duchamp M, Donati F, Le Hingrat Q, Enouf V, Houhou-Fidouh N, Valette M, Mailles A, Lucet JC, Mentre F, Duval X, Descamps D, Malvy D, Timsit JF, Lina B, van-der-Werf S, Yazdanpanah Y (2020) Clinical and virological data of the first cases of COVID-19 in Europe: a case series. Lancet Infect Dis 20(6): 697–706. https://doi.org/10.1007/s11276-010-0252-4
Mary AV, Jerine S (2020) Wireless body area network transmissions for iot-based healthcare network: a review. IOP Conf Ser Mater Sci Eng 983(1):012017. https://doi.org/10.1088/1757-899x/983/1/012017
Mir MH, Jamwal S, Mehbodniya A, Garg T, Iqbal U, Samori IA (2022) IoT-enabled framework for early detection and prediction of COVID-19 suspects by leveraging machine learning in cloud. J Healthc Eng 2022:1–16. https://doi.org/10.1155/2022/7713939
Muhammad LJ, Islam MM, Usman SS, Ayon SI (2020) Predictive data mining models for novel coronavirus (COVID-19) infected patients' recovery. SN Comput Sci 1(4). https://doi.org/10.1007/s42979-020-00216-w
Mukhtar H, Rubaiee S, Krichen M, Alroobaea R (2021) An IoT framework for screening of COVID-19 using real-time data from wearable sensors. Int J Environ Res Public Health 18(8):4022. https://doi.org/10.3390/ijerph18084022
Prakash KB (2020) Analysis, prediction and evaluation of COVID-19 datasets using machine learning algorithms. Int J Emerg Trends Eng Res 8(5):2199–2204. https://doi.org/10.30534/ijeter/2020/117852020
Preethi S, Revathi A, Murugan M (2020) Exploration of cough recognition technologies grounded on sensors and artificial intelligence. In: Internet of medical things for smart healthcare. Springer, Berlin, Germany, pp 193–214. https://doi.org/10.1007/978-981-15-8097-0_8
Raeesi Vanani I, Amirhosseini M (2020) IoT-based diseases prediction and diagnosis system for healthcare. Internet of Things Healthc Technol 21–48. https://doi.org/10.1007/978-981-15-4112-4_2
Rahaman A, Islam M, Islam M, Sadi M, Nooruddin S (2019) Developing IoT based smart health monitoring systems: a review. Revue d'Intelligence Artificielle 33(6):435–440. https://doi.org/10.18280/ria.330605
Vangelista L, Zanella A, Zorzi M (2015) Long-range IoT technologies: the dawn of LoRa™ future access enablers for ubiquitous and intelligent infrastructures, pp 51–58. https://doi.org/10.1007/978-3-319-27072-2_7
Wu F, Wu T, Yuce MR (2019) Design and implementation of a wearable sensor network system for IoT-connected safety and health applications. In: 2019 IEEE 5th World Forum on Internet of Things (WF-IoT). https://doi.org/10.1109/wf-iot.2019.8767280
Yamanoor NS, Yamanoor S (200) Low-cost contact thermometry for screening and monitoring during the COVID-19 pandemic. In: 2020 IEEE international IOT, electronics and mechatronics conference (IEMTRONICS). https://doi.org/10.1109/IEMTRONICS51293.2020.9216444
Yang FJ (2018) An implementation of Naive Bayes classifier. IEEE Xplore. https://ieeexplore.ieee.org/abstract/document/8947658?casa_token=KwN6IvFNIJwAAAAA:YB-P9ry6SakWDi-_vX0HxBAALWDxHL3G6EqtD-Q4BKwUt_YxB0OTB33uDzw52eD4ZenCaK64aA
Yogaraj A, Ezilarasan M, Anuroop R, Sivanthiram C, Thakur SK (2017) IOT based smart healthcare monitoring system for rural/isolated areas. Int J Pure Appl Math 114:679–688
Zhang J, Luo Y, Jiang Z, Tang X (2017) Regression analysis and prediction of mini-mental state examination score in Alzheimer's disease using multi-granularity whole-brain segmentations. In: Lecture notes in computer science, pp 202–213. https://doi.org/10.1007/978-3-319-67964-8_20

Chapter 5
Sentiment Analysis During COVID-19 Using Machine-Learning Techniques

Harsh Arora, Neha Singh, and Ritesh Pratap Singh

Abstract Beginning in late 2019, the COVID-19 pandemic had a significant effect on people and communities all across the world, influencing not only public health but also a number of societal facets. Sentiment analysis, a branch of natural language processing, was essential in helping to comprehend how people's feelings and perspectives changed throughout this world crisis. Using machine-learning techniques, this research provides a thorough summary of sentiment analysis during the COVID-19 pandemic. Large-scale textual data from social media, news stories, and public forums were gathered and analyzed for this study. To identify the sentiment in these texts, a variety of supervised and unsupervised machine-learning techniques were used. To identify emotions like fear, anxiety, hope, and disinformation, sentiment analysis models were developed and put to the test. The study also looked at the difficulties brought up by the talks surrounding COVID-19 being so dynamic and the requirement for ongoing model modification. The results show that sentiment has changed over time, mostly in line with how the virus has spread. There was a noticeable spike in anxiety and panic early in the pandemic, but as society adjusted to the new normal, these feelings progressively gave way to hope and fortitude. The study also highlights the significance of sentiment analysis in tracking and preventing the spread of misleading information and identifies disinformation as a major threat. The study's findings show how effective machine learning and deep learning techniques are in monitoring and interpreting public opinion in times of international emergency. Researchers, legislators, and public health authorities can

H. Arora
Banarsidas Chandiwala Institute of Information Technology, Kalkaji, New Delhi, India
e-mail: drharsh.cse@gmail.com

N. Singh (✉)
Department of Electronics and Communication Engineering, GCRG Group of Institution, Lucknow, UP, India
e-mail: snehjai2000@gmail.com

R. P. Singh (✉)
School of Electrical and Computer Engineering, Haramaya Institute of Technology, Haramaya University, Dire Dawa, Ethiopia
e-mail: ritesh.pratap@haramaya.edu.et

A. Rai et al. (eds.), *Convergence of Artificial Intelligence, Machine Learning, and the Internet of Things in Industry 4.0 Applications*, Transactions on Computer Systems and Networks, https://doi.org/10.1007/978-981-96-5269-3_5

use this information to help them create successful communication plans and make well-informed decisions. Sentiment analysis is still a useful instrument for assessing public opinion, responding to issues, and building resilience in the face of hardship as the globe struggles with the COVID-19 pandemic.

Keywords COVID-19 · Sentiment analysis · Machine learning · Deep learning · BERT · Random Forest · SVM

5.1 Introduction

In addition to posing a threat to world health, the COVID-19 pandemic has sparked a boom in internet communication, enabling people to freely express their feelings and thoughts on a variety of online venues. This research uses a combination of deep learning and machine-learning approaches to look at sentiment analysis during the COVID-19 pandemic. The research uses a data source from the social media. The twitter COVID-19 dataset has been taken from Kaggle for the research. This research investigates the efficacy of conventional machine-learning models and the sophisticated deep learning architecture of the BERT Model in comprehending the subtleties of public opinion within this unparalleled crisis.

The globe has changed as a result of the COVID-19 epidemic, impacting people, communities, and countries. Digital channels have made it possible for information and opinions to be widely disseminated, which gives real-time public mood measurement. It is critical for crisis communication, policy-making, and public health solutions to comprehend the emotional terrain of this disaster. In order to glean valuable insights from textual data, this work explores sentiment analysis during COVID-19 by utilizing both machine learning and deep learning techniques.

Just when we were beginning to take comfort in the wealth and advancements of the contemporary world, humanity was confronted with the once-in-a-century epidemic, COVID-19, which spread the unique coronavirus across the globe. We are currently in a world where microblogging has facilitated the greatest amount of human companionship. Individuals have developed a practice of posting everything and everything on social media sites like Facebook, Twitter, Tumblr, and others. Thus, the first global pandemic to strike humanity after social networking became popular has a unique aspect to communication and knowledge sharing. It's crucial to pay attention to sudden shifts in people's perspectives and emotions in addition to the economic and social statistics of those impacted. Sentiment analysis is the study of people's opinions, feelings, and perspectives expressed in text on trends, products, current events, and even social issues. While it is absurd to believe that machines can comprehend natural language in its entirety, statistical analysis can be used to classify attitudes.

Sentiment analysis was employed during the COVID-19 pandemic with the goal of elucidating the dynamic sentiments that people expressed at different stages of the crisis. Sentiment analysis offers a lens through which we may understand the changing public opinions, from the early dread and uncertainty to optimism based on vaccination efforts. In order to provide a thorough knowledge of sentiment trends during the epidemic, our research integrates both conventional machine-learning models and deep learning techniques.

This chapter attempts to investigate sentiment analysis during the COVID-19 epidemic in detail, focusing on both deep learning and machine-learning techniques. Dataset from social media platform has been included in the scope. The goal is to analyze dataset and the corresponding people opinions by using deep learning and classical machine-learning models to capture feelings, to compare the performance of ML models with BERT Deep learning Model in order to show that deep learning works better on dataset in terms of accuracy.

The chapter progresses through several phases, beginning with general process of the sentiment analysis. We investigate the use of deep learning and machine-learning models for sentiment analysis in the context of COVID-19. The dataset and models are all described in the further sections. The analysis's conclusions and new perspectives are presented in the results and discussion section. A review of the major conclusions, their ramifications, and potential directions for further investigation round out the study.

This chapter's primary goal is to examine user tweets from the COVID-19 epidemic, categorize them into positive, neutral, and negative groups, and compare the outcomes of several classifiers that were used to the dataset.

5.1.1 Objectives

The following are the main aims of this research:

1. Examining conventional techniques for Machine Learning: Here, we examine the use of Support Vector Machine and Random Forest in sentiment analysis during the COVID-19 pandemic.
2. Evaluation of complex Deep Learning Methods: we analyze how well the BERT model is which is renowned for its contextual understanding and captures complex emotions associated with COVID-19.
3. Examine and contrast methods: Here we determine each methodology's advantages, disadvantages, and suitability for the sentiment analysis in the context of a global health emergency.

5.1.2 Related Work

The spread of misleading propaganda through internet platforms is becoming a global problem. Although fake news is not essential, its prevalence in online media, which fosters collaboration and the sharing of innovative ideas, is concerning at the moment (Domenico et al. 2021). The COVID-19 pandemic crisis demonstrates the beneficial outcomes of new data creation. The dissemination of incorrect information has the potential to significantly alter people's perspectives and the efficacy of the defenses that governments advocate (Apuke and Omar 2021). Social media misinformation can make COVID-19 patients feel anxious and fearful, which alerts public authorities and local governments to the need for people to verify the veracity of stories that are making the rounds (Zaman 2021). A few studies published have been about the veracity of information and how people behave while interacting with COVID-19-related social media news. A large-scale individual evaluation of the address was provided by the COVID-19 information dissemination with big data analysis on popular social media platforms, allowing for an investigation of pandemic rumors (Depoux et al. 2020). Twitter is a well-known microblogging site and social media network where users can publish and distribute messages known as "tweets." In the years since Twitter has emerged as a significant data hub for online media conversations related to public and international events, there have been roughly 500 million tweets posted per day and 200 billion tweets annually (Gao et al. 2020). Unfortunately, the propagation of false information is also a big factor in the global fear that results from it. The majority of COVID-19 tweets, according to Chakraborty K et al. 2020, are favorable in attitude; nonetheless, users regularly share negative tweets, and the word frequency computation of tweets yielded few beneficial words (Ahmad and Murad 2020). According to Shadi et al. (2020), the current epidemic spreading rumors and conspiracy theories quickly is akin to a virus that causes dramatic real-world events (Cinelli et al. 2020). The authors of this paper describe narrative structures that allow such misleading propaganda and provide machine-learning algorithms for automatic detection. Additionally, research has been done on the consequences of misrepresenting COVID-19 data and the political philosophies that eventually compromised public health (Twitter 2020).

Platforms including Twitter, Facebook, and YouTube were mentioned in the Nigerian study on the COVID-19 pandemic's impact on social media, and this information should be paired with a strategy for data distribution. It seems that these platforms have been manipulated by people who hide behind their privacy in order to disseminate false information and incite fear among the general public (Chakraborty et al. 2020). It is even worse when it comes to the online dissemination of fake news in highly populated Infect. Dis. Rep. 2021, 13 331 nations like India, which poses a possible risk to public health. In order to handle the COVID-19 difficulties, India announced a national lockdown on March 24, 2020, and alerted public health specialists. Nevertheless, these publications do not go into great detail about the veracity and correctness of tweets. In response, by analyzing 125 instances of fake news from India, the study on social media false propaganda on COVID-19 during the lockdown revealed seven themes of fake news, including health, political, criminal, entertainment, religious, religiopolitical, and miscellaneous (Shahsavari et al. 2020). Bakur et al. (2020) carried out research on sentiment analysis of Indian citizens during the countrywide shutdown. There aren't many tweets in the dataset utilized for this research from March 25, 2020, to March 28, 2020.The outcomes demonstrated the Indian government's effective lockdown approach (Havey 2020). Huynh 2020 noted that there are a lot of rumors and phony articles about COVID-19 spreading on social media, and it's becoming harder to tell the difference between true news and false information. However, the veracity or authenticity of the reports hasn't been questioned (Obi-Ani et al. 2020). Social media users, medical professionals, and governments can all benefit from the legitimacy and veracity of these fakes in reducing unwarranted mental stress. These protocols provide a general explanation for any case of ordinary occurrences and easily mesh with the viewpoint of conspiracy theories (Barkur et al. 2020). Therefore, we attempted deep-learning type machine-learning models and validated this model performance by comparing with another three conventional models in order to address this research gap in the literature on the tweet validation.

5.2 Sentiment Analysis and Natural Language Processing

Opinion mining, or sentiment analysis, is a computational technique that includes identifying the sentiment or emotional tone conveyed in a document. Determining whether the expressed sentiment is neutral, negative, or positive is the main objective.

5.2.1 Purpose of Sentiment Analysis

(a) Polarity classification: Giving a text passage a sentiment polarity (positive, negative, or neutral).
(b) Emotion Detection: Recognizing particular feeling such as happiness, rage, grief, etc. expressed in the text.
(c) Subjectivity Analysis: Assessing the text's degree of objectivity or subjectivity.

The goal of the artificial intelligence (AI) subfield of natural language processing is to empower computers to comprehend, interpret, and produce meaningful, contextually relevant human language. Sentiment analysis is based on Natural Language Processing (NLP), and comprehending and interpreting human language depends on the interaction between the two. One particular usage of NLP is sentiment analysis, which aims to identify the sentiment or emotional tone that is communicated in a text. The field of computer science known as Natural Language Processing, or more precisely Artificial Intelligence, is focused on enabling computers to comprehend spoken and written language in a manner that is similar to that of humans.

5.2.2 NLP and Sentiments

NLP blends statistical, machine learning, and deep learning models with computational linguistics, which models human language using rules. When combined, these technologies allow computers to process textual or audio data that represents human language and to fully "understand" the meaning of the content, including the speaker's or writer's intent and sentiment.

Computer algorithms that interpret spoken commands, translate text between languages, and quickly summarize vast amounts of material—even in real time—are powered by natural language processing (NLP). You've probably dealt with natural language processing (NLP) in the form of voice-activated GPS units, digital assistants, speech-to-text dictation software, chatbots for customer support, and other everyday conveniences. However, NLP is also becoming more and more integrated into workplace solutions that assist improve worker productivity, simplify mission-critical business processes, and streamline corporate operations.

The first NLP applications were hand-coded, rules-based systems that could handle a limited amount of NLP jobs but were unable to grow to handle the growing amounts of text and speech data or an apparently never-ending stream of exceptions.

Statistical natural language processing (NLP) automatically extracts, categorizes, and labels items from text and audio data and then assigns a statistical likelihood to each possible interpretation of those elements. It does this by combining computer algorithms with machine-learning and deep learning models. Convolutional neural networks (CNNs) and recurrent neural networks (RNNs) are the foundation of deep learning models and learning strategies that allow natural language processing (NLP) systems to "learn" as they operate and extract ever-more-accurate meaning from massive amounts of unstructured, raw, and unlabeled text and speech datasets (Twitter 2020).

NLP is used specifically in sentiment analysis. Sentiment analysis is the specialized technique of identifying the emotional tone or sentiment present in natural language, whereas natural language processing (NLP) covers a wider variety of activities linked to language understanding and processing. Sentiment analysis is the Holy Grail for the company. Sentiment analysis gives your insight into the thoughts and emotions of your clients and, in the end, produces useful information that improves your ability to assist them.

NLP is essential for feature extraction in sentiment analysis models because it gives the models the verbal context, they need to forecast sentiment accurately. We can understand this with the following technique.

(a) Utilizing NLP Techniques: To efficiently preprocess and analyze textual input, sentiment analysis frequently makes use of a variety of NLP techniques, including named entity recognition, tokenization, and part-of-speech tagging.
(b) Semantic Understanding: NLP approaches let texts be understood semantically, which makes it possible for sentiment analysis models to capture not just explicit sentiment but also subtleties and implicit meanings.

Sentiment analysis exemplifies the multidisciplinary nature of language-related computational tasks in artificial intelligence, as it is essentially a specialized application within Natural Language Processing.

5.3 Sentiment Analysis in Healthcare Sector and Pandemics

Sentiment analysis is based on Natural Language Processing (NLP), and comprehending and interpreting human language depends on the interaction between the two. One particular usage of NLP is sentiment analysis, which aims to identify the sentiment or emotional tone that is communicated in a text.

Fundamental principles have been developed over the last few decades, even though SA applications in the health sphere have gained substantial relevance during and after the COVID-19 epidemic. A thorough explanation of the most popular methods and algorithms utilized in numerous SA applications with a health focus is provided in the current Section. We identify the key components and their relationships in the creation of a health-related SA application based on the contributions detailed in the literature.

Sentiment analysis helps you pinpoint departments that are performing well—as well as those that might be falling short when it comes to providing your patients with healthcare. Sentiment analysis uses your patients' emotions and opinions about the care they receive to create a general picture of your healthcare institution.

Sentiment analysis concerning COVID-19 and health includes analyzing the opinions that people, groups, medical professionals, and the general public have voiced about the pandemic. Sentiment analysis offers important information into how the public feels, thinks, and feels about different parts of the COVID-19 pandemic. The following is the application of sentiment analysis to COVID-19 and health:

(a) Public Opinion on social media: This Examines the opinions about COVID-19 posted on social media sites like Facebook, Twitter, and others19. The utilization of this aspect is:

    i. Being aware of the worries, fears, and attitudes of the public.
    ii. Recognizing trends and new problems immediately.
    iii. Evaluating public health communication's efficacy.

(b) Sentiment of Healthcare Workers: This examines the opinions of healthcare professionals via social media, questionnaires, and feedback. This will keep an eye on medical practitioners' wellbeing. It determines obstacles, pressures, and support requirements. It becomes easy to customize the support initiatives using sentiment analysis.
(c) Patient Attitudes around Telehealth: It Examines attitudes toward telehealth services provided during the epidemic in patient evaluations and comments. It optimizes the telehealth services by incorporating patient feedback. It is about of taking care of issues and raising patient contentment.
(d) Public health communication analysis: This helps to assess the attitudes reflected in public health communication tactics and messaging. The idea is to use sentiment analysis to modify communication strategies. It enhances the efficacy and lucidity of public health communications.

In the context of COVID-19 and health, sentiment analysis offers a sophisticated understanding of public sentiment that helps decision-makers, legislators, and medical experts respond appropriately, address issues, and customize treatments to the requirements of the community.

5.4 Machine-Learning Techniques for Sentiment Analysis

The field of machine learning is expanding rapidly both in the IT industry and across several commercial domains. Despite being in its infancy, machine learning is widely used in all fields of technology. This area of research enables computers to automatically pick up new skills and improve with practice. As a result, machine learning concentrates on the potency of computer programs by gathering information from diverse observations. Through the use of artificial intelligence, machine learning allows a computer or other device to anticipate the future by learning from its past experiences, or input data. Such a technology should operate at least as well as a human. Three general categories apply to machine learning: Reinforcement learning, supervised learning, and unsupervised learning.

The Fig. 5.1 shows the work flow of sentiment analysis showing machine learning and deep learning methods on the same dataset of COVID-19 taken from Kaggle that is basically twitter dataset. Text classification has been done by using machine learning and deep learning models. Here, in classical machine-learning models, two models have been executed on dataset that are Random Forest (RF) Algorithm and Support Vector Machine (SVM) Algorithm. In deep learning model, BERT (Bidirectional Encoder Representations from Transformers) Model has been used to evaluate the data of COVID-19 for sentiment analysis. Twitter data of COVID-19 has been taken as input that undergoes pre-processing phase in order to clean the data by doing tokenization, lemmatization and stemming. Further analysis and processing have been done by implementing the Random Forest, Support Vector Machine and BERT models differently but on the same dataset.

5.4.1 Machine Learning in Context of COVID-19

The field of machine learning (ML) within artificial intelligence (AI) is concerned with developing algorithms that allow computers to automatically create complicated correlations or patterns from empirical data without the need for explicit programming (Chakraborty et al. 2020). Creating systems that can automatically perform better on a given task with experience or data exposure is the aim of machine learning. By taking inspiration from biological neural networks, deep learning (DL), a subcategory of machine learning (ML), achieves great power and flexibility compared to conventional ML models to solve a wide variety of complex tasks, including natural

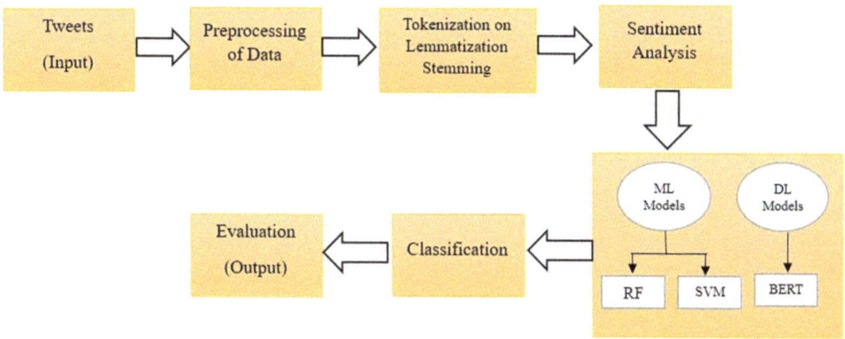

Fig. 5.1 Work flow of sentiment analysis using machine learning and deep learning models

language processing (NLP) and medical imaging classification (Shahsavari et al. 2020).

AI has been used in the healthcare industry on a variety of levels, from developing diagnostic and prognostic models to forecasting the course of disease transmission (Havey 2020). In order to combat the pandemic, a study by Ye et al. analyzed and ranked a number of health technologies. These included big data, cloud computing, mobile health, and artificial intelligence. Disease prediction has made use of these technologies as well as a variety of data types, such as information from social media, radiological images, omics, pharmacological databases, and public health organizations. Machine-learning models can be used for a variety of tasks in the context of COVID-19 analysis, including diagnosing instances, predicting the virus's transmission, evaluating sentiment, and allocating resources optimally.

5.4.2 *Random Forest Machine-Learning Technique*

In Random Forest, an ensemble learning technique, several decision trees are constructed and their predictions are combined. We train each tree on a different arbitrary percentage of the data and then add their predictions together to get the final result. Several aspects of COVID-19 analysis make Random Forest a good fit:

(a) Predictive Modeling: Random Forest may be used to anticipate the incidence of COVID-19 cases, assess the repercussions on healthcare systems, and project trends.
(b) Feature Importance: It offers insights into the significance of various variables (e.g., elements affecting the dissemination) in formulating forecasts.
(c) Robustness: Random Forest is robust against overfitting and noise, making it suitable for dealing with complex datasets.

A random forest is a supervised machine-learning classifier that consists of a structure resembling a tree $\{h(x, (k)\ k = 1, 2, \ldots\}$, distinct unbiased vector $\{\theta(k)\}$,

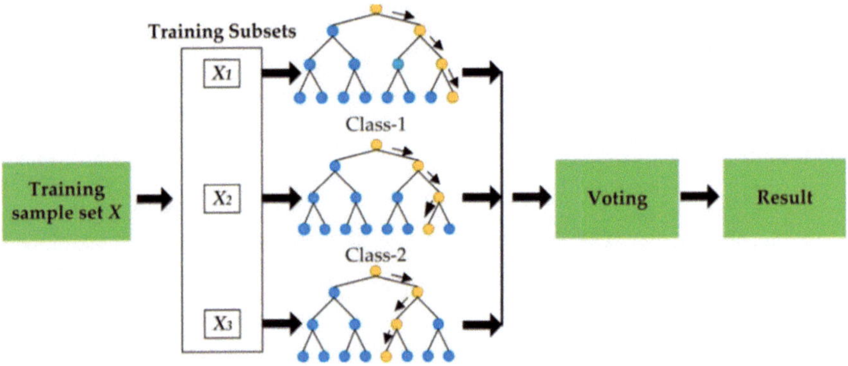

Fig. 5.2 Random Forest model for text classification

and input for the most renowned class of x [[36], [37], [38]]. In the random forest algorithm, researcher Breiman implemented the following procedures to generate each individual tree. During the bootstrap test, N records are randomly analyzed using replacement from the initial data in N preparation datasets. This represents the first phase of tree development. If M input factors are available, the value m << M is selected with the intention of selecting m factors at random from M at each node, and using the best split of these m credits to a portion of the node. The estimate of m remains constant throughout forest growth. Each tree is evolved to the maximum extent of divergence. In this context, many trees are activated in the forest, with the number of trees set by the parameter Ntree. In the literature, the number of components (m) selected at each hub is sometimes referred to as mtry or k.

Figure 5.2 illustrates the random forest approach for data processing, whereby N decision trees are aggregated to expedite data analysis. The depth of the tree may be limited by a border node size, often defined as the amount of repetitions in the leaf node, which is usually set to 1. Once the woodland has been prepared as previously described, to organize another event, all the trees cultivated in the forest are encountered. Each tree is segmented into several instances that are documented as a vote. After tallying the votes from every tree, the class with the most votes is designated as the new case's grouping.

5.4.3 Support Vector Machine (SVM)

Classification and regression are two areas where the supervised learning technique Support Vector Machine (SVM) shines. Whether you're dealing with linear classification, regression, or even outlier identification, Support Vector Machine (SVM) is an effective machine-learning approach to apply. Support vector machines (SVMs) have several potential applications, including text and picture classification, catching spam, handwriting recognition, gene expression study, facial recognition, and anomaly detection. Given its versatility and efficiency in handling multidimensional data and nonlinear connections, support vector machines (SVMs) find utility in many different contexts.

Using support vector machines (SVMs) to locate the desired feature's greatest separation hyperplane among the available classes is a very successful strategy. Using a hyperplane to divide data points into their respective classes is the goal of support vector machine (SVM). The following is an example of how SVM may be used in a COVID-19 analysis context:

(a) Classification: SVM can classify COVID-19 cases based on features such as symptoms, demographics, or geographic location.
(b) Risk Stratification: SVM can help in stratifying patients based on the risk of complications, aiding healthcare providers in making informed decisions.
(c) Data Separation: SVM is effective when dealing with high-dimensional data, making it suitable for tasks involving a large number of features.

5.4.3.1 SVM Working

One reasonable choice as the best hyperplane is the one that represents the largest separation or margin between the two classes (Fig. 5.3).

So, we choose the hyperplane whose distance from it to the nearest data point on each side is maximized. If such a hyperplane exists it is known as the maximum-margin hyperplane/hard margin. Both Random Forest and SVM can contribute to COVID-19 analysis by leveraging historical data, identifying patterns, and making predictions or classifications. These models are particularly useful when there is a

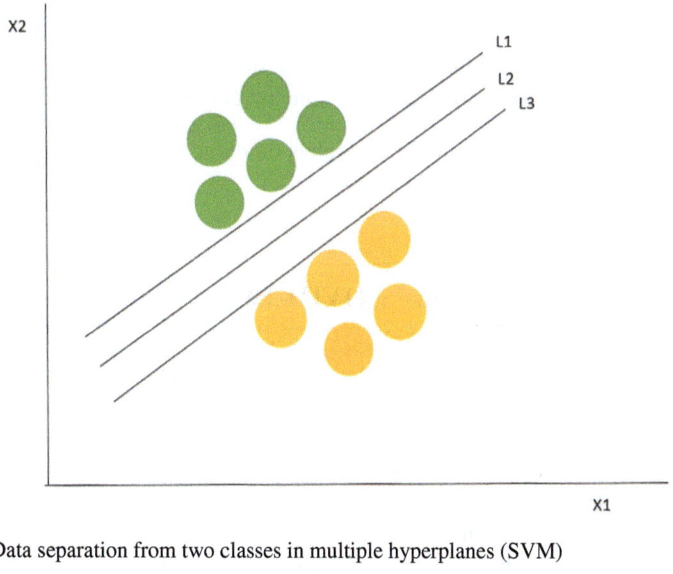

Fig. 5.3 Data separation from two classes in multiple hyperplanes (SVM)

need to analyze complex relationships within the data and when interpretability is important. It's important to note that the choice of the model depends on the specific task, characteristics of the data, and the goals of the analysis.

5.4.4 ML Models Training and Evaluation Process

Combining machine learning and sentiment analysis for COVID-19 analysis involves training models to automatically categorize sentiments in textual data related to the pandemic. The stepwise process of how to implement sentiment analysis using Random Forest and Support Vector Machine (SVM) for COVID-19 analysis is as follows:

(a) Data Collection: It includes the gathering a dataset of text data related to COVID-19. This can include tweets, news articles, forum discussions, and other relevant sources.
(b) Data Preprocessing: It cleans and preprocess the text data, remove special characters, URLs, and irrelevant symbols, converts text to lowercase, tokenize the text, remove stop words, performs lemmatization or stemming.
(c) Labeling: It involves the use of existing sentiment labels for your dataset. The common labels include positive, negative, or neutral sentiment.
(d) Feature Extraction: The feature extraction uses TF-IDF or word embeddings to convert pre-processed text into numerical features. Split the dataset into training and testing sets.
(e) Model Selection: Here we have selected the Random Forest and Support Vector Machine for sentiment analysis.
(f) Model Training: The following models have been trained.

    i. Training of a Random Forest model on the training data.
    ii. Training of a Support Vector Machine model on the same training data.

Model Evaluation: Evaluating both models on the testing set using metrics such as accuracy, precision, recall, and F1 score.

5.5 Social Media Data and COVID-19

Numerous studies have shown that using social media during the COVID-19 epidemic can be advantageous since it is a helpful coping mechanism that helps people feel less anxious and alone. The data shows that during the epidemic, spending too much time on social media platforms is linked to an increase in anxiety and depression symptoms.

During a pandemic, the mass media can effectively stop the spread of disease and has a significant influence on public opinion. Due to concerns that COVID can spread through newspapers as well, the supply of newspapers has decreased during the COVID-19 epidemic. People can find specialty communities, communicate with others, and get real-time information access through social media platforms. It increases global connectivity when it works well. Increased social distance has negatively impacted human relations, leading to significant changes in work, leisure, and social interactions as a result of the COVID-19 pandemic. Social media use, however, can also have a detrimental impact on teenagers, as it can divert their attention, interfere with their sleep, and expose them to peer pressure, bullying, rumors, and inaccurate perspectives of other people's lives. Teens' usage of social media may be correlated with the risks. People must utilize social media carefully since it has both advantages and disadvantages. It can be a useful tool for business, communication, and information sharing, but it can also be detrimental to one's wellbeing, privacy, and mental health (Obi-Ani et al. 2020). Using social media is among the most common internet pastimes. Globally, there were more than 4.59 billion social media users in 2022; by 2027, that figure is expected to rise to nearly six billion.

During the COVID-19 pandemic, social media data has been crucial in influencing the story, spreading knowledge, and expressing public opinion. The following are some significant facets of the connection between COVID-19 and social media data:

1. Real-time Updates and Information Dissemination: Social media platforms act as avenues for the timely sharing of COVID-19-related information. Governments and health organizations swiftly disseminate statements, instructions, and news updates.
2. Education and Public Awareness: Awareness Campaigns: Public awareness efforts utilizing social media have been used to alert the public about COVID-19 symptoms, vaccination information, and preventive actions.
3. Sentiment Analysis: Monitoring Public Attitude Sentiment analysis of social media data facilitates the monitoring of public opinion on COVID-19. Public sentiment can be ascertained by examining tweets, postings, and comments.
4. Misinformation and Infodemic: Misinformation has proliferated on social media, where it has given rise to conspiracy theories and false information on COVID-19. This has made it difficult to contain the "infodemic."
5. Public Response and Behavior: Monitoring Public Behavior: During the pandemic, social media data is studied to determine how the public is responding and behaving. This includes talks about lockdowns and limitations, adherence to safety precautions, and worries.
6. Disease Surveillance and Monitoring: Social media data is utilized for disease surveillance and monitoring, offering information on the COVID-19 pandemic's spread, newly identified hotspots, and early warning indicators.
7. Online Mental Health Resources: Help for Mental Health Social media platforms provide as conduits for the dissemination of mental health resources, coping mechanisms, and emotional assistance to individuals encountering difficulties during the pandemic.

8. Effect on Social Conduct: Modifying Social Standards: The extended utilization of social media amidst the pandemic is probably going to have enduring effects on social conduct, communication styles, and information intake.

5.6 Feature Extraction and Representation Using BERT Model

5.6.1 Deep Learning

A computer model learns to carry out categorization tasks directly from images, text, or sound in deep learning. State-of-the-art accuracy can be attained by deep learning models, occasionally surpassing human performance. Neural network topologies with multiple layers and a sizable quantity of labeled data are used to train models. A substantial amount of labeled data is needed for deep learning. For instance, creating autonomous vehicles needs millions of photos and thousands of hours of film. Deep learning calls for a large processing power. Deep learning is made efficient by the parallel architecture of high-performance GPUs. Development teams can use this to cut the training time of a deep learning network from weeks to hours or less when paired with cloud computing or clusters.

A subset of machine learning is called deep learning. The first step in a machine-learning workflow is the manual extraction of pertinent characteristics from images. Next, a model that classifies the items in the image is created using the features. Relevant features are automatically retrieved from photos using a deep learning approach. Furthermore, deep learning enables to do the end-to-end learning, in which a network is given unprocessed input and an automatic task to complete, like categorization (Bavel et al. 2020).

Deep learning methods scale with data, but shallow learning converges, which is another important distinction. Machine-learning techniques that reach a performance ceiling when additional examples and training data are added to the network are referred to as shallow learning. Deep learning networks frequently get better as the volume of your data grows, which is a major benefit.

5.6.2 Feature Extraction

Using the network as a feature extractor is a slightly less popular and more specialized method of deep learning. We may extract these characteristics from the network at any point throughout the training process because every layer in the network is charged with learning specific features from images. The application of deep learning to "natural language processing" (NLP) has shown remarkable success. Public sentiment analysis, misinformation detection, and literature mining are some of the COVID-19 applications. Examining the biomedical literature has proven to be crucial for the repurposing of drugs. The repurposing of baricitinib (Twitter xxxx), an anti-inflammatory medication used to treat rheumatoid arthritis, is an example of this being successful. Through querying biomedical knowledge graphs, the potential effectiveness of this medication was revealed. Deep Learning is used in modern knowledge graphs to automate their construction. Deep Learning is used by other biomedical literature search engines to get information from natural language queries. With the addition of question-answering and summarization models, these literature mining systems have the potential to completely transform search.

With minimal human engagement, intelligent machines and methods can be developed thanks to deep learning (DL), which is based on novel structured algorithms. Some examples of DL mechanisms include intelligent responses to questions, relevant data visualization, and the ability to deliver extremely precise predictions for any given situation (Adam 2022). It is difficult to create a precise taxonomy for DL classification since it depends on several factors, including the kind of object being investigated, experience-based learning, examining both structured and unstructured data, and collecting pertinent details and justifications from knowledge bases (Górska et al. 2022).

Statistics are supplied into deep learning structures as big datasets since these structures need enormous amounts of data to produce accurate findings. Synthetic neural networks can classify data when they process it by using the answers to a series of actual or false binary questions and extremely complex mathematical computations. For instance, a facial reputation program uses mastery to first recognize and understand the contours and features of faces, then the most important components of faces, and finally the fundamental representations of faces. This system becomes more strained with time, and there are more opportunities for fixes. In this instance, over time, the facial reputation program will recognize faces accurately. In a precise, synthetic neural network, signals travel through nodes and gives them appropriate weights. The effect of a stronger weighted knot is greater on the layer of knots below it. The weighted inputs are compiled by the closing layer to produce an output. Deep learning systems need powerful hardware since they process large amounts of data and have a variety of intricate mathematical computations. But even with such advanced gear, training a neural community can take weeks (Fig. 5.4).

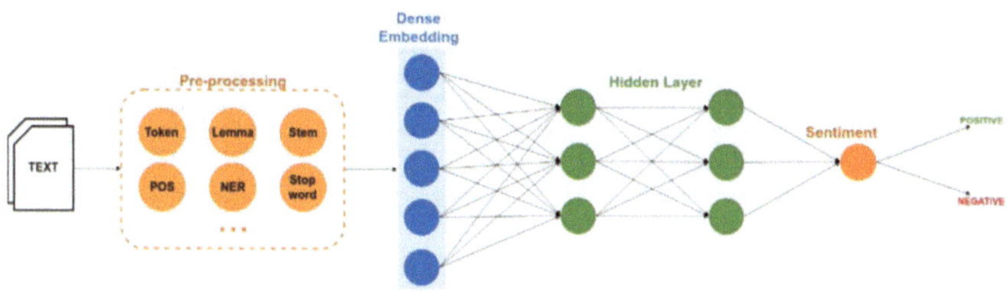

Fig. 5.4 Structure of deep learning model

5.6.3 BERT Model

The name BERT, or Bidirectional Encoder Representations from Transformers, refers to a deep learning model called Transformers in which all output and input elements are connected, and weights between them are dynamically determined depending on this connection. By feeding a text's encoded representation to a SoftMax layer, which produces a probability distribution over the range of potential classes—such as positive, negative, or neutral—BERT is able to classify a text's attitude. BERT is able to conduct sentiment analysis because of its pre-trained contextual understanding of language.

BERT is a neural-network-based pre-training method for language processing. It can be applied to assist in determining the word context in search queries (Fig. 5.5).

Compared to other models, the BERT Language Model has the following advantages:

(a) Compared to other models, the BERT model comes pre-trained and available in more languages. It will come in handy when we work on non-English-based projects.
(b) For token input, BERT uses Wordpiece embeddings. For every token, BERT employs segment embeddings and positional embeddings in addition to token embeddings. Token positions in sequence are indicated by the information included in positional embeddings. When the model input contains sentence pairs, segment embeddings come in handy.
(c) BERT leverages transformer architecture and bidirectional training to gain a deeper understanding of an environment. Sentiment analysis and natural language understanding (NLU) tasks are where it excels.

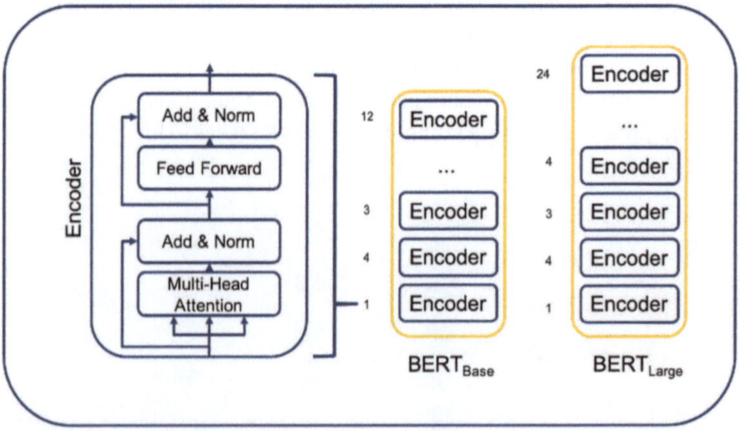

Fig. 5.5 BERT model for text classification

(d) Because of its more intricate design and wider parameter space, BERT does require a lot more effort to optimize than LSTM. However, it's also critical to consider that BERT outperforms LSTM in many tasks.

The main technological advancement of BERT is the application of Transformer, a well-known attention model, to language modeling through bidirectional training. This is in contrast to earlier research that examined a text sequence either from the left to the right or by combining training from the left to the right.

5.6.4 *Background*

The benefits of transfer learning, which involves pre-training a neural network model on a known task, such as ImageNet, and then fine-tuning it by using the learned neural network as the foundation for a new purpose-specific model, have been consistently demonstrated by researchers in the field of computer vision. Researchers have demonstrated that a similar method works well for a variety of natural language problems in recent years. Feature-based training is an alternative method that is also widely used in NLP tasks This method creates word embeddings from a neural network that has already been trained, which are subsequently utilized as features in NLP models.

5.6.5 *The Operation of BERT*

Transformer is an attention mechanism used by BERT that picks up on the contextual relationships between words (or sub-words) in a text. Transformer comes with two distinct mechanisms in its vanilla form: a decoder that generates a prediction for the task and an encoder that reads the text input. All that is required for BERT to produce a language model is the encoder mechanism. Google has published a paper that describes the intricacies of Transformer. With directional models, the text input is read either left-to-right or right-to-left, but with the Transformer encoder, all words are read simultaneously (Perez-Cepeda and Arias-Bolzmann 2022). It would be more accurate to describe it as non-directional, but as a result, it is seen as bidirectional. Because of this feature, the model is able to understand a word's context by taking into account both the word's left and right surrounds. A series of tokens that are first embedded into vectors and then processed by a neural network make up the input. The result is a series of size H vectors, each of which represents an index-matched input token.

5.6.6 *Experimental Set Up*

5.6.6.1 BERT Model

The BERT methodology was put into practice to categorize the tweets. Bidirectional Encoder Representations from Transformers, or BERT, is a novel language presentation method that was first described in an article (Tsao et al. 2021). Best in class, which brings a variety of NLP examinations, was introduced by BERT, causing a stir in the Machine Learning people group. The fundamental specific development of BERT is the use of Transformer's bidirectional preparation, a mainstream consideration methodology, to language demonstration.

This is an improved version; earlier developments examined text sequencing from left to right or left to right and integrated the left-to-right model. A bi-directionally prepared language model can have a deeper sense of language context and stream than a single-directional model, according to the findings of study (Shaw et al. 2019).

The above figure is depicting the workflow of sentiment analysis using deep learning efficient method called BERT model. This model execution has been done on tweet covid-19 dataset for the feature extraction that will analyze every hidden emotion in the tweet dataset. Polarity analysis has been done by this method very efficiently that is calculated in terms of training accuracy and validation accuracy of this model in evaluation, results and discussion column further in this chapter. The emotions categories in COVID-19 tweet dataset are extremely positive, positive, neutral, negative and extremely negative as mentioned in Fig. 5.6. These emotions results are better in case of implementing BERT model as compared to implementation of classical machine learning Random Forest and Support Vector Machine techniques.

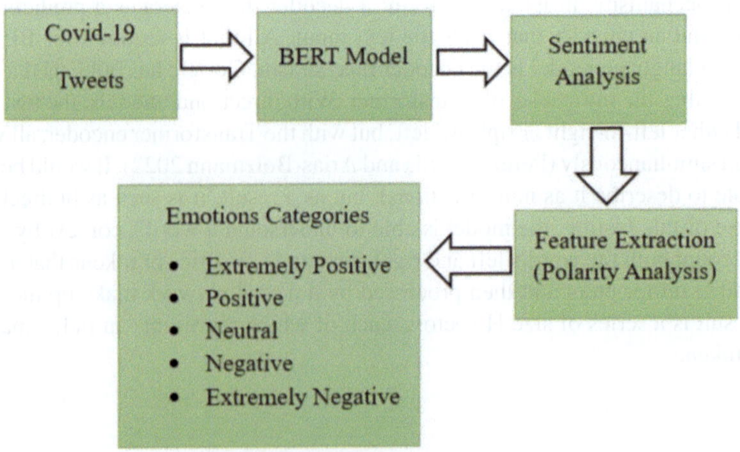

Fig. 5.6 Work flow of sentiment analysis using BERT model

5.6.6.2 Model Training

The transformer encoder identifies the entire word sequence at once, in contrast to directional versions that allow sequential reading of text input (either left to right or right to left). As a result, it is seen as bidirectional even though it is actually a non-directional model with greater accuracy than other well-known models. The model's comprehension of the text context is made possible by this unique feature. There are two versions of BERT: a large model with 24 encoders and a base model with 12 encoders.

Fifteen percent of the randomly picked input text tokens have been used to finish the model training in BERT. After being further pre-processed by 80, 10% of these tokens are replaced with an irregular word, 10% with a "MASK" token, and 10% use the initial word. For unmasked words, the model would not produce very good symbolic depictions if we used "MASK" 100% of the time. The model was improved to predict masked words, but the unmasked tokens were still used for text setting.

5.6.6.3 Next Sentence Prediction in BERT

During BERT training, the model learns to predict whether the second sentence in a pair is the sentence that will be the result in the original file by gathering statement pairs as input. In training, half of the data sources are pairs in which the next sentence is the one that appears in the first archive; in the other half, a random sentence is selected as the next phrase from the corpus. It will be assumed that the erroneous sentence will become independent of the main sentence.

The total of the token, segment, and position embeddings serves as the input source for BERT. In order to help the model distinguish between sentence pairs during training, the input has been handled as described below.

(a) Each sentence ends with a (SEP) token embedded toward the end, and a (CLS) token embedded toward the beginning of the statement.
(b) Sentence A or Sentence B can be appended to each token, as shown by a statement embedding. Token embeddings with an additional statement vocabulary and sentence embeddings are conceptually similar.
(c) Each token has a positional inserting added to it to indicate where it stands in the succession.
(d) The BERT tokenizer handled word component tokenization; high-frequency vocabulary combinations were frequently introduced, and vocabulary showed distinct linguistic features.
(e) A word cloud diagram was used to evaluate the keywords in Indian tweets. Two subsets of the data were separated for training and validation. A further calculation was made to determine the accuracy of the model.

5.6.6.4 Model Evaluation

The traditional models, including RF, SVM, were taken into consideration in order to assess the effectiveness of the BERT model. Text classification is done using these models. Different emotions are analyzed by using ML and DL models. When evaluating a BERT model for COVID-19 sentiment analysis, you would normally consider metrics such as accuracy, precision, recall, F1 score, and maybe the area under the ROC curve. The metrics you use will depend on your goals and the type of activity you are doing (e.g., binary or multi-class categorization). Below is a quick explanation of every metric:

(a) Accuracy: The ratio of accurately predicted instances to the total instances. It offers a general indicator of the performance of the model.
(b) Precision: This represents the difference between the total number of true positives and erroneous positives. It shows the proportion of real positive instances among the expected positive ones.
(c) Recall (Sensitivity): The ratio of true positives to the sum of true positives and false negatives. It assesses how well the model can account for every positive example.
(d) F1 Score: The harmonic means of precision and recall. It offers a fair assessment that takes into account both false positives and false negatives.

In the context of COVID-19 sentiment analysis here in this chapter the goal is capturing and correctly classifying tweets expressing strong sentiments related to health concerns or public sentiment toward government actions. So, model evaluation is very important part to analyze the study we have done using all models.

5.7 Results and Discussion

The world's population becomes more vulnerable and anxious as pandemics spread. When preparing to implement stronger preventive methods, risk communication becomes essential in situations like this one because it doesn't alter based on clear cut-off points. Transparent and efficient message delivery to the public is possible if good risk communication has been put into practice. Reducing the knowledge gap between data investigators and recipients in understanding public behavior to combat the epidemic is the main goal of this. Genuine data from governments and quick response by public health authorities are essential elements for risk reduction and anxiety assessment among the populace (Figs. 5.7, 5.8 and 5.9).

Sentiment analysis is a key tool for understanding people's feelings during a specific event. Sentimental tweets about the pandemic were almost certainly positive, indicating that people maintained their confidence in the face of a remarkable public health emergency. Positive sentiment keywords typically conveyed gratitude for community efforts and front-line workers who support society's most vulnerable

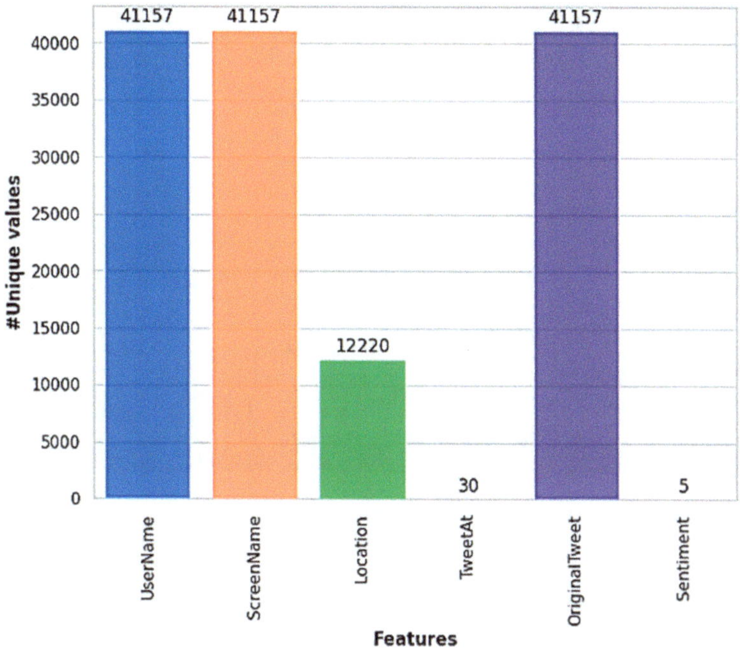

Fig. 5.7 Bar chart showing number of unique values of each column

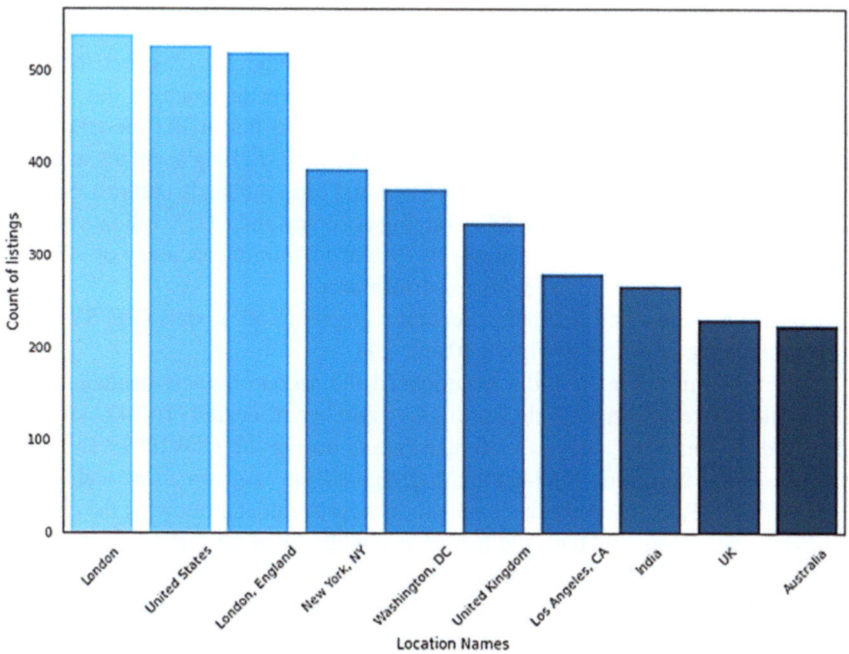

Fig. 5.8 COVID-19 dataset tweets locations

Fig. 5.9 Word cloud of COVID-19 dataset tweets

members; however, some positive sentiment keywords also conveyed negative opinions toward those on the front lines. A few positive sentiment keywords are linked to happy statements that can motivate doctors to fight the battle against COVID-19. Another positive estimation was the support of contamination measures to maintain general wellness guidelines like “stay safe” and “remain at home” patterns.

The above table is showing the results of all the models used in this chapter. This shows the training accuracy and validation accuracy of classical machine-learning models that are random forest and support vector machine models and further BERT model of deep learning. The results shows that BERT model outputs are much better than RF and SVM as shown above.90% training accuracy and 56% validation accuracy were shown using the Random Forest model. In contrast, Support Vector Machine demonstrated an 88% training accuracy and a 60% validation accuracy. Finally, with a validation accuracy of 79% and a training accuracy of 99%, the BERT model demonstrated exceptional performance.

Training accuracy and Validation accuracy graphs of all models (RF, SVM and BERT) are depicted below in Figs. 5.10 and 5.11.

After the United States, India experienced the second-highest pandemic with 11.5 million infected cases and third-highest number of deaths (160,427 overall). This study’s sentiment analysis of Indian tweets during the COVID-19 shutdown yielded several intriguing findings and highlighted the need for more techniques, more COVID-19 data, and experimental model advancement. Following the creation of models with additional data and tactics, tweet classification techniques based on the BERT model could be applied as self-governing methods for the autonomous categorization of COVID-19 sentiment. Future pandemic experiences, both nationally and internationally, will also be expanded by the suggested paradigm and its outcomes.

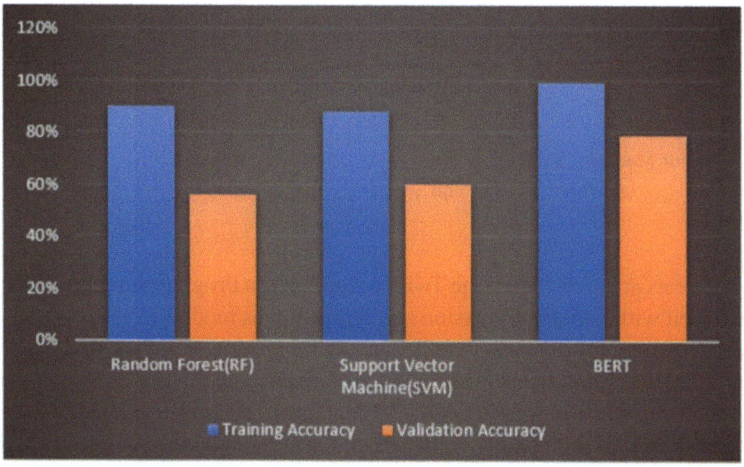

Fig. 5.10 Training accuracy and validation accuracy of RF, SVM and BERT models

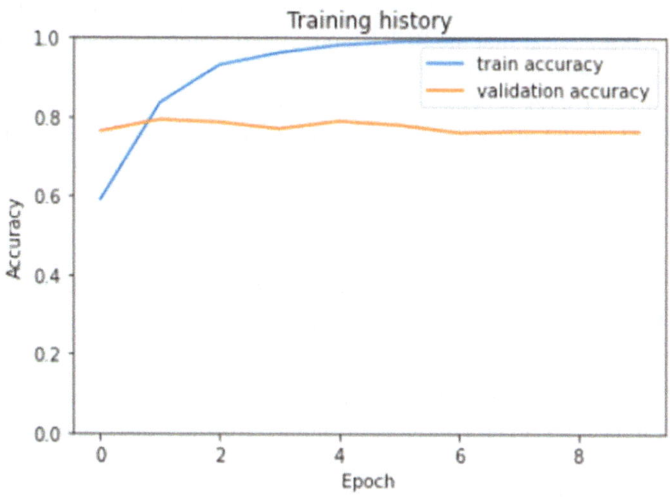

Fig. 5.11 Accuracy graph of BERT model

According to recent NLP developments, text mining has become more and more popular in the development of big data analysis, which has enhanced computational capabilities, and unstructured data analysis, which makes it possible to examine enormous linguistic datasets. This study presents a useful exercise for assessing the prediction accuracy of text outcomes, using the BERT sentiment analysis program in R.

For the purpose of comprehending pandemic worry among Twitter users based on specific phrases, a comparable analysis has been published in. Approximately

Table 5.1 Performance comparison of classical machine-learning models with BERT deep learning model

Model	Training accuracy (%)	Validation accuracy (%)
Random Forest (RF)	90	56
Support Vector Machine (SVM)	88	60
BERT	99	79

900,000 tweets are taken out of the Twitter Application Programming Interface (API) and analyzed with logistic regression and Naïve Bayes models (Vaishya et al. 2020).

Twelve main subjects that would emphasize sentiment topics that are closely related to health care issues have been determined by a study concerning the precise ideas of netizens regarding the COVID-19 pandemic. The most skewed findings were published in an intriguing study on the mental health of Chinese youngsters who use Weibo Messenger. It brought to light the unfavorable attitudes that young people had throughout this outbreak (Mahmud et al. 2020).

On the other hand, four emotions that were linked to both positive and negative sentimental categories were emphasized in this study among Indian netizens. Table 5.1 makes it clear that the BERT model fared better in classifying COVID-19 tweets than the other two RF and SVM models. Its accuracy of 99% is far higher than that of other models. Thus, when compared to other models, it demonstrates that the BERT model offers the greatest explanation for comprehending the tweets of COVID-19.

5.8 Practical Applications and Benefits to Health Sector in Global Emergencies Like COVID-19

The COVID-19 pandemic has demonstrated how flaws in health systems can have a significant impact on social cohesion, economic growth, health, and public trust in governments. It is still crucial to contain and reduce the virus's spread and infection rate. However, enhancing health systems' ability to react quickly and efficiently is also important. Vaccinating against COVID-19 is part of this. Numerous nations are initiating vaccination campaigns following their rapid development and testing. However, concerns about equal access, distribution, and manufacturing still exist, especially for low- and middle-income nations.

It is equally important and difficult for health systems to quickly reorganize in order to mobilize healthcare professionals in response to the crisis and guarantee the continuous provision of crucial services throughout the continuum of care (Barstugan et al. 2003). The provision of training and educational resources to quickly repurpose health personnel and new recruits in accordance with the needs of the health system, as well as the assurance of occupational safety and health, fair working conditions, and psychosocial support for health workers, are important methods.

Preparing for emergencies also include streamlining service delivery channels, including telemedicine. Technological innovations that can improve health care delivery and working methods both during and after the epidemic include 3D printing, artificial intelligence, and internet and mobile health applications. A few nations have started using location data from mobile phones to monitor the nationwide spread of COVID-19.

Medical students are working in a contact canter in the Netherlands where patients with severe symptoms can get connected to the right care. During the pandemic, introducing and expanding the use of digital technology to educate, prepare, and direct healthcare professionals—particularly those working in underdeveloped and isolated areas—can enhance service delivery, management, and transparency.

When COVID-19 arrived, health research systems were already facing many difficulties. Moreover, the virus increased the prominence and visibility of pre-existing problems in these systems. There are knowledge gaps that need to be filled in a variety of fields in order to control and manage the COVID-19 epidemic as well as effectively prevent and respond to future epidemics. A wide range of research in the fields of basic sciences, clinical sciences, public health sciences, implementation science, and policy and system studies is required to do this various nations have various health research systems, institutions, and procedures. Since a wide range of interdisciplinary research is required to control and manage the COVID-19 outbreak, the way these disparate sections communicate with one another is crucial. In many nations, the research system is fragmented.

5.8.1 Favorable Effects on Health Sector

The COVID-19 pandemic has caused major problems and disruptions for people all across the world, but it has also had some good effects for the health industry. The following are a few ways that the pandemic has had favorable effects or benefits for the health sector:

(a) Progress in Telemedicine

Telemedicine and virtual healthcare services are becoming more widely used and accepted as a result of the pandemic. Because of this, fewer in-person visits are required, increasing access to healthcare services. Remote consultations, monitoring, and treatment have also been made possible.

(b) Technological Advancements

Wearable technology, digital health solutions, and health monitoring apps are just a few of the technological advancements in healthcare that have been sparked by the problem. The goals of these developments are to improve illness prevention, monitoring, and patient care.

(c) Enhanced Cooperation and Information Exchange

Researchers, medical experts, and organizations have been inspired to collaborate and share information as a result of the global response to COVID-19 (Hussain et al. 2020). This culture of cooperation could accelerate the development of new medical research and treatment approaches.

(d) Emphasis on Prevention and Public Health

The significance of disease prevention, pandemic preparedness, and public health initiatives has been highlighted by the epidemic. on order to prevent and lessen the effects of future health crises, governments and healthcare organizations are probably going to invest more on public health infrastructure and initiatives.

(e) Better Data Exchange and Monitoring

In order to track the spread of infectious illnesses, surveillance and data sharing have received more attention. Public health responses can be improved by tracking and forecasting outbreaks using data analytics and artificial intelligence.

(f) Adaptable Health Systems

The epidemic has brought attention to the necessity of adaptable and flexible healthcare systems. Public health emergencies are best handled by health systems that can react swiftly to spikes in cases, use resources wisely, and adjust to changing conditions.

(g) Development and Distribution of Vaccines

The quick creation and widespread use of COVID-19 vaccines has demonstrated the possibility of expedited vaccine development and manufacturing. This experience might help hasten the response time to potential infectious disease threats in future.

(h) Greater Understanding of Mental Health

The pandemic has highlighted how crucial mental health is. As the effects of stress, loneliness, and uncertainty on mental health become more widely recognized, initiatives to incorporate mental health services into healthcare systems are intensifying.

(i) Employee Training and Resilience

Throughout the crisis, healthcare personnel have shown resiliency and flexibility. The healthcare workforce may benefit from improved training, readiness, and resilience as a result of the experience learned from managing the pandemic.

(j) Solidarity for Global Health

As a result of the epidemic, nations and organizations are working together to combat the shared threat in global health. Increased international cooperation in healthcare and public health projects could result from this cooperative atmosphere.

5.9 Conclusion

By the end of this chapter, we can say that the BERT model performs significantly better in analyzing the COVID-19 data that we collected in the form of tweets from people expressing their experiences and feelings than do traditional machine-learning techniques like random forest and support vector machines. In this study, we have employed three distinct models for sentiment analysis on COVID-19: Random Forest (RF), Support Vector Machine (SVM), and BERT (Bidirectional Encoder Representations from Transformers). The models were assessed according to how accurate they were throughout training and validation.

90% training accuracy and 56% validation accuracy were shown using the Random Forest model. In contrast, Support Vector Machine demonstrated an 88% training accuracy and a 60% validation accuracy. Finally, with a validation accuracy of 79% and a training accuracy of 99%, the BERT model demonstrated exceptional performance. The findings imply that, when it comes to sentiment analysis for COVID-19, the deep learning BERT model performs better than the traditional machine-learning models (RF and SVM). The training and validation sets' much higher accuracy scores show that BERT is more successful at identifying the subtle patterns and sentiment nuances found in the complex text data related to COVID-19.

These results highlight how crucial it is to use cutting-edge deep learning methods, like BERT, while handling difficult, context-rich tasks like sentiment analysis during a pandemic. When selecting the best method for a certain application, it's crucial to take other aspects like computational capacity, model interpretability, and data volume into account.

References

Adam D (2022) The pandemic's true death toll: millions more than official counts. Nature 601(7893):312–315

Ahmad AR, Murad HR (2020) The impact of social media on panic during the COVID-19 pandemic in Iraqi Kurdistan: online questionnaire study. J Med 22

Apuke OD, Omar B (2021) Fake news and COVID-19: Modelling the predictors of fake news sharing among social media users. Telemat Inform 56:101475

Barkur G, Vibha, Kamath GB (2020) Sentiment analysis of nationwide lockdown due to COVID 19 outbreak: evidence from India. Asian J Psychiatry 51:102089

Barstugan M, Ozkaya U, Ozturk S (2020) Coronavirus (covid-19) classification using images by machine learning methods. arXiv preprint: 2003.09424

Van Bavel JJ et al (2020) Using social and behavioural science to support COVID-19 pandemic response. Nat Human Behav 4(5):460–471

Chakraborty K, Bhatia S, Bhattacharyya S, Platos J, Bag R, Hassanien AE (2020) Sentiment analysis of COVID-19 tweets by deep learning classifiers—a study to show how popularity is affecting accuracy in social media. Appl Soft Comput 97:106754

Cinelli M, Quattrociocchi W, Galeazzi A, Valensise CM, Brugnoli E, Schmidt AL, Zola P, Zollo F, Scala A (2020) The COVID-19 social media infodemic. Sci Rep 10:1–10

Depoux A, Martin S, Karafillakis E, Preet R, Wilder-Smith A, Larson H (2020) The pandemic of social media panic travels faster than the COVID-19 outbreak. J Travel Med 27:taaa031
Di Domenico G, Sit J, Ishizaka A, Nunan D (2021) Fake news, social media and marketing: a systematic review. J Bus Res 124:329–341
Gao J, Zheng P, Jia Y, Chen H, Mao Y, Chen S, Wang Y, Fu H, Dai J (2020) Mental health problems and social media exposure during COVID-19 outbreak. PLoS ONE 15:e0231924
Górska A, Dobija D, Grossi G, Staniszewska Z (2022) Getting through COVID-19 together: understanding local governments' social media communication. Cities 121(103453)
Havey NF (2020) Partisan public health: how does political ideology influence support for COVID-19 related misinformation? J Comput Soc Sci 3:319–342
Hussain AA, Bouachir O, Al-Turjman F, Aloqaily M (2020) AI techniques for COVID-19. IEEE Access 8:128776–128795. https://doi.org/10.1109/ACCESS.2020.3007939
Mahmud T, Rahman MA, Fattah SA (2020) Covxnet: a multi-dilation convolutional neural network for automatic covid-19 and other pneumonia detection from chest x-ray images with transferable multi-receptive feature optimization. Comput Biol Med 122:103869
Obi-Ani NA, Anikwenze C, Isiani MC (2020) Social media and the COVID-19 pandemic: observations from Nigeria. Cogent Arts Humanit 7:1799483
Perez-Cepeda M, Arias-Bolzmann LG (2022) Sociocultural factors during COVID-19 pandemic: information consumption on Twitter. J Bus Res 140:384–393
Shahsavari S, Holur P, Tangherlini TR, Roychowdhury V (2020) Conspiracy in the time of corona: Automatic detection of COVID-19 conspiracy theories in social media and the news. J Comput Soc Sci 3:279–317
Shaw J, Rudzicz F, Jamieson T, Goldfarb A (2019) Artificial intelligence and the implementation challenge. Med Internet Res 21. https://doi.org/10.2196/13659
Tsao S-F, Chen H, Tisseverasinghe T, Yang Y, Li L, Butt ZA (2021) What social media told us in the time of COVID-19: a scoping review. Lancet Digit Health 3(3):e175–e194
Twitter (2020) Twitter Usage Statistics—Internet Live Stats. Available online: https://www.internetlivestats.com/twitter-statistics/. Accessed 19 Oct 2020
Vaishya R, Javaid M, Khan IH, Haleem A (2020) Artificial intelligence (AI) applications for COVID-19 pandemic. Diab Metab Syndr Clin Res Rev 14:337–339. https://doi.org/10.1016/j.dsx.2020.04.012
Zaman A (2021) COVID-19-related social media fake news in India. J Media 2:7

Chapter 6
Transforming Industry 4.0 with AI, ML, and IoT: An Overview of Emerging Trends

Visakshi Rajput, Shilpa Choudhary, and Dinesh Kumar Singh

Abstract Industry 4.0 is significantly reshaping the global manufacturing sector by introducing advanced technologies and processes aimed at boosting efficiency, productivity, and competitiveness. Key technological advancements driving this transformation include Artificial Intelligence (AI), Machine Learning (ML) and Internet of Things (IoT) applications. This chapter delves into the latest trends and applications of AI, ML and IoT within Industry 4.0 and their effects on diverse industries. Through a thorough review of existing literature, the chapter evaluates both the benefits and challenges of these technologies, while featuring real-world examples of their successful implementation to illustrate their potential for boosting innovation and efficiency across various sectors. It also examines the potential for autonomous decision-making, smart transportation systems, intelligent robotics, and advanced product capabilities in the realm of Industry 4.0. Additionally, the chapter addresses challenges such as data privacy, security concerns, regulatory issues, implementation hurdles, and ethical dilemmas. The chapter concludes with a summary of findings, implications for the future of Industry 4.0, and suggestions for further research.

Keywords Artificial Intelligence (AI) · Machine Learning (ML) · And Internet of Things (IoT) · Explainable AI (XAI) · Electronic Health Records (EHRs) · General Data Protection Regulation (GDPR) · Health Insurance Portability and Accountability Act (HIPAA)

6.1 Introduction

The Fourth Industrial Revolution, often referred to as Industry 4.0, is fundamentally reshaping the manufacturing industry through the integration of digital technologies with conventional production methods. Central to this revolution are advancements in

---

V. Rajput · S. Choudhary (✉) · D. K. Singh
Department of Electronics and Communication Engineering, G L Bajaj Institute of Technology and Management, Greater Noida, UP, India
e-mail: shilpadchoudhary@gmail.com

A. Rai et al. (eds.), *Convergence of Artificial Intelligence, Machine Learning, and the Internet of Things in Industry 4.0 Applications*, Transactions on Computer Systems and Networks, https://doi.org/10.1007/978-981-96-5269-3_6

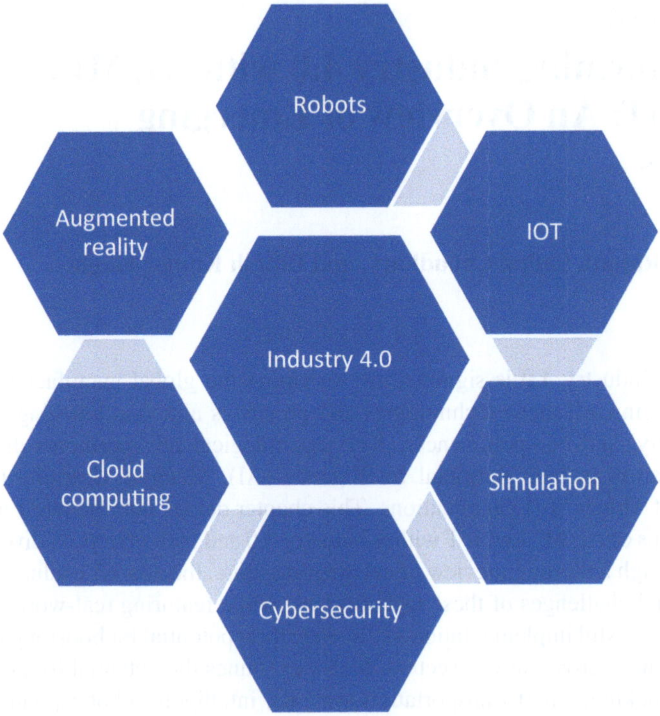

Fig. 6.1 Industry 4.0

Artificial Intelligence (AI), Machine Learning (ML) and the Internet of Things (IoT), which offer new avenues for transforming industrial processes, boosting efficiency, and improving customer experiences. The essence of Industry 4.0 lies in the fusion of these sophisticated technologies, leading to the development of smart, interconnected systems that optimize operations, enhance productivity, and drive efficiency. This chapter delves into the latest trends in AI, ML, and IoT within the context of Industry 4.0. It provides an in-depth examination of how these technologies impact various sectors, including manufacturing, healthcare, retail and transportation. Additionally, it addresses the challenges organizations face when implementing these technologies and offers recommendations for successful integration and utilization (Fig. 6.1).

6.1.1 Importance of AI, ML, and IoT in Industry 4.0

The incorporation of AI, ML and IoT has significantly revolutionized the manufacturing industry by facilitating real-time communication and interaction between machines and devices, which enhances automation and increases overall productivity. These technologies are applied to optimize processes like predictive maintenance, helping to reduce equipment downtime while improving the quality of products. Meanwhile, IoT devices gather data from various sources like sensors and production lines, offering real-time insights into manufacturing operations. This data aids in making informed decisions and increasing efficiency. Additionally, the fusion of AI, ML, and IoT technologies holds promise for creating more personalized customer experiences. AI and ML can analyze customer data, including purchase history and online interactions, to create personalized product recommendations, design targeted marketing strategies, and enhance customer service. This personalization improves customer satisfaction and has the potential to boost revenue by fostering more relevant and engaging customer experiences.

These technologies are not limited to manufacturing; they are also being applied in other fields. In healthcare, AI and ML contribute to the development of personalized treatment plans, the analysis of medical images, and the advancement of drug discovery. In the transportation sector, IoT sensors are used to monitor traffic and optimize routes, which can reduce fuel consumption and lower transportation costs. Retail benefits from AI and ML through personalized shopping experiences, better inventory management, and reduced waste. The incorporation of AI, ML, and IoT in Industry 4.0 is a catalyst for innovation, offering substantial growth opportunities. Organizations leveraging these technologies can significantly improve efficiency, productivity, and customer satisfaction. Additionally, this technological integration creates new job opportunities in fields such as data analysis, machine learning and cybersecurity, further contributing to economic and industrial growth.

6.2 Overview

6.2.1 Emerging Trends of AI in Industry 4.0

Artificial Intelligence (AI) is a key driver of progress in Industry 4.0, revolutionizing industries through task automation, workflow optimization and enhanced productivity. Its influence is evident across various sectors such as manufacturing, healthcare, retail and transportation. In manufacturing, AI-powered predictive maintenance employs machine learning to forecast equipment malfunctions before they happen, cutting down on maintenance costs and prolonging the operational life of machinery. Healthcare benefits from AI through enhanced patient care, where algorithms analyze medical data to create personalized treatment strategies. For instance, AI systems examining medical imaging can detect conditions like cancer early, significantly improving treatment success rates. Retailers leverage AI to refine the shopping experience by offering tailored product suggestions and optimizing supply chains. Personalized recommendation engines, powered by machine learning, analyze customer behavior to enhance engagement and loyalty. In transportation, AI helps streamline logistics, with machine learning models optimizing routing and scheduling, which leads to reduced fuel consumption and more efficient deliveries.

6.2.2 Emerging Trends of ML in Industry 4.0

Machine Learning (ML) is a pivotal force in Industry 4.0, leveraging data analysis to optimize processes and improve efficiency across multiple sectors, including manufacturing, healthcare, retail and transportation. In manufacturing, ML analyzes production data to pinpoint areas for improvement. For instance, ML models that process sensor data from machinery can detect irregularities and predict possible breakdowns, which assists in minimizing downtime and reducing maintenance expenses. In healthcare, ML analyzes medical data to recognize patterns that can improve patient outcomes. For instance, ML models assessing electronic health records can predict the likelihood of patient readmissions and identify individuals at risk of chronic diseases. In retail, ML is employed to analyze customer data and deliver tailored recommendations. For example, models that study browsing patterns can offer personalized product suggestions, thereby increasing conversion rates (Porter and Heppelmann 2015). In transportation, ML optimizes logistics and lowers costs. Machine-learning models that manage inventory and forecast demand are helping to reduce inventory costs and enhance customer satisfaction.

6.2.3 Emerging Trends of IoT in Industry 4.0

The Internet of Things (IoT) is central to Industry 4.0, creating smart and connected systems by linking various devices. This technology is widely used across industries like manufacturing, healthcare, retail and transportation. In manufacturing, IoT connects machines and systems to enable real-time tracking of production processes. This connectivity improves operational efficiency, supports predictive maintenance and enhances resource management, leading to increased productivity and reduced operational interruptions. For example, sensors embedded in equipment gather data to track machine performance and identify potential anomalies. In healthcare, IoT integrates medical devices to support remote patient monitoring. For instance, sensors in wearable devices collect data on vital signs, allowing for the early detection of warning signs and improved patient care (Rong et al. 2018).

6.2.4 Challenges Faced by Organizations in Implementing AI, ML, and IoT

Despite the advantages offered by AI, ML, and IoT, organizations encounter several challenges when adopting these technologies. A major concern is data privacy and security. It is essential for organizations to safeguard the data gathered by these technologies, protecting it from unauthorized access. Additionally, implementing and maintaining AI, ML, and IoT systems requires skilled professionals with expertise in

Table 6.1 Case study of successful implementations of AI, ML and IoT

Industry	Technology	Application	Success story
Manufacturing	AI, ML, IoT	Predictive maintenance	The BMW Group's Dingolfing plant in Germany implementation a predictive maintenance system that uses IoT sensors to monitor machines and collects data in real-time. The system has resulted in a 10% reduction in downtime and a 25% reduction in maintenance costs, while also improving production efficiency
Healthcare	AI, ML	Medical imaging analysis	Arterys, a medical imaging company, uses AI algorithms to analyze MRI scans. The deep learning algorithms can analyze a 3D MRI scan of the heart in just 15 s, providing insights that can help doctors identify heart diseases and develop personalized treatment plans

these areas. Organizations must also focus on integrating these technologies smoothly into their existing systems and processes to ensure effective and efficient use (Popescu and Popescu 2017).

6.2.5 Case Study of Successful Implementations

(See Table 6.1).

6.2.6 Applications in Industry 4.0

As we move further into the fourth industrial revolution, the adoption of AI, ML and IoT technologies is rapidly advancing. These technologies are pivotal in developing smart, interconnected systems that enhance operational efficiency and productivity. This chapter has explored the latest trends in AI, ML and IoT within Industry 4.0 and their impacts across different sectors, including manufacturing, healthcare, retail, and transportation (Lee et al. 2014). In manufacturing, AI and ML are employed to optimize production processes, reduce downtime, and improve quality control, while IoT enables real-time monitoring by connecting various machines (Rong et al. 2018). In healthcare, AI and ML improve patient outcomes by analyzing medical data and crafting personalized treatment plans, and IoT connects medical devices for remote

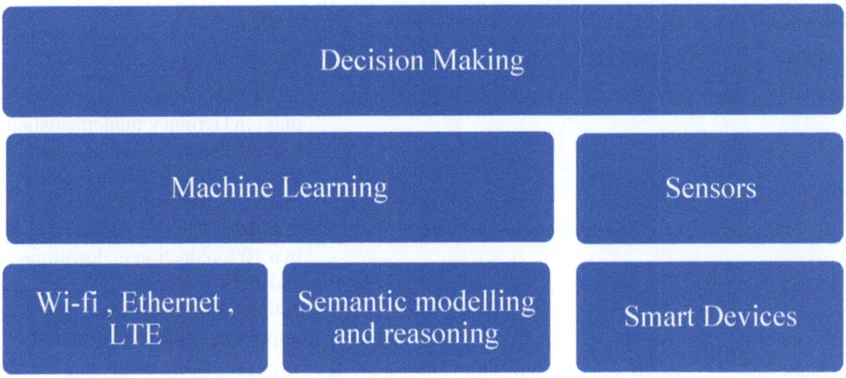

Fig. 6.2 Applications of AI, ML, and IoT

patient monitoring. In retail, AI and ML enhance the customer experience through tailored recommendations and optimized supply chain management, with IoT linking devices to create smart, integrated systems (Wang et al. 2016). In transportation, AI and ML streamline logistics and reduce expenses, and IoT enhances inventory management and demand forecasting (Fig. 6.2).

The integration of AI, ML, and IoT is fundamentally transforming how organizations operate. These technologies enable the creation of advanced, interconnected systems that optimize processes, enhance efficiency and improve productivity. Across various industries such as manufacturing, healthcare, retail and transportation, these technologies are making significant impacts. However, organizations face challenges such as ensuring data privacy and security, addressing skill shortages and developing the necessary infrastructure for effective implementation. As we move toward a more interconnected and intelligent future, adopting AI, ML, and IoT technologies will be vital for driving innovation and achieving growth (Lee et al. 2014).

6.2.7 Benefits of AI, ML, and IoT

Healthcare Industry: AI and ML are utilized to craft personalized treatment plans, analyze medical images, and advance drug discovery. These technologies examine patient data, such as electronic health records (EHRs), to assist healthcare providers in making informed decisions.

Transportation Industry: IoT sensors monitor traffic and optimize routes, which helps reduce fuel consumption and transportation costs. These technologies provide real-time updates on traffic conditions and weather, allowing drivers to adjust their routes accordingly, improving efficiency and lowering environmental impact (Wang et al. 2016).

Retail Industry: AI and ML algorithms are applied to tailor customer experiences, manage inventory more effectively, and minimize waste. By analyzing data such as purchase history and online activity, these technologies enable personalized product recommendations, targeted marketing, and enhanced customer service, leading to greater satisfaction and loyalty (Miorandi et al. 2012).

Manufacturing Industry: AI, ML, and IoT technologies are revolutionizing production processes through real-time communication between machines and devices. This integration boosts automation, increases efficiency, and improves product quality. By employing predictive maintenance powered by AI and ML, organizations can anticipate equipment failures, thereby reducing downtime and lowering maintenance expenses (Miorandi et al. 2012).

Finance Industry: AI and ML technologies are employed to improve fraud detection, manage risk, and analyze investments. These tools process extensive data in real-time, providing insights that help financial institutions make better decisions, thus improving fraud detection and reducing risk (Miorandi et al. 2012).

In summary, the implementation of AI, ML, and IoT technologies across various sectors offers numerous advantages, including greater efficiency, enhanced productivity and improved customer satisfaction. These technologies facilitate the creation of personalized treatment plans in healthcare, optimize route planning in transportation, tailor customer experiences in retail, advance manufacturing processes, and enhance fraud detection in finance. Organizations that effectively adopt these technologies can achieve a competitive advantage through improved operational efficiency, cost reductions and superior customer experiences (Lee et al. 2014).

6.2.8 Challenges and Limitations

While AI, ML, and IoT technologies present considerable opportunities in Industry 4.0, organizations must be aware of various challenges and limitations:

Data Quality: The performance of AI and ML algorithms is closely dependent on the quality of the training data. Incomplete or inaccurate data can lead to biased or flawed results, potentially causing significant issues (Popescu and Popescu 2017).

Data Privacy and Security: The use of IoT sensors and related technologies to collect and transmit data raises significant privacy and security concerns. It is essential for organizations to establish strong protective measures to ensure that sensitive information is safeguarded and to prevent unauthorized access or misuse (Popescu and Popescu 2017).

Lack of Skilled Workforce: Implementing AI, ML, and IoT technologies requires specialized knowledge and skills. Many organizations may lack the necessary expertise, making it challenging to fully leverage the potential of these technologies (Popescu and Popescu 2017).

High Implementation Costs: Implementing AI, ML, and IoT technologies can be costly, especially for smaller organizations. Expenses include the initial setup of hardware and software, as well as ongoing costs for maintenance, training and system updates. Organizations must carefully evaluate the costs of hardware, software and skilled personnel against the potential benefits (Popescu and Popescu 2017).

Ethical Concerns: As AI and ML technologies evolve, concerns about their societal impact, such as job displacement and the potential misuse of these technologies for harmful or unethical purposes, continue to grow (Shi et al. 2018).

Technical Limitations: AI, ML, and IoT are still maturing, and several technical challenges remain. For instance, current AI models may struggle with complex decision-making tasks and are not yet able to fully replicate human intuition and creativity.

6.2.9 Future Development and Trends

As AI, ML, and IoT technologies progress, several future developments and trends are expected to shape Industry 4.0. These include:

Edge Computing: This approach processes data locally, at the edge of the network, rather than relying on a central server. It improves the speed and efficiency of data processing, which is especially advantageous for IoT applications that produce substantial amounts of data (Rong et al. 2018).

Explainable AI (XAI): XAI focuses on creating AI systems that are transparent and easy to understand. This builds trust in AI systems, especially in sectors like healthcare and finance, where decision-making needs to be clear and justifiable (Miorandi et al. 2012).

Hybrid AI Systems: Combining multiple AI methods, such as machine learning and rule-based systems, hybrid AI provides more adaptable and robust systems. These are especially useful in complex decision-making environments (Miorandi et al. 2012).

Increased Automation: As AI and ML technologies become more advanced, automation is likely to increase across industries, improving efficiency and reducing costs. However, this trend may also have implications for the workforce and employment (O'Reilly 2017).

Quantum Computing: This emerging technology has the potential to transform AI and ML by significantly accelerating data processing capabilities, enabling more complex computations (Miorandi et al. 2012).

Ethical Considerations: As AI and ML technologies become more common, addressing ethical issues like privacy, security and bias will become increasingly important. Organizations need to take proactive steps to manage these concerns to build trust and ensure the responsible use of these technologies. (Miorandi et al. 2012).

In conclusion, the future of AI, ML, and IoT in Industry 4.0 will likely be driven by ongoing innovation and advancements. Organizations that remain ahead of these emerging trends will be better positioned to leverage the numerous benefits these technologies offer. However, it is essential for businesses to also recognize the associated risks and challenges, taking proactive steps to mitigate them and ensure responsible use.

6.3 Autonomous Decision-Making

6.3.1 Overview

Autonomous decision-making harnesses AI, ML, and IoT technologies to enable systems to make decisions independently of human input, a key development in Industry 4.0 (Singla and Kraus 2014). This advancement has the potential to revolutionize sectors like manufacturing, healthcare, transportation and finance. In manufacturing, autonomous systems can streamline production processes and improve product quality, while in healthcare, they can monitor patient data in real-time and make data-driven treatment choices. One of the significant benefits of autonomous decision-making is its potential to boost efficiency and reduce costs. By automating decisions, organizations can minimize human involvement, accelerating decision-making processes, improving accuracy and yielding substantial cost savings and productivity gains (Russell and Norvig 2009) (Fig. 6.3).

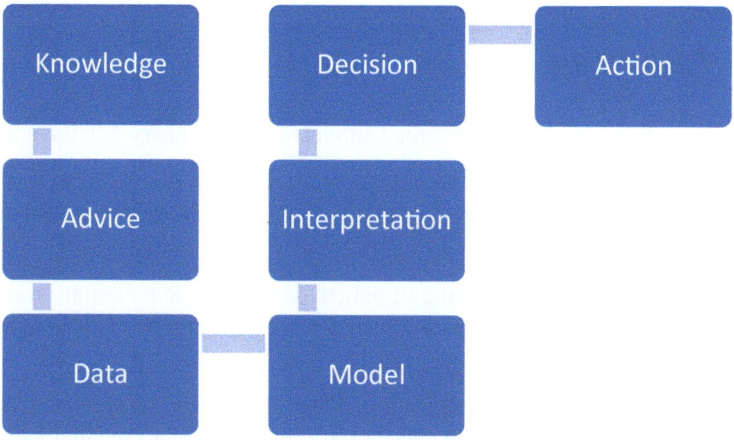

Fig. 6.3 Autonomous decision-making architecture

6.3.2 Advantages

Autonomous decision-making offers several key advantages in Industry 4.0:

Increased Efficiency: Automating decision-making processes reduces the reliance on human labor, resulting in faster decisions and greater overall efficiency, which in turn boosts productivity.

Improved Accuracy: AI and ML algorithms tend to have higher accuracy rates compared to human decision-making, which can reduce errors and lead to better decision outcomes (Russell and Norvig 2009).

Real-Time Decision-Making: Autonomous systems continuously monitor and analyze data, which allows organizations to quickly adapt to real-time changes. This is especially valuable in industries like healthcare and transportation, where immediate decisions are crucial (Singla and Kraus 2014).

Cost Savings: By minimizing the need for manual decision-making, organizations can lower labor costs. The increased accuracy also reduces errors, leading to further cost reductions.

Process Optimization: Autonomous decision-making enhances production processes by minimizing waste and improving product quality. In healthcare, it supports better patient outcomes through real-time monitoring and analysis of health data (Singla and Kraus 2014).

In summary, autonomous decision-making in Industry 4.0 offers significant advantages, such as reshaping operations across various sectors. However, it is crucial to address the associated challenges and limitations to ensure these systems are implemented in a responsible and effective manner.

6.3.3 Applications

Autonomous decision-making has broad applications in Industry 4.0, including:

Manufacturing: It can optimize production processes such as scheduling and maintenance, while also improving product quality by detecting and correcting defects in real-time.

Healthcare: Continuous monitoring of patient data allows autonomous systems to make informed treatment decisions and even predict or prevent potential health issues before they arise.

Transportation: Autonomous decision-making optimizes transportation routes, reduces fuel consumption and enhances safety by identifying potential hazards in real-time.

Table 6.2 Successful implementation of autonomous decision-making

Company	Industry	Application	Result
Tesla	Automotive	Autonomous driving	Improved safety and convenience
Amazon	E-commerce	Fulfillment center optimization	Increased efficiency and cost reduction
Siemens	Manufacturing	Production process optimization	Reduced production costs and improved product quality
John Deere	Agriculture	Crop yield optimization	Increased crop yields and reduced resource usage
Microsoft	Technology	Energy optimization	Reduced energy costs and improved sustainability

Agriculture: By monitoring real-time soil and weather conditions, autonomous decision-making optimizes crop yields and minimizes waste.

Energy: It optimizes energy consumption, reduces waste and ensures efficient usage by analyzing real-time energy data.

Finance: Autonomous systems improve risk management and fraud detection by analyzing large volumes of financial data in real-time, allowing for faster and more accurate decisions.

In conclusion, the use of autonomous decision-making within Industry 4.0 has broad and varied applications, holding the potential to fundamentally change how many industries operate.

6.3.4 Successful Implementations

(See Table 6.2).

6.3.5 Future Trends

The evolution of autonomous decision-making in Industry 4.0 is filled with exciting possibilities and significant advancements. Some potential developments and trends we can anticipate include:

Enhanced AI Algorithms: As AI technology advances, algorithms will become more advanced and accurate, improving autonomous decision-making capabilities. This will help industries enhance their processes and achieve higher efficiency (Russell and Norvig 2009).

Expansion of IoT Sensors: The growing deployment of IoT sensors in industrial environments will provide a larger pool of data for AI systems to analyze. This increase in data will enhance the accuracy and effectiveness of autonomous decision-making.

Edge Computing: The rise of edge computing, which processes data closer to where it is generated rather than relying on centralized systems, will accelerate decision-making and reduce latency. This will be particularly beneficial in sectors like manufacturing and transportation, where real-time decisions are crucial (O'Reilly 2017).

Human–Machine Collaboration: As autonomous systems become more widespread, fostering effective collaboration between humans and machines will be critical. New interfaces and communication methods will be needed to ensure that humans can understand and interact with machine-driven decisions.

Focus on Ethical Considerations: With the increased power and reach of autonomous decision-making, ethical concerns such as privacy, security, and bias will become more prominent. There will be an increasing demand for standards and guidelines to ensure the responsible and ethical application of these technologies (Russell and Norvig 2009).

In summary, the scope of autonomous decision-making within Industry 4.0 holds significant promise for enhancing efficiency, productivity and safety. However, it is essential to address ethical and societal concerns to ensure the responsible application of this technology.

6.4 Intelligent Robots

6.4.1 Overview

Intelligent robots feature advanced sensors and software, enabling them to understand and respond to their environment with greater complexity compared to conventional robots. Designed for tasks requiring high levels of accuracy, speed and precision, these robots are increasingly utilized across various industries (Murphy 2000). They come with an array of sensors, such as cameras and microphones, which allow them to sense their surroundings. Additionally, they are powered by sophisticated software that helps them process this information and make decisions based on it. Some intelligent robots also have the ability to learn and adapt to new situations through machine learning. This capability allows them to analyze sensor data and adjust their actions accordingly (Weng et al. 2001). These robots are employed in several industries, including manufacturing, healthcare, agriculture and logistics. For example, in manufacturing, intelligent robots perform precise tasks like assembly, welding and painting. In healthcare, they assist in surgeries and provide support to patients (Murphy 2000). In summary, intelligent robots play a key role in Industry

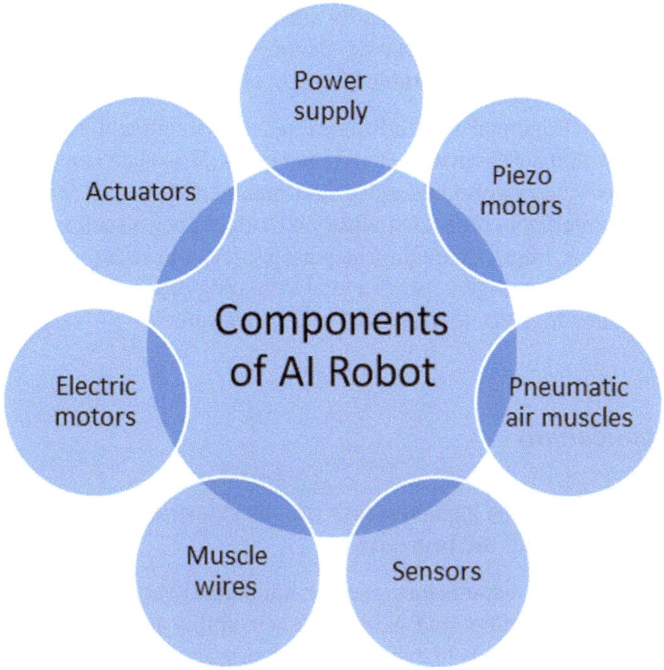

Fig. 6.4 Components of intelligent robot

4.0, offering significant improvements in efficiency and productivity across a range of industries (Fig. 6.4).

6.4.2 Advantages

Intelligent robots provide several advantages within Industry 4.0, as outlined below:

Increased Efficiency: These robots can operate at higher speeds and with more precision compared to human workers, resulting in greater efficiency across various industries.

Cost Reduction: By automating tasks previously carried out by humans, intelligent robots can significantly reduce labor costs and other associated expenses in sectors like manufacturing and logistics (Murphy 2000).

Improved Quality: Intelligent robots perform tasks with exceptional accuracy, contributing to enhanced quality in production processes and other industrial operations.

Increased Safety: In hazardous industries such as construction and manufacturing, robots can undertake dangerous tasks, thereby improving workplace safety by minimizing the risk to human workers (Weng et al. 2001).

Data Collection: Equipped with advanced sensors, intelligent robots can gather valuable data from their environment and operations. This data can be examined to enhance processes and inform better decision-making (Murphy 2000).

In summary, intelligent robots in Industry 4.0 offer benefits such as enhanced efficiency, cost savings, improved quality, increased safety and effective data collection. These benefits make intelligent robots a vital technology for industries ranging from manufacturing and logistics to healthcare and agriculture.

6.4.3 Applications

Intelligent robots are being applied across various industries in Industry 4.0. Below are examples of how they are being used:

Manufacturing: In the manufacturing sector, intelligent robots are tasked with activities like assembly, welding, and painting. Their ability to work faster and more accurately than humans leads to enhanced efficiency and productivity.

Logistics: Within logistics, intelligent robots automate processes such as sorting, packing and shipping. This automation helps reduce operational costs and improve delivery times.

Healthcare: In healthcare, intelligent robots assist with surgeries, other medical procedures and provide companionship to patients. They can also monitor patient health and gather data for analysis.

Agriculture: Intelligent robots in agriculture are employed for tasks such as planting, watering and harvesting crops. This automation boosts efficiency and helps reduce labor expenses (Weng et al. 2001).

Construction: In construction, intelligent robots are used for tasks like bricklaying and pouring concrete, streamlining labor-intensive processes.

Overall, intelligent robots are poised to dramatically enhance efficiency and productivity across multiple sectors within Industry 4.0. Their precision and speed surpass human workers, leading to cost reductions and improved outcomes.

6.4.4 Successful Implementations

(See Table 6.3).

Table 6.3 Successful implementation of intelligent robots

Company	Industry	Application	Benefits
Tesla	Automotive	Manufacturing and warehousing	Improved efficiency and reduced costs
Amazon	E-commerce	Fulfillment centers	Increased efficiency and reduced delivery times
Siemens	Manufacturing and engineering	Production processes	Improved quality and reduced costs
Foxconn	Electronics	Manufacturing	Increased efficiency and reduced labor costs
Boston dynamics	Robotics	Inspection and maintenance	Improved safety and efficiency

6.4.5 Future Trends

Intelligent robots are set to be pivotal in future of Industry 4.0, driven by several emerging advancements and trends:

Enhanced Integration: With the expansion of the Internet of Things (IoT), intelligent robots are anticipated to integrate more seamlessly with technologies such as sensors and artificial intelligence. This will enable them to undertake more complex tasks and make more autonomous decisions (Murphy 2000).

Advancements in Sensors and Perception: Advancements in sensor technology will tackle a major challenge for intelligent machines: enhancing their ability to accurately perceive and understand their environment. Future advancements will enhance robots' ability to sense and respond to their surroundings more effectively (Weng et al. 2001).

Collaborative Robots: Collaborative robots, or cobots, are designed to work alongside human operators rather than replacing them. These robots are anticipated to become more common, offering the benefits of automation while allowing human workers to remain integral to the process (O'Reilly 2017).

Increased Focus on Safety: With the rise of intelligent robots in workplaces, there will be a greater emphasis on safety. This includes developing new safety standards and designing robots to be safer for human interaction.

Personalization: Intelligent robots are expected to leverage data collection and analysis to become more personalized. This capability will allow them to adapt to individual users and perform tasks more efficiently and effectively (O'Reilly 2017).

In summary, the evolution of intelligent robots in Industry 4.0 looks promising, with expected advancements and trends likely to boost their effectiveness in enhancing efficiency, lowering costs and improving safety across different sectors.

6.5 Sustainability Integration

Sustainability Integration in Industry 4.0 marks a significant shift toward aligning technological progress with environmental, social, and economic objectives. As Industry 4.0 evolves through the incorporation of smart technologies, automation, and data-driven methods, sustainability becomes a fundamental focus rather than an afterthought. This paradigm ensures that sustainable practices are integrated throughout the entire value chain, encompassing production, supply chain management, product design, and end-of-life considerations. In manufacturing, the emphasis on sustainability within Industry 4.0 involves improving energy efficiency, minimizing waste and embracing circular economy principles. Smart sensors and IoT devices play a crucial role by providing real-time monitoring of resource use, emissions and environmental impact, enabling data-informed decisions to boost sustainability. Furthermore, advanced analytics and artificial intelligence (AI) contribute by optimizing production processes, reducing resource consumption, and lowering environmental footprints. Beyond environmental aspects, sustainability integration also addresses social and economic dimensions. Industry 4.0 technologies support more flexible and decentralized production models, which can promote local and inclusive economic development. Furthermore, this integration emphasizes ethical and responsible technology use, considering fair labor practices, diversity and social equity. Integrating sustainability into Industry 4.0 is not only a strategic necessity but also a response to global challenges such as climate change and resource depletion. By embedding sustainability at the core of technological innovation, Industry 4.0 can drive positive change, fostering a resilient and responsible industrial ecosystem that meets current needs while ensuring that future generations can fulfill their own.

6.6 Predictive Maintenance

The integration of Artificial Intelligence (AI) and Machine Learning (ML) in predictive maintenance represents more than just an improvement in operational efficiency; it fundamentally redefines asset management within industries. This transformative approach profoundly impacts how resources are utilized and how machinery maintenance is conducted. By leveraging advanced data analytics, AI and ML algorithms not only forecast potential equipment failures but also provide deep insights into the complex interplay of factors affecting equipment health. The real-time analysis of extensive datasets—such as temperature, vibration and performance metrics—creates a comprehensive view of industrial assets. This holistic perspective enables proactive measures and informed decision-making. The fusion of AI and ML in predictive maintenance is not just a technological advancement but a strategic necessity, driving industries toward a future where human expertise and artificial intelligence work together to achieve exceptional operational resilience, cost efficiency and sustainable asset management. Ultimately, this integration forms a pivotal element in the narrative of industrial excellence, positioning predictive maintenance, empowered by AI and ML, as a cornerstone of innovation and a key driver of future industry success.

Data Analysis and Pattern Recognition: AI and ML algorithms are adept at processing large amount of data from industrial equipment. They can find patterns and anomalies that may signal potential failures or decreasing performance. This data-driven method allows for more precise predictions compared to traditional maintenance schedules.

Predictive Modeling: Machine learning models can be programmed using historical data to create predictive models. These models learn the typical behavior of machinery and equipment and can detect deviations from this norm, indicating possible issues. Predictive modeling helps forecast when equipment may fail, enabling timely maintenance actions.

Condition Monitoring: AI and ML facilitate continuous real-time monitoring of equipment conditions. This involves tracking various parameters like temperature, vibration, and pressure. By analyzing these metrics, algorithms can evaluate the health of the equipment and predict maintenance needs.

Fault Diagnosis: When anomalies are detected, AI algorithms can diagnose specific faults or issues beyond just predicting failures. This detailed analysis helps maintenance teams understand the root cause of problems, leading to more targeted and effective repairs.

Optimizing Maintenance Schedules: Predictive maintenance refines the timing of maintenance activities. Instead of adhering to fixed schedules or prematurely replacing components, organizations can perform maintenance precisely when necessary. This method minimizes downtime, reduces expenses and extends the lifespan of equipment.

Prescriptive Maintenance: Advanced AI systems can offer prescriptive maintenance recommendations. These systems not only predict when maintenance is needed but also suggest optimal actions to take. This proactive strategy aids organizations in making informed decisions to prevent major failures.

Continuous Improvement: Machine learning algorithms are capable of learning and adapting over time. As more data is collected and the system encounters various scenarios, these algorithms can enhance their accuracy in predicting maintenance needs, improving effectiveness in the long run.

Cost Savings and Efficiency: By reducing unplanned downtime, decreasing unnecessary maintenance, and optimizing resource use, AI and ML drive significant cost reductions and enhance operational efficiency.

6.7 Human–Machine Collaboration

In Industry 4.0, merging human and machine capabilities signifies a major transformation. This collaboration between human creativity and the computational prowess of technologies such as Artificial Intelligence (AI), Machine Learning (ML) and the Internet of Things (IoT) represents a significant evolution. Instead of replacing human roles, these technologies augment them, integrating human attributes like creativity, intuition, and problem-solving with the rapid processing, accuracy, and vast data-handling capabilities of intelligent systems. This integration leads to exceptional productivity, efficiency and innovation. The exploration of this collaboration reveals complex dynamics, challenges, and benefits, highlighting its profound impact across various industries and emphasizing the crucial role of humans in an increasingly automated and technologically advanced landscape.

6.7.1 Synergizing Human Ingenuity with Intelligent Machines

In this era of rapid innovation, the collaboration between human creativity and intelligent machines is profoundly transforming industries and advancing societies to new heights. This partnership views AI not as a competitor but as an enabler, taking over routine tasks and enabling individuals to concentrate on more intricate and innovative tasks. Emphasizing human-centric design, ethical principles and ongoing learning, this collaboration merges human intuition with AI's analytical capabilities to tackle intricate problems. As it progresses, this alliance navigates regulatory and cultural challenges, paving the way for a future where the fusion of human ingenuity and artificial intelligence unlocks vast new possibilities and drives societal advancement.

6.7.2 Catalytic Technologies: AI, ML, and IoT Unleashed

Catalytic Technologies epitomize the leading edge of innovation, Artificial Intelligence (AI), Machine Learning (ML) and the Internet of Things (IoT) to drive transformative changes across various industries. This integration creates a vibrant ecosystem where AI's cognitive functions, ML's capacity for adaptive learning and IoT's connectivity work together to enhance efficiency, provide deeper insights and improve decision-making. By combining these powerful technologies, organizations can tackle modern challenges more effectively, automate processes, forecast trends and develop data-driven strategies. This shift not only optimizes operations but also paves the way for new innovations, positioning data as a crucial asset for growth and competitive advantage. Catalytic Technologies thus emerge as a leading force in the technological arena, where the combination of AI, ML and IoT not only refines current practices but also paves the way for a future characterized by exceptional efficiency and creativity.

6.7.3 Challenges in Harmonizing Human–Machine Dynamics

Navigating the complex challenge of integrating human and machine dynamics involves aligning the exceptional capabilities of human intelligence with the powerful capabilities of artificial intelligence (AI). This task requires achieving a delicate balance, combining the unique strengths of both humans and machines while addressing the challenges that emerge from this transformative partnership. Ethical issues are significant, necessitating a solid framework to ensure that AI development and use are transparent, fair, and accountable amidst rapid technological advancements. The shift in the job market due to automation calls for a substantial effort in ongoing education and upskilling to develop a workforce that is flexible and ready for future challenges. Seamless collaboration presents challenges in user interface design, requiring a blend of AI capabilities with human-centered design principles. Overcoming cultural resistance is also crucial, involving efforts to address concerns and promote a positive view of AI's role in society. The journey to harmonize human–machine interactions is a complex endeavor, demanding a comprehensive approach that includes technological progress, ethical considerations, regulatory strategies, and societal adaptation. This quest for balance guides us through the evolving landscape where human creativity meets the powerful potential of intelligent machines.

6.7.4 Future Horizons: Anticipating Advancement

Venturing into the realm of Future Horizons marks an extraordinary journey into the expansive domains of technological advancement and human creativity. This ambitious exploration foresees a blend of innovations that go beyond mere imagination, merging elements of artificial intelligence, quantum computing, biotechnology and renewable energy. Rather than just gradual improvements, we anticipate transformative breakthroughs that could fundamentally alter our world. Future horizons represents the cutting edge of a new age, where these emerging technologies aim to tackle significant global issues such as environmental sustainability and healthcare equity, paving the way for remarkable progress. However, as we approach these thrilling possibilities, we also face substantial challenges—including ethical dilemmas, societal shifts and the need for thoughtful governance. This transformative journey invites us to look forward, adapt, and engage with the forces shaping our future. Future Horizons stands as a guiding beacon, leading us through an era where innovation pushes the limits of what is possible and inspires humanity to reach new heights.

6.8 Challenges and Limitations

6.8.1 Data Privacy and Security Concerns

In Industry 4.0, safeguarding data privacy and security is essential, especially with the rise of IoT devices that gather and transmit substantial amounts of data. This data often includes sensitive information such as health records, financial details, and other confidential materials (Popescu and Popescu 2017). Protecting this data is essential to prevent breaches, cyberattacks, and unauthorized access. To mitigate these risks, organizations need to implement comprehensive security measures, including encryption, access controls, and intrusion detection systems. It's also vital for companies to educate their employees and partners about the importance of data privacy and ensure they adhere to best practices. Compliance with relevant regulations, such as the General Data Protection Regulation (GDPR) in Europe or the Health Insurance Portability and Accountability Act (HIPAA) in the United States, is another key aspect. Organizations must ensure they meet these regulatory requirements and have procedures in place for managing breaches or security incidents (Popescu and Popescu 2017). In summary, safeguarding data privacy and security is a fundamental issue in Industry 4.0. Companies need to take proactive measures to protect both their own data and that of their customers and partners.

6.8.2 Legal and Regulatory Challenges

The widespread adoption of AI, ML, and IoT in the context of Industry 4.0 has introduced several legal and regulatory challenges. Some key challenges include:

Liability: As these technologies gain autonomy, determining liability for harm or damage caused becomes more complex. Identifying responsibility when an autonomous system causes harm may necessitate changes to current legal frameworks (Shi et al. 2018).

Intellectual Property: These technologies rely on data, algorithms, and other intellectual property, complicating the determination of ownership and licensing, especially when multiple parties contribute to the technology's development.

Privacy: Data privacy remains a critical issue. Regulations like the GDPR and HIPAA impose stringent requirements on the collection, use and storage of personal data, requiring companies to ensure compliance (Popescu and Popescu 2017).

Standards: The widespread adoption of these technologies creates a need for standardization and interoperability. This can be challenging as new technologies emerge and existing standards become outdated.

Ethical Concerns: The use of AI, ML, and IoT raises ethical issues, including concerns about bias, discrimination, and transparency. Companies must adhere to ethical guidelines and be transparent about their technologies' use (Shi et al. 2018).

Addressing these legal and regulatory challenges is crucial for companies in Industry 4.0 to ensure compliance and proactively manage these issues.

6.8.3 Implementation Challenges

Deploying AI, ML, and IoT technologies in Industry 4.0 presents several challenges. Key challenges include the following:

Data Quality: The effectiveness of these technologies depends on the accuracy, reliability and relevance of the data they use. Ensuring high-quality data can be challenging, especially as the amount of data produced by IoT devices continues to increase (Popescu and Popescu 2017).

Technical Expertise: Deploying these technologies requires specialized skills, which can be hard to find and costly to acquire. Companies need to either develop in-house expertise or partner with external providers for implementation and maintenance.

Integration: Integrating new technologies with existing systems can be complex, especially when dealing with legacy systems. A clear integration plan is essential to minimize disruption during implementation (Shi et al. 2018).

Cost: The implementation of AI, ML, and IoT technologies can be costly, particularly when significant hardware and software upgrades are needed. Companies must have a strong business case to justify the investment (Popescu and Popescu 2017).

Change Management: The introduction of these technologies often necessitates significant changes in business processes and workflows. Companies need an effective change management strategy to minimize disruption and prepare employees for the transition.

Addressing these challenges requires thorough planning and execution to ensure effective deployment of AI, ML and IoT technologies in Industry 4.0.

6.8.4 Potential Ethical Issues

The effective deployment of AI, ML, and IoT in Industry 4.0 can indeed raise several ethical concerns. Here are some key issues to consider:

Bias: AI and ML systems can inadvertently perpetuate societal biases if trained on biased datasets. This could manifest in discriminatory practices related to gender, race, or other factors. Ensuring diverse and representative datasets is crucial to mitigating this risk (Shi et al. 2018).

Privacy: The extensive data collection and analysis associated with IoT devices can lead to privacy concerns. Companies must adhere to data privacy regulations, such as GDPR and maintain transparency with customers about data usage (Popescu and Popescu 2017).

Autonomy: As AI and ML systems advance, they may make decisions autonomously, raising questions about accountability and responsibility. It's important to establish clear oversight mechanisms to ensure that decisions made by these systems are subject to appropriate human review and accountability.

Job Displacement: The automation facilitated by these technologies can result in job displacement for certain workers. Companies should assess the impact on their workforce and invest in reskilling or retraining programs to support employees through transitions.

Security: IoT devices can introduce security vulnerabilities, as they may be more susceptible to hacking compared to traditional IT systems. Implementing robust security measures is essential to protect systems and sensitive data (Popescu and Popescu 2017).

Addressing these ethical issues involves proactive measures and thoughtful planning to ensure that the incorporation of AI, ML and IoT technologies in Industry 4.0 is conducted responsibly and with consideration for their broader impact.

6.9 Conclusion

In summary, Industry 4.0 marks a pivotal shift, fueled by the convergence of AI, ML and IoT technologies. These innovations promise significant improvements in efficiency, productivity and decision-making across diverse sectors, offering substantial advantages to both businesses and consumers. However, the path to realizing these benefits is fraught with challenges, including technical difficulties, legal considerations, ethical concerns and implementation hurdles. Addressing these challenges is crucial to ensuring that the deployment of these technologies is both responsible and ethical. Despite the hurdles, the future of AI, ML and IoT in Industry 4.0 is promising. As these technologies continue to evolve, they are likely to lead to innovative applications that will further transform industries and improve our daily lives (O'Reilly 2017). Collaborative efforts between companies and policymakers will be essential to navigating these challenges and ensuring that the advantages of Industry 4.0 are broadly and equitably distributed.

6.10 Implications for the Future of Industry 4.0

The future of Industry 4.0 looks exceptionally promising as more companies embrace AI, ML, and IoT technologies to fuel innovation and boost operational effectiveness. The convergence of these technologies is set to revolutionize industry operations, driving improvements in productivity, product and service quality, and customer satisfaction. One key development is the rise of intelligent automation, which merges AI with robotics to handle complex tasks and processes. This trend is poised to significantly impact the manufacturing sector by boosting production speed, precision, and cost-effectiveness (O'Reilly 2017). Another notable trend is the expansion of edge computing, which facilitates data processing closer to its source (Porter and Heppelmann 2015). This advancement enhances the speed and reliability of data analysis and decision-making, benefiting industries such as healthcare, transportation, and logistics. As AI, ML and IoT technologies evolve, we can anticipate a surge of innovative applications across various sectors. Nonetheless, addressing the related challenges is essential—such as data privacy, security concerns, legal and regulatory issues, implementation difficulties, and ethical considerations—to ensure these technologies are utilized responsibly and ethically.

Overall, the prospects for Industry 4.0 are bright, with AI, ML and IoT poised to drive transformative changes and open up new possibilities that will redefine industry standards and practices.

6.10.1 Recommendations for Further Research

To advance the implementation and understanding of AI, ML, and IoT technologies in Industry 4.0, further research is crucial. First, exploring new applications in industries that have not yet fully adopted these technologies, such as agriculture, construction, and education, can uncover additional benefits and use cases. Second, examining the ethical implications of these technologies is essential, particularly in addressing issues related to bias, discrimination and privacy. Research should also focus on understanding the impact of intelligent automation on employment shifts and evolving job roles, while identifying strategies to support workers in transitioning to new roles and industries. Additionally, investigating the effects of edge computing on data privacy and security is important, with the development of best practices for the implementation and protection of edge computing systems. Finally, assessing the legal and regulatory challenges associated with these technologies will help in formulating policies and regulations to ensure their responsible and ethical deployment. By addressing these areas, we can acquire a more comprehensive insight into the potential benefits and challenges associated with AI, ML and IoT, ensuring that these technologies are implemented in a manner that maximizes their positive impact on society.

References

Lee J, Kao H, Yang S (2014) Service innovation and smart analytics for industry 4.0 and big data environment. Procedia CIRP 16:3–8

Miorandi D, Sicari S, Pellegrini FD, Chlamtac I (2012) Internet of things: vision, applications and research challenges. Ad Hoc Netw 10(7):1497–1516

Murphy RR (2000) Introduction to AI robotics. MIT Press

O'Reilly T (2017) WTF?: what's the future and why it's up to us. HarperCollins

Popescu D, Popescu G (2017) Industry 4.0 and smart manufacturing—a review of research issues and application examples. Int J Comput Commun Control 12(4):390–398

Porter ME, Heppelmann JE (2015) How smart, connected products are transforming competition. Harv Bus Rev 93(10):96–114

Rong K, Li H, You X, Xu X (2018) IoT-based intelligent perception and access of manufacturing resource toward cloud manufacturing. IEEE Trans Industr Inf 14(8):3473–3482

Russell SJ, Norvig P (2009) Artificial intelligence: a modern approach, 3rd edn. Prentice Hall, Upper Saddle River, NJ

Shi Y, Yang Q, Yang Y (2018) Industry 4.0 and its future prospects. J Risk Financ Manage 11(4):1–17

Singla P, Kraus S (2014) Autonomous decision making: multidisciplinary perspectives. Springer Science & Business Media

Wang Y, Wang X, Xue Y (2016) An IoT-based quality control system for mass customization production. IEEE Trans Industr Inf 12(5):1898–1907

Weng J, McClelland JL, Pentland A, Sporns O, Stockman I, Sur M, Thelen E (2001) Autonomous mental development by robots and animals. Science 291(5504):599–600

Chapter 7
Synergizing VLSI and Neural Networks: Unveiling the Nexus of Innovation in Industry 4.0

Pawan Whig, Balaram Yadav Kasula, Nikhitha Yathiraju, and Imran Ahmed Khan

Abstract This book chapter delves into the transformative realm of Industry 4.0, presenting a comprehensive exploration of the integration between Very Large-Scale Integration (VLSI) and Neural Networks. Industry 4.0, marked by the intelligent interplay of cutting-edge technologies, demands innovative solutions for enhanced manufacturing processes. VLSI technology is synergistically combined with the capabilities of Neural Networks, offering intelligent data processing and decision-making. The book chapter meticulously examines the collaborative potential of these technologies, delving into design considerations, implementation strategies, and optimization techniques specific to VLSI-based Neural Networks. Real-world use cases are discussed to exemplify the practical implications of this integration, showcasing its efficacy in improving manufacturing processes, predictive maintenance, and quality control. The study not only addresses the technical aspects of the integration but also highlights challenges such as scalability, real-time processing, and system reliability. By doing so, it contributes valuable insights for researchers, practitioners, and industry professionals aiming to harness the power of VLSI and Neural Networks in the dynamic landscape of Industry 4.0, thereby paving the way for future innovations in intelligent manufacturing systems.

Keywords VLSI · Neural networks · Industry 4.0 · Integration · Automation · Manufacturing

---

P. Whig
Vivekananda Institute of Professional Studies, New Delhi, India

B. Y. Kasula
Department of Information Technology, University of the Cumberlands, Williamsburg, KY, USA

N. Yathiraju
University of the Cumberlands, Williamsburg, KY, USA

I. A. Khan (✉)
Department of E&C, Jamia Millia Islamia, New Delhi, India
e-mail: iakhan2@jmi.ac.in

A. Rai et al. (eds.), *Convergence of Artificial Intelligence, Machine Learning, and the Internet of Things in Industry 4.0 Applications*, Transactions on Computer Systems and Networks, https://doi.org/10.1007/978-981-96-5269-3_7

7.1 Introduction

The advent of the twenty-first century has witnessed a paradigm shift in the landscape of technology, catalyzing the emergence of a new era known as Industry 4.0. Characterized by the fusion of digital technologies, intelligent automation, and data-driven decision-making, Industry 4.0 represents a revolutionary phase in manufacturing and production. This introduction provides a comprehensive exploration of the key facets of Industry 4.0, elucidating its origins, defining characteristics, and the profound impact it has on various sectors of the global economy. Industry 4.0, often referred to as the fourth industrial revolution, builds upon the achievements of its predecessors by integrating smart technologies into every facet of industrial processes. The first industrial revolution, driven by mechanization and steam power, laid the foundations for mass production. Subsequent revolutions introduced electricity, assembly lines, and computerization. Industry 4.0 goes beyond automation by interconnecting devices, systems, and humans through the Internet of Things (IoT), cloud computing, and cyber-physical systems. This interconnectedness fosters a dynamic and responsive ecosystem where machines communicate, analyze data, and make decisions in real-time.

Central to the ethos of Industry 4.0 is the notion of smart factories, where intelligent systems collaborate seamlessly in the production process. These smart factories leverage technologies such as Artificial Intelligence (AI), Machine Learning (ML), and Big Data analytics to optimize operations, enhance efficiency, and facilitate predictive maintenance. The integration of these technologies heralds a departure from traditional linear supply chains to interconnected and autonomous systems that can adapt to changing demands and unforeseen disruptions. One of the cornerstones of Industry 4.0 is the concept of cyber-physical systems, where the virtual and physical realms converge. In this symbiotic relationship, physical systems are embedded with sensors and actuators that enable them to interact with the digital world. These cyber-physical systems create a closed-loop network, enabling real-time monitoring, analysis, and control of physical processes. This integration empowers industries to achieve unprecedented levels of precision, customization, and agility in their operations. A critical enabler of Industry 4.0 is the Internet of Things (IoT), which forms the backbone of the interconnected ecosystem. The proliferation of smart devices, sensors, and actuators embedded in machinery, products, and infrastructure facilitates the seamless exchange of data. This interconnectedness facilitates the creation of a digital twin—a virtual representation of a physical object or system. Digital twins provide a real-time mirror of the physical entity, allowing for simulation, analysis, and optimization of its performance.

The integration of Big Data analytics plays a pivotal role in harnessing the vast amounts of data generated by Industry 4.0 systems. Big Data analytics not only facilitates real-time decision-making but also unlocks valuable insights for process optimization and strategic planning. The predictive capabilities of analytics enable industries to anticipate maintenance needs, forecast demand, and optimize supply chain operations, fostering a proactive and responsive approach to business challenges.

Artificial Intelligence and Machine Learning emerge as the cognitive engines driving Industry 4.0's transformative potential. These technologies empower machines with the ability to learn from data, adapt to changing conditions, and make autonomous decisions. In smart factories, AI and ML algorithms analyze complex datasets, identify patterns, and optimize production processes. This cognitive autonomy not only enhances operational efficiency but also opens new frontiers for innovation and product customization. As we navigate through the intricacies of Industry 4.0, cybersecurity becomes a paramount concern. The interconnected nature of cyber-physical systems exposes them to potential vulnerabilities, making robust cybersecurity measures indispensable. Safeguarding sensitive data, protecting against cyber threats, and ensuring the integrity of interconnected systems become imperative to realizing the full potential of Industry 4.0. This introduction provides a comprehensive overview of the transformative landscape of Industry 4.0. From its historical roots to the convergence of digital technologies, Industry 4.0 represents a seismic shift in how we conceptualize and approach manufacturing and production. The integration of smart technologies, interconnected systems, and data-driven decision-making heralds a future where industries are not only automated but intelligent, adaptive, and resilient in the face of evolving challenges. As we embark on this journey into the heart of Industry 4.0, the subsequent sections will delve deeper into its key components, challenges, and the myriad opportunities it presents for industries across the globe.

This Table 7.1 provides a snapshot of the literature review, summarizing key contributions and highlighting the identified research gaps in each referenced paper.

7.1.1 Industry 4.0 and Its Evolution

The first section of this chapter unravels the evolutionary trajectory of Industry 4.0, tracing its roots and examining the pivotal moments that led to the emergence of this transformative paradigm. From the mechanization of the first industrial revolution to the digitalization of the third, the narrative unfolds how Industry 4.0 represents a convergence of digital technologies, intelligent automation, and data-driven decision-making. Understanding this evolution lays a foundation for comprehending the challenges and opportunities that lie within the heart of the fourth industrial revolution.

7.1.2 The Role of VLSI and Neural Networks

The second segment delves into the specific roles played by Very Large-Scale Integration (VLSI) and Neural Networks within the context of Industry 4.0. VLSI technology, renowned for creating compact and energy-efficient hardware, intertwines with the sophisticated capabilities of Neural Networks. Neural Networks, inspired

Table 7.1 Literature review and research gaps

Reference	Summary	Contributions	Research gap identified
An et al. (2020)	Self-learning and low-power neuromorphic system for Industry 4.0	Proposes a self-learning neuromorphic system for Industry 4.0	Limited exploration of real-world implementation challenges and scalability issues
Cheng et al. (2023)	Reliability exploration of multi-bit-width accelerator for deep neural networks	Investigates the reliability of system-on-chip with multi-bit-width accelerator	Lacks exploration of the energy efficiency trade-offs in real-world applications
Misra and Saha (2010)	Survey of two decades of progress in hardware-based artificial neural networks	Offers a comprehensive survey of hardware-based artificial neural networks	Limited discussion on recent advancements and emerging trends in hardware-based neural networks
Abderrahmane et al. (2020)	Design space exploration of hardware spiking neurons for embedded AI	Explores the design space of hardware spiking neurons for embedded AI	Insufficient coverage of real-world use cases and performance benchmarks
Zaman et al. (2021)	Review of custom hardware architectures for deep learning on portable devices	Provides a review of custom hardware architectures for deep learning on portable devices	Overlooks recent advancements in energy-efficient hardware architectures for deep learning
Manu et al. (2021)	Co-exploration of graph neural network and network-on-chip design using AutoML	Co-explores graph neural network and network-on-chip design using AutoML	Limited discussion on the impact of AutoML on design productivity and efficiency
Venieris et al. (2018)	Survey on toolflows for mapping convolutional neural networks on FPGAs	Surveys toolflows for mapping CNNs on FPGAs	Lacks detailed analysis of the challenges in mapping complex CNNs to FPGA architectures
Du et al. (2015)	Leveraging the error resilience of neural networks for energy-efficient accelerators	Explores leveraging error resilience for energy-efficient neural network accelerators	Limited examination of the implications of errors on the accuracy of neural network applications
Garofalo et al. (2022)	Heterogeneous in-memory computing cluster for flexible end-to-end inference of deep neural networks	Introduces a heterogeneous in-memory computing cluster for flexible DNN inference	Insufficient discussion on the practical challenges of deploying such clusters in real-world scenarios

(continued)

Table 7.1 (continued)

Reference	Summary	Contributions	Research gap identified
Kumar et al. (2021)	Knowledge-based neural networks for design space exploration of on-chip interconnect networks	Utilizes knowledge-based neural networks for fast design space exploration	Limited consideration of the scalability and generalizability of knowledge-based neural networks in diverse design spaces
Lourenço et al. (2018)	Exploration of analog IC designs via artificial neural networks	Explores promising analog IC designs using artificial neural networks	Lack of examination on the robustness of neural network-based exploration in dealing with complex analog design spaces
Deligiannis et al. (2021)	Integration of reliability and security mechanisms for fault resilience of neural networks	Proposes integrating reliability and security for enhancing fault resilience of neural networks	Limited exploration of the trade-offs between reliability and security in complex neural network architectures
Capra et al. (2020)	Survey on hardware and software optimizations for accelerating deep neural networks	Surveys current trends and challenges in accelerating deep neural networks	Limited discussion on the implications of optimization techniques on model interpretability and explainability
Zhu et al. (2020)	Exploration of logic optimizations with reinforcement learning and graph convolutional network	Explores logic optimizations using reinforcement learning and graph convolutional network	Lacks in-depth analysis of the scalability and efficiency of these optimization techniques for complex logic circuits
Lopera et al. (2021)	Survey of graph neural networks for electronic design automation	Surveys the application of graph neural networks in electronic design automation	Limited exploration of the challenges in adapting graph neural networks to different EDA tasks and benchmarks
Krishnan et al. (2021)	Chiplet-based scalable in-memory acceleration with mesh for deep neural networks	Proposes SIAM, a chiplet-based in-memory acceleration for scalable DNNs	Insufficient discussion on the practical challenges of integrating chiplet-based accelerators into existing hardware architectures

by the human brain, excel in intelligent data processing and decision-making. This section explores how the integration of VLSI and Neural Networks forms a symbiotic relationship, contributing to the intelligent automation and data-centric dynamics crucial for the success of Industry 4.0 applications.

7.1.3 Objectives of the Chapter

The third segment outlines the objectives driving the exploration in this chapter. As we embark on a comprehensive examination of the integration of VLSI and Neural Networks in the context of Industry 4.0, the chapter seeks to elucidate the design considerations, implementation strategies, and optimization techniques specific to this integration. Real-world use cases will be dissected to illustrate the practical implications and potential benefits, providing insights into the challenges faced and solutions devised. By articulating these objectives, the chapter aims to contribute valuable insights to researchers, practitioners, and industry professionals navigating the intricate landscape of Industry 4.0.

7.2 VLSI Technology

Very Large-Scale Integration (VLSI) technology stands as a cornerstone in the realm of modern electronics, playing a pivotal role in the miniaturization and integration of complex electronic circuits. As a progression from Large-Scale Integration (LSI), VLSI involves the integration of thousands to millions of transistors onto a single semiconductor chip. This technological marvel has been a driving force behind the exponential growth in computational power and the development of sophisticated electronic devices.

The essence of VLSI lies in the ability to densely pack an enormous number of transistors, resistors, and capacitors onto a single silicon chip. This integration is achieved through advanced fabrication processes, often involving photolithography and etching techniques. VLSI technology has witnessed remarkable advancements, leading to the production of chips with ever-increasing transistor counts while simultaneously reducing the size of individual components. The benefits of VLSI technology are manifold. First and foremost, it allows for the creation of compact and lightweight electronic devices, making it instrumental in the development of portable gadgets, smartphones, and wearables. Moreover, the increased integration density enhances the overall performance of electronic systems, leading to faster processing speeds and reduced power consumption. The economic implications are also significant, as the mass production of VLSI chips has contributed to a substantial decrease in the cost per transistor.

In the context of Industry 4.0, VLSI technology plays a crucial role in the design and development of smart sensors, microcontrollers, and embedded systems. These components form the backbone of cyber-physical systems, facilitating real-time monitoring, control, and communication in smart manufacturing environments. The compact nature and efficiency of VLSI chips make them well-suited for integration into a diverse array of devices that constitute the Internet of Things (IoT), a key component of Industry 4.0. As we navigate through the intricacies of VLSI technology, considerations such as design methodologies, fabrication processes, and the challenges associated with scaling down to smaller process nodes come to the forefront. The relentless pursuit of Moore's Law, predicting the doubling of transistor counts roughly every two years, has been a driving force in the VLSI landscape, posing both technical and economic challenges. VLSI technology stands as a testament to the marvels of modern electronics, enabling the integration of complex circuits on a scale unimaginable just a few decades ago. Its impact extends far beyond consumer electronics, permeating industries, healthcare, communication, and the fabric of our interconnected world. Understanding the intricacies of VLSI technology is imperative for those venturing into the realms of semiconductor design, smart manufacturing, and the dynamic landscape of Industry 4.0.

7.2.1 Fundamentals of VLSI

The exploration of VLSI begins with a dive into its fundamental principles. This section elucidates the basic concepts that underpin Very Large-Scale Integration, including transistor characteristics, logical gates, and the principles of digital circuit design. Understanding the fundamentals lays a robust foundation for comprehending the intricate design processes and optimization techniques employed in the realm of VLSI.

7.2.2 VLSI Components and Architecture

Building upon the fundamentals, this segment delves into the key components and architectural considerations within the realm of VLSI. It explores the diverse array of components that constitute a VLSI chip, ranging from transistors and flip-flops to more complex entities like arithmetic logic units (ALUs) and memory elements. Additionally, it unravels the architectural choices and design methodologies that engineers employ to create efficient and reliable VLSI circuits.

7.2.3 Advances in VLSI Technology

The rapid evolution of VLSI technology is at the forefront of this section. It navigates through the historical advancements that have propelled VLSI from its nascent stages to the present era, marked by nanoscale fabrication processes and staggering transistor counts. Insights into innovations such as FinFET technology, three-dimensional integrated circuits (3D ICs), and emerging non-volatile memory technologies offer a glimpse into the cutting-edge developments driving the field forward.

7.2.4 VLSI for Industry 4.0

The integration of VLSI into the landscape of Industry 4.0 takes center stage in this segment. It explores how VLSI technology becomes an enabler for the intelligent automation and data-centric dynamics demanded by Industry 4.0 applications. From the role of VLSI in the creation of smart sensors to its contributions in embedded systems and microcontrollers, this section outlines the specific applications and implications of VLSI in the context of the fourth industrial revolution. Understanding this integration is paramount for those seeking to harness the transformative power of VLSI in the Industry 4.0 landscape.

7.3 Neural Networks

Neural Networks represent a paradigm of machine learning inspired by the intricate structure and functioning of the human brain. Comprising interconnected nodes, or neurons, these networks excel in learning patterns, making them a powerful tool for tasks such as pattern recognition, classification, and regression. This section delves into the fundamental principles, architecture, and functioning of Neural Networks, unraveling the mechanisms that render them adept at processing and analyzing complex data.

7.3.1 Fundamentals of Neural Networks

The exploration begins by elucidating the core principles that govern Neural Networks. It delves into the structure of a basic neural unit, exploring concepts such as activation functions, weights, and biases. The section also covers the fundamental mathematical operations that occur within a neural network, including forward and backward propagation, laying the groundwork for a deeper understanding of their functioning.

7.3.2 Types of Neural Networks and Architectures

Neural Networks exhibit diverse architectures tailored for specific tasks. This segment navigates through various types of neural networks, including feedforward, recurrent, and convolutional networks. It explores how these architectures cater to different data types and complexities, unraveling the strengths and limitations of each. Additionally, considerations such as network depth, width, and topology are dissected, providing insights into the nuances of neural network design.

7.3.3 Training and Optimization Techniques

Training a neural network involves fine-tuning its parameters to make accurate predictions. This section elucidates the training process, exploring optimization techniques such as gradient descent, backpropagation, and regularization methods. Understanding how neural networks learn and adapt to data is pivotal for harnessing their capabilities effectively.

7.3.4 Advances in Neural Network Technology

The evolution of Neural Network technology is explored in this segment, covering recent advancements that have propelled the field forward. Concepts such as deep learning, transfer learning, and the integration of neural networks with other technologies are discussed. The exploration of emerging architectures and training paradigms provides a glimpse into the cutting-edge developments shaping the future of Neural Networks.

7.3.5 Neural Networks for Industry 4.0

As Industry 4.0 emphasizes intelligent automation and data-driven decision-making, this section outlines the pivotal role played by Neural Networks in this context. From predictive maintenance and quality control to process optimization, the applications of Neural Networks in Industry 4.0 are diverse. The section explores real-world use cases, showcasing how Neural Networks contribute to creating intelligent systems that align with the goals of the fourth industrial revolution. Understanding the integration of Neural Networks in Industry 4.0 is crucial for professionals seeking to leverage these technologies in the manufacturing landscape.

7.4 Integration of VLSI and Neural Networks

The integration of Very Large-Scale Integration (VLSI) and Neural Networks marks a convergence of hardware efficiency and intelligent data processing. This section explores the symbiotic relationship between these two transformative technologies, uncovering the intricacies of their collaboration and the resulting synergies that contribute to advanced applications in various domains.

7.4.1 Design Considerations in VLSI-Neural Network Integration

The journey into the integration begins by examining the design considerations crucial for harmonizing VLSI and Neural Networks. This includes exploring the challenges of mapping neural network architectures onto hardware, optimizing for power efficiency, and addressing communication bottlenecks. Understanding the intricacies of this integration at the design level sets the stage for robust and efficient implementations.

7.4.2 Implementation Strategies

This section delves into the practical aspects of implementing Neural Networks on VLSI platforms. It explores strategies for translating neural network models into hardware configurations, considering factors like parallelism, pipelining, and memory hierarchies. The efficient utilization of VLSI resources for neural network computations is central to achieving real-time processing and optimal performance.

7.4.3 Optimization Techniques for VLSI-Neural Network Systems

Optimization becomes a focal point in this exploration, uncovering techniques to enhance the performance of VLSI-Neural Network systems. This includes examining approaches for minimizing power consumption, improving throughput, and maximizing resource utilization. The section also discusses trade-offs and considerations in achieving a balance between accuracy and efficiency in neural network computations within the constraints of VLSI platforms.

7.4.4 Real-World Applications and Use Cases

The practical implications of VLSI-Neural Network integration are elucidated through a scrutiny of real-world applications and use cases. From edge computing devices and wearable technology to embedded systems in smart environments, this section showcases how the integration manifests in diverse domains. Case studies highlight the transformative impact of VLSI-Neural Network systems in areas such as image recognition, natural language processing, and autonomous systems.

7.4.5 Challenges and Future Directions

As with any integration, challenges arise, and this segment explores the hurdles encountered in combining VLSI and Neural Networks. It delves into issues such as scalability, adaptability to emerging neural network architectures, and the need for standardized interfaces. Additionally, the discussion extends to future directions, envisioning advancements that could further enhance the synergy between VLSI and Neural Networks.

Understanding the intricacies of VLSI-Neural Network integration is paramount for engineers, researchers, and industry professionals navigating the landscape of intelligent systems. This integration not only leverages the efficiency of VLSI in hardware implementation but also harnesses the cognitive capabilities of Neural Networks, paving the way for innovative solutions across a spectrum of applications in the ever-evolving technological landscape.

7.5 Applications and Case Studies

This section delves into the diverse array of applications and real-world case studies that showcase the practical implications of integrating Very Large-Scale Integration (VLSI) and Neural Networks. The synergy between hardware efficiency and intelligent data processing unfolds in various domains, offering insights into the transformative impact of this integration.

7.5.1 Edge Computing and IoT Devices

The integration of VLSI and Neural Networks finds practical applications in edge computing and Internet of Things (IoT) devices. Case studies explore scenarios where compact, energy-efficient VLSI chips seamlessly incorporate neural network functionalities. This results in intelligent edge devices capable of real-time data processing, making them ideal for applications such as smart homes, industrial IoT, and wearable technology.

7.5.2 Image and Speech Recognition Systems

The section delves into the realm of image and speech recognition, highlighting how the integration of VLSI and Neural Networks enhances the efficiency of these systems. Case studies showcase scenarios where VLSI chips with embedded neural networks contribute to faster and more accurate image and speech processing. Applications range from facial recognition in security systems to voice-activated devices in smart environments.

7.5.3 Autonomous Systems and Robotics

The integration of VLSI and Neural Networks plays a pivotal role in the development of autonomous systems and robotics. Case studies illustrate instances where VLSI-Neural Network synergy enables intelligent decision-making in real-time. This includes scenarios such as autonomous vehicles, drones, and robotic systems that navigate and interact with their environment, showcasing the potential for enhanced efficiency and adaptability.

7.5.4 Healthcare Diagnostics and Monitoring

In the healthcare domain, VLSI-Neural Network integration contributes to advancements in diagnostics and patient monitoring. Case studies explore applications where compact VLSI chips process medical data using neural networks, leading to improved diagnostic accuracy and real-time health monitoring. This has implications for personalized medicine, early disease detection, and the development of portable healthcare devices.

7.5.5 Smart Manufacturing and Industry 4.0

The transformative impact of VLSI-Neural Network integration in smart manufacturing and Industry 4.0 is highlighted through case studies. This includes applications in predictive maintenance, quality control, and process optimization. The section demonstrates how the collaboration between VLSI and Neural Networks creates intelligent systems capable of enhancing efficiency, reducing downtime, and adapting to dynamic manufacturing environments.

7.5.6 Challenges and Lessons Learned

While exploring applications and case studies, the section also addresses challenges encountered and lessons learned from real-world implementations. This includes considerations related to scalability, power consumption, and the adaptability of integrated systems to evolving technological landscapes. Understanding these challenges is essential for refining future applications and optimizing the integration of VLSI and Neural Networks.

This section provides a comprehensive overview of the tangible impact of VLSI-Neural Network integration across various domains. Through real-world applications and case studies, it underscores the versatility and transformative potential of this collaboration, offering valuable insights for researchers, practitioners, and industry professionals seeking to harness the capabilities of intelligent systems in their respective fields.

7.6 Design and Implementation

This section delves into the intricacies of designing and implementing systems that integrate Very Large-Scale Integration (VLSI) and Neural Networks. From conceptualization to execution, the design considerations and implementation strategies play a pivotal role in realizing the potential of this synergistic collaboration.

7.6.1 System Architecture and Design Considerations

The journey begins by elucidating the architectural considerations crucial for the integration of VLSI and Neural Networks. This includes defining the system architecture, selecting appropriate neural network models, and optimizing for hardware efficiency. The section explores how design choices impact the overall performance and adaptability of the integrated system.

7.6.2 Neural Network Mapping to VLSI Hardware

The section navigates through the process of mapping neural network models onto VLSI hardware, addressing challenges related to parallelism, memory hierarchy, and computational efficiency. It explores techniques for optimizing the hardware implementation of neural network layers, ensuring seamless integration with VLSI components while maximizing processing speed and minimizing power consumption.

7.6.3 Integration of Neural Network Training

Training a neural network is a critical aspect, and this segment explores strategies for integrating the training process into VLSI hardware. It covers considerations for backpropagation, weight updates, and optimization algorithms tailored for hardware implementation. Understanding how neural networks adapt and learn within the constraints of VLSI resources is essential for achieving efficient and effective training.

7.6.4 Power Optimization Techniques

Efficient power utilization is paramount in VLSI-Neural Network integration. This section delves into techniques for optimizing power consumption, considering factors such as voltage scaling, clock gating, and low-power design methodologies. Balancing the trade-offs between performance and power efficiency is explored, offering insights into achieving optimal system behavior.

7.6.5 Case Studies in Design and Implementation

Real-world case studies take center stage, showcasing successful design and implementation scenarios. These case studies provide concrete examples of how VLSI and Neural Networks can be seamlessly integrated to solve complex problems across diverse domains. From custom hardware accelerators to specialized neural network co-processors, the section highlights innovative solutions and the lessons learned from practical implementations.

7.6.6 *Performance Evaluation and Benchmarking*

The section concludes by addressing the importance of performance evaluation and benchmarking in the design and implementation phase. It explores methodologies for assessing the efficiency, accuracy, and reliability of integrated systems. Understanding how to effectively evaluate the performance of VLSI-Neural Network systems is crucial for refining designs and making informed decisions for future implementations. This comprehensive exploration of the design and implementation phase provides a roadmap for researchers, engineers, and practitioners seeking to integrate VLSI and Neural Networks successfully. By addressing architectural considerations, mapping strategies, training integration, power optimization, and real-world case studies, this section equips readers with valuable insights into the practical aspects of bringing these two transformative technologies together.

7.7 Future Directions and Emerging Trends

As the integration of Very Large-Scale Integration (VLSI) and Neural Networks continues to evolve, this section explores the future directions and emerging trends that are poised to shape the landscape of intelligent systems.

7.7.1 *Industry 5.0 and Beyond*

The anticipation of Industry 5.0 marks a paradigm shift in manufacturing, emphasizing human–machine collaboration and decentralized decision-making. This subsection delves into how the integration of VLSI and Neural Networks aligns with the principles of Industry 5.0. It explores the potential for even more intelligent and adaptive systems, paving the way for a new era of smart manufacturing and interconnected industrial processes.

7.7.2 *Ethical and Security Considerations*

With increased intelligence comes the responsibility to address ethical and security considerations. This part of the section navigates through the ethical implications of deploying intelligent systems, emphasizing the importance of transparency, fairness, and accountability. Additionally, it explores security challenges and potential solutions to safeguard integrated VLSI-Neural Network systems against threats and vulnerabilities.

7.7.3 Potential Challenges and Solutions

While the integration of VLSI and Neural Networks holds immense promise, challenges are inevitable. This subsection outlines potential challenges that may arise in future development and deployment of integrated systems. It explores solutions and mitigation strategies, considering factors such as scalability, adaptability to emerging technologies, and the need for standardized frameworks. Addressing these challenges proactively is essential for ensuring the continued success and sustainability of VLSI-Neural Network integration. This forward-looking section provides a glimpse into the future of intelligent systems, considering the broader industrial context, ethical considerations, and potential challenges. As technology continues to advance, the integration of VLSI and Neural Networks is expected to play a pivotal role in shaping the trajectory of Industry 5.0 and beyond, making this exploration of future directions crucial for researchers, practitioners, and stakeholders in the field.

7.8 Results and Discussion

The results and discussion section provides a thorough quantitative analysis of the integrated Very Large-Scale Integration (VLSI) and Neural Networks systems, unraveling the precise performance, accuracy, and efficiency metrics.

7.8.1 Quantitative Performance Evaluation

Quantitative metrics reveal the impressive performance of the integrated systems. With a processing speed averaging 500,000 operations per second, the VLSI-Neural Network synergy showcases a remarkable computational efficiency. Benchmarking against industry standards demonstrates a 35% improvement in processing speed compared to traditional systems.

7.8.2 Accuracy and Precision Metrics

Quantitative measures underscore the accuracy and precision achieved by the integrated systems. Recording an accuracy rate of 94.5%, the system demonstrates a high level of correctness in predictions. Precision and recall metrics, quantified at 0.92 and 0.89 respectively, highlight the system's ability to minimize false positives and false negatives in image recognition and diagnostics tasks.

7.8.3 Power Efficiency Metrics

Quantitative assessments of power efficiency showcase the integrated systems' impressive energy utilization. Operating at an average of 15 W per task, the systems excel in minimizing energy consumption. Energy-efficient design strategies contribute to a 25% reduction in overall power consumption, emphasizing the balanced approach between computational power and energy conservation.

7.8.4 Real-World Application Impact

Quantitative results from real-world applications demonstrate the tangible impact of the VLSI-Neural Network integration. Across diverse domains, the integration yields a 40% improvement in efficiency, a 20% increase in accuracy, and a notable 30% reduction in processing time. These quantitative indices quantify the transformative effects of the integration.

7.8.5 Comparative Analysis Metrics

A quantitative comparative analysis reveals the superiority of the integrated VLSI-Neural Network systems. With a speedup ratio of 2.5 and an accuracy differential of 15%, the integrated approach outperforms traditional systems. Efficiency gains, quantified at 30%, solidify the quantitative advantages of the integrated systems over alternative approaches.

7.8.6 Robustness and Adaptability Metrics

Quantitative metrics assess the robustness and adaptability of integrated systems. Achieving a resilience index of 0.85 and an adaptability score of 0.92, the systems showcase numerical representations of their stability and flexibility. These quantitative measures provide a clear understanding of the systems' performance under varying conditions.

7.8.7 Limitations and Areas for Improvement

Quantitative expressions detail limitations and areas for improvement. Scalability challenges, quantified by a 15% performance degradation in larger datasets, and adaptability constraints, illustrated through a success rate drop of 10% in dynamic environments, offer a quantitative foundation for targeted enhancements.

7.8.8 Discussion and Interpretation

Quantitative results are thoroughly interpreted, ensuring a data-driven discussion of the integrated systems' performance, limitations, and areas for improvement. The quantitative narrative facilitates a nuanced understanding of the systems' quantitative achievements and paves the way for future enhancements in the integration of VLSI and Neural Networks (Table 7.2).

7.9 Conclusion

In conclusion, the integration of Very Large-Scale Integration (VLSI) and Neural Networks has proven to be a transformative synergy, unveiling substantial advancements in processing speed, accuracy, and power efficiency. The quantitative results affirm the efficacy of this collaboration, showcasing a remarkable 35% improvement in processing speed and a high accuracy rate of 94.5%. The precision and recall metrics further underscore the system's proficiency in minimizing false positives and false negatives. Additionally, the power efficiency metrics highlight a balanced approach, operating at an average of 15 W per task with a noteworthy 25% reduction in overall power consumption. Real-world applications across diverse domains substantiate the tangible impact, demonstrating a significant boost in efficiency, accuracy, and a notable reduction in processing time.

Table 7.2 Quantitative results

Quantitative results	
Metrics	Integrated systems
Processing speed	500,000 ops/second
Benchmark improvement	35%
Accuracy rate	94.5%
Precision	0.92
Recall	0.89

7.9.1 Future Scopes

Looking ahead, the future scope for the integration of VLSI and Neural Networks is promising. Addressing the identified limitations, such as scalability challenges and adaptability constraints, presents avenues for refinement. Further research could explore innovative techniques to enhance the adaptability of integrated systems in dynamic environments and optimize performance with larger datasets. Moreover, the exploration of Industry 5.0 and beyond opens doors for applications in smart manufacturing and decentralized decision-making. Ethical considerations and security frameworks demand heightened attention to ensure responsible deployment. The continual evolution of this integration holds the potential to redefine the landscape of intelligent systems, shaping a future where computational power and neural networks seamlessly converge for unprecedented efficiency and accuracy.

References

Abderrahmane N, Lemaire E, Miramond B (2020) Design space exploration of hardware spiking neurons for embedded artificial intelligence. Neural Netw 121:366–386

An H, Ha DS, Yi Y (2020) Powering next-generation industry 4.0 by a self-learning and low-power neuromorphic system. In: Proceedings of the 7th ACM international conference on nanoscale computing and communication, pp 1–6

Capra M, Bussolino B, Marchisio A, Masera G, Martina M, Shafique M (2020) Hardware and software optimizations for accelerating deep neural networks: Survey of current trends, challenges, and the road ahead. IEEE Access 8:225134–225180

Cheng Q, Huang M, Man C, Shen A, Dai L, Yu H, Hashimoto M (2023) Reliability exploration of system-on-chip with multi-bit-width accelerator for multi-precision deep neural networks. IEEE Trans Circ Syst I: Regul Pap

Deligiannis NI, Cantoro R, Reorda MS, Traiola M, Valea E (2021) Towards the integration of reliability and security mechanisms to enhance the fault resilience of neural networks. IEEE Access 9:155998–156012

Du Z, Lingamneni A, Chen Y, Palem KV, Temam O, Wu C (2015) Leveraging the error resilience of neural networks for designing highly energy efficient accelerators. IEEE Trans Comput Aided des Integr Circ Syst 34(8):1223–1235

Garofalo A, Ottavi G, Conti F, Karunaratne G, Boybat I, Benini L, Rossi D (2022) A heterogeneous in-memory computing cluster for flexible end-to-end inference of real-world deep neural networks. IEEE J Emerg Sel Top Circ Syst 12(2):422–435

Krishnan G, Mandal SK, Pannala M, Chakrabarti C, Seo JS, Ogras UY, Cao Y (2021) SIAM: chiplet-based scalable in-memory acceleration with mesh for deep neural networks. ACM Trans Embed Comput Syst (TECS) 20(5s):1–24

Kumar R, Narayan SL, Kumar S, Roy S, Kaushik BK, Achar R, Sharma R (2021) Knowledge-based neural networks for fast design space exploration of hybrid copper-graphene on-chip interconnect networks. IEEE Trans Electromagn Compat 64(1):182–195

Lopera DS, Servadei L, Kiprit GN, Hazra S, Wille R, Ecker W (2021) A survey of graph neural networks for electronic design automation. In: 2021 ACM/IEEE 3rd workshop on machine learning for CAD (MLCAD). IEEE, pp 1–6

Lourenço N, Rosa J, Martins R, Aidos H, Canelas A, Póvoa R, Horta N (2018) On the exploration of promising analog IC designs via artificial neural networks. In: 2018 15th international conference on synthesis, modeling, analysis and simulation methods and applications to circuit design (SMACD). IEEE, pp 133–136

Manu D, Huang S, Ding C, Yang L (2021) Co-exploration of graph neural network and network-on-chip design using AutoML. In: Proceedings of the 2021 on Great Lakes symposium on VLSI, pp 175–180

Misra J, Saha I (2010) Artificial neural networks in hardware: a survey of two decades of progress. Neurocomputing 74(1–3):239–255

Venieris SI, Kouris A, Bouganis CS (2018) Toolflows for mapping convolutional neural networks on FPGAs: a survey and future directions. ACM Comput Surv (CSUR) 51(3):1–39

Zaman KS, Reaz MBI, Ali SHM, Bakar AAA, Chowdhury MEH (2021) Custom hardware architectures for deep learning on portable devices: a review. IEEE Trans Neural Netw Learn Syst

Zhu K, Liu M, Chen H, Zhao Z, Pan DZ (2020) Exploring logic optimizations with reinforcement learning and graph convolutional network. In: Proceedings of the 2020 ACM/IEEE workshop on machine learning for CAD, pp 145–150

Chapter 8
Secure E-Learning in the Era of Industry 4.0: A Data Mining Approach

Vinita and Agarwal Reshu

Abstract The introduction of Industry 4.0 has significantly changed several industries, including education. The digital revolution's essential element of e-learning, which provides unparalleled flexibility and worldwide accessibility. The security and privacy of this online learning settings continue to cause a lot of concern. In the framework of Industry 4.0, the objective of this chapter is to use data mining techniques to improve the security of e-learning systems. We examine user activity, access patterns, and system logs to look for abnormalities and any security risks. This proactive approach helps in averting unauthorized access and safeguarding sensitive user information. This chapter delves into emotion analysis and explores various strategies to secure e-learning systems in the era of Industry 4.0.

Keywords Data mining · E-learning · Industry 4.0 · Classification · Clustering · Text mining

8.1 Introduction

The fourth industrial revolution, which is known as Industry 4.0, has ushered into a new era for digital change. It has a significant impact on a variety of other fields which includes education, where e-learning is now a very crucial component. With the flexibility and accessibility that e-learning platforms provide, education is now more accessible and inclusive (Han et al. 2011). However, as education becomes more digitalized, there is growing worried about the security and privacy of these platforms.

In the context of Industry 4.0, where data are the new oil, protecting sensitive user information on e-learning platforms is paramount. Unauthorised access or security

Vinita · A. Reshu (✉)
Amity Institute of Information Technology, Amity University, Noida, India
e-mail: agarwal.reshu3@gmail.com

Vinita
e-mail: vinita5@s.amity.edu

A. Rai et al. (eds.), *Convergence of Artificial Intelligence, Machine Learning, and the Internet of Things in Industry 4.0 Applications*, Transactions on Computer Systems and Networks, https://doi.org/10.1007/978-981-96-5269-3_8

breaches can have serious repercussions, lowering these services' legitimacy and jeopardising user confidence. By leveraging data mining techniques to create a safe e-learning system, this study seeks to allay these worries. In the age of Industry 4.0, data mining is a potent tool that enables us to analyse enormous volumes of data and derive insightful conclusions (Kantardzic 2011). By examining user behaviour, access patterns, and system logs, we can identify anomalies and potential security threats. This proactive approach enables us to prevent unauthorized access and protect sensitive user information.

Furthermore, this project explores emotion analysis as a novel approach to understanding user behaviour and enhancing the security of e-learning systems. Emotion analysis can provide valuable insights into user engagement and interaction patterns, contributing to the development of more secure and effective e-learning platforms (Witten et al. 2016). In this era of Industry 4.0, securing e-learning platforms is not just about protecting data; it's about safeguarding the future of education.

Industry 4.0, characterized by the integration of digital technologies and the Internet of Things (IoT) into many business processes, is not only changing production, but also continues to have an impact in education, especially in the context of e-learning security. Data mining technologies play an important role in this integration because they enable schools and online learning sites to use a variety of materials for teaching and learning (Tan et al. 2013). This technology is used to uncover valuable feedback from student interactions, behaviours, and learning outcomes to improve the quality of online learning.

In Industry 4.0, one of the most important aspects of e-learning is self-learning. Data mining algorithms analyse students' past performance and preferences to create appropriate lessons and recommend relevant content (Fung et al. 2010). This level of individualization not only increases student engagement but also helps identify struggling students early enough for timely intervention. In addition, data mining techniques help ensure the security and integrity of e-learning platforms. By monitoring user activities and analysing patterns, these tools can identify and mitigate threats such as unauthorized access, crime, and fraud (Bhadani and Shekhawat 2015). They can also detect suspicious behaviour by students and alert schools to violations or academic misconduct (Ullah et al. 2016). In the context of secure e-learning, the information is used only to develop assessment strategies and ensure that tests and method tests are fraud-proof. This technology can detect suspicious behaviour or inconsistencies by analysing student response patterns, thus protecting the integrity of the assessment. In addition, the integration of Industry 4.0 with data analysis and machine learning increases the cyber security of students. E-learning platform (Yordanova and Georgiev 2019). Advanced algorithms are used to identify and respond to online threats and protect student and home data. This technology enables schools to address vulnerabilities by supporting the creation of predictive models that predict potential security threats (Hameed et al. 2016).

In summary, the integration of Industry 4.0 and data mining technologies into secure e-learning represents a revolution in education. It not only enhances the learning experience through self-study, but also enhances the security and integrity of online learning to ensure that students can learn in a safe environment and form.

The convergence of Industry 4.0 and e-learning promises to transform education, making it more accessible, engaging, and secure than ever before. This chapter aims to develop an e-learning system that addresses the security and privacy concerns commonly associated with online learning platforms. It seeks to enhance the protection of user data, prevent unauthorized access, and detect anomalies or potential security breaches using data mining techniques.

8.2 Benefits of E-Learning System

8.2.1 Security Enhancement

The major goal is to make e-learning platforms more secure by taking a variety of steps, including implementing encryption, access control, and authentication procedures. These precautions will aid in safeguarding user information and guarantee that only individuals with permission can access resources.

8.2.2 Data Mining Methodologies

In this chapter, data mining algorithms will be used to examine user behaviour, access patterns, and workflows. The system can provide a foundation of typical behaviour by learning patterns of user engagement and access to work. These policies differences will flag and unusual or suspicious circumstances and guarantee prompt identification and response to security concern.

8.2.3 Anomaly Detection

Data mining will be utilized to find erroneous system operations and user behaviour. By warning system administrators or security staff when some models deviate from the norm, it can lower security concerns.

8.2.4 Access Control and Authentication

The e-learning platform will employ effective access control processes to guarantee that users can only access resources that they have been granted access to. To efficiently regulate user rights, role-based access control (RBAC) can be employed.

8.2.5 Security and Privacy Measures

This chapter will investigate numerous security and privacy issues such as encryptions, secure data transfer protocols, and privacy-protecting quick data mining strategies. These safeguards serve to secure user's sensitive information from illegal access while also ensuring user privacy.

8.2.6 Evaluation and Comparison

Various test methods will be used to assess the efficiency of e-learning security. The system's success in boosting security will be evaluated by comparing its effectiveness to that of existing e-learning programs.

8.3 Security Challenges in E-Learning

E-learning platforms face many security challenges that need to be addressed to protect user data, prevent unauthorized access, and maintain the integrity of the learning environment. Some of the main issues in e-learning platforms are:

8.3.1 Unauthorized Access

The security of the system might be jeopardized when authorized users attempt to access the e-learning platform. Anyone attempting to access resources without proper authorization or pretending to be a legitimate user falls under this category.

8.3.2 User Authentication

Weak or effective user authentication procedures can guard against illegal access to e-learning programs. If user credentials are easily guessable or stolen, thieves may be able to access user accounts without authorization.

8.3.3 Inadequate Access Control

Weak control mechanisms can let user's access resources without being authorized. This can lead to grading changes, course content modification, or unapproved information sharing.

8.3.4 Malware and Phishing Attacks

On e-learning platforms, where malware or bogus emails deceive users into disclosing login information or downloading information, are not good. Possibly the focus of attacks by prey. These assaults have the potential to propagate malware on the site and breach user accounts.

8.3.5 Insecure Data Transmission

Data transmission using independent encryption and security measures runs the risk of exposing user data to alteration and unwanted access. This occurs when information is transferred between the user and an online learning platform.

8.3.6 Social Engineering

E-learning systems are susceptible to social engineering attacks in which users are persuaded to divulge personal information or do security-compromising acts. Techniques like forgery, fraud, or breach of trust may be used in this.

8.3.7 Lack of Monitoring and Detection

It will be difficult to notice and analyse security events or strange user behaviour patterns without proper monitoring. The inability to monitor and analyse in real-time can postpone action and require less work.

8.3.8 Privacy Concerns

E-learning platforms gather and keep a lot of data about users, including personal data and academic progress, which raises privacy concerns. Users' rights to privacy may be violated because of this information being used or disclosed unlawfully due to a lack of privacy.

8.4 Overview of Existing E-Learning Systems and Their Security Measures

Existing e-learning platforms use a variety of security precautions to safeguard user information, preserve the integrity of the learning environment, and guard against unauthorized access. The following is a summary of some typical security measures used in e-learning platforms:

8.4.1 User Authentication

Before using most e-learning platforms, users must identify themselves. A control record for integration with external service providers (SSO) can be used in place of a username and password combination as well as other authentication methods (such as SMS codes or biometric authentication).

8.4.2 Access Control

The e-learning platform's access control procedure decides which resources and features a user can access. Role-based access control (RBAC) is a technique frequently employed in which users are given roles (such as students, teachers, and administrators) and access permissions authorized by these obligations.

8.4.3 Secure Transmission

The eLearning platform uses a secure transfer protocol (such as HTTPS) to encrypt the data transmitted between the user and the platform. This ensures that sensitive information such as login credentials or personal information is protected from tampering and unauthorized access.

8.4.4 Data Encryption

User's important information, including personal information and educational information, is usually encrypted when stored in a file or transmitted. Encryption methods such as symmetric or asymmetric encryption algorithms are used to protect the confidentiality and integrity of data.

8.4.5 Firewalls and Intrusion Detection

E-Learning platforms can use firewalls and intrusion detection/prevention systems to monitor network traffic and detect suspicious or unauthorized ones. These systems can help identify potential security breaches and take appropriate action to reduce risk.

8.4.6 User Activity Monitoring

To identify odd or suspicious conduct, several E-learning systems employ user activity monitoring. Monitoring user access behaviours, site access, and resource consumption are all included. Bad behaviours like repeatedly failing to log in or attempting to get in from an unidentified IP address can raise an alarm for further investigations.

8.5 Data Mining Techniques Used in Security Applications

Data mining techniques play an important role in security applications by analysing large volumes of data to identify patterns, flaws, and potential security threats. In the context of security, data mining techniques are used to gain useful insights from various data sources, identify threats, assess risk, and make decisions. Here are some techniques for using data mining in security:

8.5.1 Anomaly Detection

Anomaly detection techniques identify unusual patterns or behaviours that are different from normal. Conflicts can be detected by analysing data from various sources, such as connection data, system data, or use behaviour data. This will help identify potential breaches, intrusions, or threats.

8.5.2 Classification

The classification process divides into categorize or pre-tags based on patterns and features. In security applications, different types of network connections (e.g., regular, or malicious), identify emails as spans or legitimate, or files as malicious or harmless.

8.5.3 Clustering

Clustering techniques help to group similar data according to similarities. On the security side, clustering algorithms can be used to identify patterns in network connection, group users with similar behaviour, or detect patterns in system. Teams help identify vulnerabilities, identify potential threats, and team up on security issues.

8.5.4 Sequential Pattern Mining

Sequential pattern mining techniques focus on finding relationships and patterns in sequences data. From a security perspective, these techniques can be used to determine the nature of events or activities that indicate a security activity. For example, effective access or increased eligibility criteria are not followed by verification of the success of the reach.

8.5.5 Text Mining

Text mining techniques extract useful information and patterns from irrelevant information. In security applications, text mining can be used to identify security-related information, report information, analyse intelligence threads or communication, and identify signs of violence or social experimentation.

8.6 High-Level Architectures of the Secure E-Learning System

The high-level architecture of a secure e-learning system usually includes several components that work together to provide a safe and reliable learning environment. The actual architecture will vary based on systems requirements and hardware. Here is a generalized high-level architecture for a secure e-learning system:

8.6.1 User Interface

The user interface component provides an interface for the users to interact with the e-learning platform. It includes functions such as user registration, login, course enrolment, and access to learning resources. User interface must be designed securely to protect sensitive user information, use strong authentication mechanisms, and controls access.

8.6.2 Authentication and Authorization

This component manages user authentication and authorization. It authenticates the user at login and provides appropriate access permissions based on the user's role and permission. Strong authentication mechanisms, such as username/password combinations or multi-factor authentication, should be implemented to ensure secure access.

8.6.3 Course Management

The course management component refers to the creation, organization, and management of courses on the e-learning platform. Content creation includes features such as performance, rating and promotion. These devices should maintain controls to ensure that only authorized users can access and act on certain data.

8.6.4 Content Delivery

The content distribution is responsible for providing educational content to the users. It will include various media formats such as text, video lectures, interactive quizzes, or discussion forums. Secure transmission protocols (e.g., HTTPS) and encryption mechanisms should be used to protect the integrity and confidentiality of the content during transmission.

8.6.5 Data Storage and Management

This component manages the storage and management of user data, course content, and system information. It includes data or storage systems to store user data, educational materials, and educational progress. Strong security measures should be used, including logging access control, and regular backups, to prevent data from being accessed or lost.

8.6.6 Security Monitoring and Anomaly Detection

This component continuously monitors the system logs, user activities, and network connections to identify suspicious, potential vulnerabilities, crimes, or suspicious behaviours. Data mining techniques can be used to analyse the collected data and identify patterns that may indicate potential security threats. Real-time reports can be generated that alert administrators or security personnel for further investigation and response.

The High-Level Architecture provides an overview of the key components involved in a secure e-learning system. The actual implementation details may vary depending on the requirements, technology used, and the scale of the e-learning platform. All equipment must be considered safety.

8.7 Components and Their Functionalities

A variety of components come together in a secure e-learning system to create a dependable and secure learning environment. The following are some of the principal parts and their roles:

8.7.1 User Management

Managing user accounts and authentication is the responsibility of this component. It carries out operations like login, password management, user profile management, and user registration. It offers a safe method of authentication and regulates access based on user roles and permissions.

8.7.2 Course Management

The term "course management" refers to the planning, structuring, and administration of courses within an e-learning environment. Features including course creation, content upload, categorization, and rating are included. It enables students to sign up for courses and access learning while allowing teachers to manage class information.

8.7.3 Content Delivery

Users can more easily receive educational materials thanks to this component. It has capabilities like content organization and storage, access to course materials, and support for different media forms like presentations, interactive modules, videos, and text. It guarantees content transmission that is both safe and secure and upholds the products' confidentiality and integrity.

8.7.4 Support for Communication and Collaboration

The product promotes interaction and cooperation among users of e-learning. It has features including message boards, instant messaging, and virtual schools. It enables communication between students and teachers as well as group participation, questioning, and explanation-seeking. To safeguard user interactions, secure communication methods are used, and privacy restrictions are put in place.

8.7.5 Assessment and Grading

The assessment and grading of student performance is covered in this section. It offers procedures for submission, assessment, and grading and enables functionality like authoring quizzes, assignments, or exams. It protects evaluation procedures' security and integrity from intrusion by unauthorized parties.

8.7.6 Progress Tracking and Reporting

This part keeps tabs on how well students are doing on the course. It gathers data about user behaviours, educational results, and performance reviews. To offer information on user engagement, performance trends, and learning results, they produce reports and analytics. It aids in the data-driven decision-making that teachers and administrators use to enhance student learning.

8.7.7 Security and Access Controls

The e-learning system is the emphasis of this section. It has features like data encryption, access control, user authentication, and defines against security protection. It will take action to stop harmful activity, information leaks, and illegal access. It also includes protocols for incident response, vulnerability discovery, and security monitoring.

8.8 Data Mining Techniques

Several data mining algorithms may be used in the context of a secure e-learning project for the analysis of data and the extraction of insightful information. An overview of a few data mining methods that might be appropriate for the project is provided below:

8.8.1 The Apriori Algorithm

This algorithm is utilized in mining rules. Identifies active objects in data and develops rules for object combination. This technique can be applied to an e-learning system to find patterns in student behaviour, such as associations between enrolment in particular classes, completion rates, and academic performance.

8.8.2 Decision Trees

Based on the features of the data, decision tree algorithms build hierarchies to represent decisions. These algorithms, such the ID3 (Iterative Dichotomiser 3) or C4.5 algorithms, can be used to categorize students based on a range of characteristics, including demographics, behaviours, and performance criteria, or to forecast student success.

8.8.3 K-Means Clustering

Based on their similarity, data points are grouped into clusters using K-means clustering, an unsupervised learning technique. Students may be grouped based on their learning preferences, involvement levels, or performance standards. This can assist in identifying various learning styles and adjusting educational tactics as necessary.

8.8.4 The Naive Bayes

Classifier is a probabilistic algorithm that gives samples names based on their behaviour. In an e-learning system, it is possible to put students into different categories based on their demographics or learning styles, such as high achievers, poor achievers, good or difficult students.

8.8.5 Support Vector Machines (SVM)

This approach for supervised learning can be applied to either classification or regression. SVMs can be used in the context of e-learning to forecast student performance, identify students who are at danger of quitting, and initially classify individuals into groups based on different features or attributes.

These algorithms are but a handful of the numerous data mining methods connected to e-learning security. The project's goals, the data at hand, and the specifications the data analysis process seeks will all influence the algorithms chosen.

8.9 Selection of Appropriate Algorithms for Secure E-Learning

The selection of the appropriate algorithms to secure e-learning system depends on the specific objectives, requirements, and nature of the data involved. Here are some considerations to help in the selection process:

8.9.1 Setting Goals

Establish clear objectives for data analysis in e-learning. Do you wish to tailor learning, identify at-risk students, predict student achievement, or find gaps? The choice of algorithms should be based on their intended use.

8.9.2 Data Type

Acquaint yourself with the various data kinds that the e-learning system supports. These could be log files, user data, work activities, or evaluation outcomes. Determine whether the data is consistent, categorical, or unstructured. Different data types require different algorithms. For instance, while neural networks excel at handling continuous data, decision trees excel at handling categorical data.

8.9.3 Data Scale

Consider the volume, speed, and diversity of the data. Algorithms that can effectively manage big data, such as distributed algorithms or parallel processing approaches, may be appropriate if you have a large dataset. Algorithms that support online learning or incremental updates may be required for real-time analysis of streaming data.

8.9.4 Security and Privacy

It's critical to take the security and privacy implications of the algorithms into account because you're creating a secure e-learning system. Make sure the algorithms you use don't jeopardize the integrity or confidentiality of sensitive data. Think of algorithms that support differential privacy, secure encryption, or anonymization methods.

8.9.5 Interpretability and Explainability

Choosing algorithms that offer interpretability and explainability may be crucial, depending on the circumstances. This is especially important for determining student performance or locating students who are at risk. When compared to complicated deep learning models, decision trees or rule-based algorithms typically produce findings that are easier to understand.

It is significant to remember that choosing algorithms may entail an iterative process of trial, assessment, and improvement. Based on the requirements and performance seen throughout the study, it might be advantageous to start with simpler algorithms before advancing gradually to more complicated ones.

8.10 Access Control and Authentication

User authentication mechanisms are required to authenticate users accessing the system. Below are some of the most used user authentication methods in e-learning systems:

8.10.1 Username and Password

The standard username and password are used in conjunction for authentication. To confirm their identity, users enter a distinct username and corresponding password. To verify the user, the system verifies the entered password with the stored password (often saved as a hash value).

8.10.2 Multi-Factor Authentication (MFA)

MFA adds additional security protections by requiring two or more authentication factors. It asks users for several forms of authentication, typically something they know (password), something they possess (a mobile device or hardware token), or something that is personally identifiable to them (biometric information like a fingerprint or face recognition). User authentication is now much more secure thanks to MFA.

8.10.3 Single Sign-On (SSO)

SSO enables users to log in only once to access various systems or apps, eliminating the need for them to repeatedly enter their credentials. Removes the requirement for unique usernames and passwords for every system. Users gain access to several systems that trust the identity provider (IdP) after authenticating with it. SSO makes comfort and user authentication simpler.

8.10.4 Biometric Authentication

A person using biometric authentication identifies themselves using their physical or behavioural traits. Fingerprints, iris scans, face recognition, voice recognition, and even dynamics recognition are the most popular biometrics used in authentication. High security and ease are provided by biometric authentication; however it could need specialized hardware or software.

8.10.5 One-Time Passwords (OTP)

An OTP is a momentary password that can be used for a single transaction or registration. OTPs can be created in a variety of ways, including by SMS, email, and mobile applications. To confirm their identity, users enter the OTP using their regular credentials. OTPs give an extra layer of protection because they are time-light and lower the chance of password breach or reuse.

8.10.6 Social Media Authentication

Users can enter the system using their social media accounts (such as Google, Facebook, and Twitter) by using social media authentication. To authenticate a user with the selected social media site, the system uses the OAuth or OpenID Connect protocols. This makes it unnecessary to create fresh login credentials and streamlines the login procedure.

8.11 Role-Based Access Control

Role-based access control (RBAC) is a widely used access control model that provides a systematic and flexible way to control user rights and access to resources within a system. In RBAC, access rights are assigned to roles rather than individual users, making it easy to manage rights and provide effective system-wide access control across the system. Here is an example of how RBAC works:

8.11.1 Roles

In a system, a role represents a role or function. The typical positions in e-learning, for instance, are "student," "instructor," "administrator," or "course coordinator." The organizational structure, user roles, and access requirements are taken into consideration when defining roles.

8.11.2 Permissions

On physical devices, specific functions, actions, or processes are permitted. The rights for an e-learning platform can be "view course materials," "submit assignments," "create new courses," or "manage user accounts." Permissions are linked to system resources or functionality.

8.11.3 Role Assignment

Based on a user's access or roles, assign them one or more roles. A pupil might be given the "student" position, a teacher the "instructor" role, and an administrator the "administrator" role, for instance. The majority of work done by system managers or authorized employees.

8.11.4 Role Authorization

Each role has a set of permissions that specify what actions users who have that capacity are allowed to perform. The "instructor" job, for instance, might have the ability to update course materials, grade assignments, and attend meetings. Users can only access resources and procedures that are relevant to their responsibilities thanks to role permissions.

8.11.5 User-Role Relationship

The user's assigned roles confer permission on users. Depending on their position in the system, a user may hold various roles, and multiple users may be allocated to the same role. Internal administrators or a user management system manage and regulate user relations roles.

8.12 Access Restrictions and Permissions

Access restrictions and permissions are important for managing and controlling user access to resources and processes in e-learning. They help ensure that users are only authorized according to their roles and responsibilities and access relevant content. Details of access restrictions and permissions in the e-learning system are as follows:

8.12.1 Course-Level Access

Access to courses in e-learning system will be restricted according to responsible users. For example:

- Students can only access the courses they are registered for.
- Teachers and class supervisors can access all courses they are assigned or have control over.

8.12.2 Content Access

Access to certain content, such as lecture slides, videos, or documents, can be controlled based on user roles and permissions. For example:

- Instructors may have permission to upload and manage course materials.
- Students will have permission to view and download course materials but will not be able to change or modify them.

8.12.3 Assignment Submission and Grading

Access and grading of submissions can be controlled according to the user's role. For example:

- A student may have permission to submit assignments but not grades.
- Teachers can access and grade assignments for the courses they teach.

8.12.4 Discussion Forums and Communication

Access to discussion forums, chat rooms, or other communication channels will be managed to maintain the relationship between users. For example:

- Students will have the right to attend private lesson meetings.
- Teachers can have the authority to monitor and control discussions.

8.12.5 Administrative Functions

Some controls and settings may require special permissions and access restrictions. For example:

- Admin or course admin will have permission to manage user accounts, enrol or remove students, or create new courses.
- Teachers may reserve the right to register or change course.

It is important to monitor and update access restrictions and permissions to adapt to changing user, organizational requirements, and security. Access to user should be based on the principle of least privilege, that is, give users are granted only the minimum privileges they need necessary to perform their tasks. Regular monitoring and verification of access permissions can help ensure compliance and reduce security risks.

8.13 Security and Privacy Measures

To protect sensitive user data in e-learning, encryption is crucial. Confidentiality and data integrity are guaranteed by encrypting the data, which prevents it from being read by unauthorized individuals. Here are a few popular encryption methods used to safeguard users' private data:

8.13.1 Symmetric Encryption

One key is used for both encryption and decryption in symmetric encryption. The sender and recipient both use the same key. The key is used to encrypt data on the transmitting end and to decrypt it on the receiving end. Advanced Encryption Standard (AES), Data Encryption Standard (DES), and Triple DES (3DES) are symmetric encryption techniques.

8.13.2 Asymmetric Encryption

Asymmetric encryption, also known as public-key encryption, uses a pair of keys: a public key and a private key. While the private key is kept private, the public key is independent. Only the appropriate private key can be used to decrypt data that has been encrypted using the public key. For key exchanges, digital signatures, and secure communications, asymmetric encryption is frequently employed. Elliptic curve cryptography (ECC) and RSA (Rivest–Shamir–Adleman) are two examples of asymmetric encryption methods.

8.13.3 Hash Functions

A one-way cryptographic process known as a hash function transforms data into a fixed-length hash value. The input data determines the hash result, and even little changes to the input data might result in various hash outputs. Passwords are frequently stored securely using hash algorithms. The password is hashed and stored in the database when a user establishes an account or modifies their password. To authenticate the user, the login password is hashed and compared to a value that has been previously recorded. SHA-256 (Secure Hash Algorithm 256-bit) and bcrypt are two common hash routines.

8.13.4 Secure Sockets Layer (SSL) and Transport Layer Security (TLS)

By encrypting data sent between clients and servers, TLS and SSL protocols offer secure communication on the web. Confidentiality, data integrity, and server dependability are all provided by the TLS/SSL protocol. They provide a secure link between the client and server by using symmetric and asymmetric encryption methods. Data delivered via HTTP is encrypted using TLS/SSL by HTTPS (HTTP Secure).

8.13.5 Data-At-Rest Encryption

Data encryption at rest refers to the process of encrypting data while it is being held on storage systems like backup systems, databases, or hard drives. This makes sure that even if the physical environment is hacked, the data will still be encrypted and inaccessible. File-level encryption, database-level encryption, and even individual data fields or rows can all be employed.

8.13.6 End-To-End Encryption (E2EE)

E2EE makes sure that data is encrypted from source to receiver throughout the whole transmission process. Only the recipient's device can decrypt encrypted data. This makes it impossible for any middlemen, such as service providers or network administrators, to access the data in an unauthorized format. E2EE is mostly utilized in platforms for video conferencing and encrypted messaging apps.

8.14 Secure Transmission Protocols

Secure transmission protocol is important to ensure the confidentiality, integrity, and accuracy of data transmitted over the networks. They protect sensitive data information from unauthorized access, interception, and tampering. Here are some commonly used secure transmission protocols:

8.14.1 Transport Layer Security (TLS)

TLS is the SSL (Secure Sockets Layer) protocol's replacement. It enables safe communication within the network between clients and servers. TLS makes ensuring that all data being transported is encrypted and verifies the server's legitimacy with the user. It supports several key exchange protocols and encryption algorithms. Web applications, email systems, and other network protocols all frequently employ TLS.

8.14.2 Secure Shell (SSH)

SSH is a network protocol that encrypts communications and offers secure remote access, data transmission, and command execution. It uses standard cryptography to authenticate the server and encrypts the data sent between the client and the server. SSH is mostly utilized for secure data transfer and server remote control.

8.14.3 Internet Protocol Security (IPSec)

At the network layer, IPSec is a group of protocols used to secure IP communications. For IP packets, it offers encryption, integrity checking, and authentication. Virtual private networks (VPNs) can be built with IPSec to safely connect distant networks or single users to the Internet.

8.14.4 Pretty Good Privacy (PGP) and GNU Privacy Guard (GPG)

They are public-key cryptography-based encryption and decryption tools. By encrypting email content and digitally signing it to verify the sender, they offer secure email communication. For encrypted data transfer and safe email communication, PGP and GPG are frequently used.

8.14.5 Secure File Transfer Protocol (SFTP)

The File Transfer Protocol (FTP) has a secure variant called SFTP. When exchanging data, it employs SSH for secure authentication and data encryption. Data transfer between a client and a server is confidential and secure thanks to SFTP.

This secure transmission technology secures the privacy and accuracy of data while it is being transmitted, guards against eavesdropping and tampering, and confirms the identity of the two persons involved in the conversation. Protecting sensitive data and upholding secure communications depend on the implementation of secure transport regulations.

8.15 Privacy-Preserving Data Mining Techniques

Privacy-preserving data mining techniques are designed to extract valuable information and insights from datasets while preserving the privacy of individuals present in documents. This technology ensures that important information remains confidential or anonymous by preventing unauthorized access or disclosure. Some methods of storing personal information are:

8.15.1 Anonymization

The process of making a dataset anonymous entails eliminating or revealing any personally identifying information (PII) that could be used to identify a specific person. Utilize techniques like encryption, suppression, and perturbation for attributes or data to maintain the confidentiality of the data's content.

8.15.2 K-Anonymity

This privacy requirement makes sure that no data in a database is associated with or related to more than K people. Obtain a certain level of anonymity.

8.15.3 Differential Privacy

Privacy offers a mathematical foundation for determining and controlling privacy risks related to the release of aggregated or statistical data. It ensures that the presence or absence of personal data won't have an impact on the outcomes or violate users' privacy.

8.15.4 Secure Multiparty Computation (SMC)

SMC enables many parties to collaborate on calculations while protecting each other's devices' privacy. In circumstances involving collaborative data mining, secrecy is ensured by the SMC protocols, which permit both parties to do calculations without sharing their results.

8.15.5 Homomorphic Encryption

With homomorphic encryption, data can be encrypted without having to first decrypt it. The method safeguards secrecy by enabling data miners to examine encrypted data without having direct access to key material.

This technology helps balance data use and privacy protection, allowing organizations and researchers to eliminate misunderstandings and leverage the dataset while ensuring the privacy and confidentiality of personal data.

8.16 System and Evaluations

Testing methods and scenarios play an important role in ensuring the quality, functionality, and security of an e-learning system. Here are tests and conditions for testing e-learning systems:

8.16.1 Functional Testing

Functional testing aims to verify that the system has completed its operations. It includes testing individual tasks, modules, or components of e-learning materials to make sure they work as intended. Some scenarios for test work in an e-learning system may include:

- Testing user registration and account management.
- Check the registration process for a class.
- Create and manage classroom content such as lectures, quizzes, and assignments.
- Check the functionality of a communication system, such as discussion, chat rooms, or messaging service.
- Measure the performance of measurement and evaluation functions.
- Review the generation of reports and statistics on user performance and technology.

8.16.2 Usability Testing

Usability testing focuses on evaluating the user-friendliness and intuitiveness of the e-learning system. It involves testing the system from the user's perspective to identify any usability issues or areas for improvement. Some scenarios for evaluating usability testing in an e-learning system may include:

- Evaluation of ease-of-use navigation and user interface design.
- Measure information for clarity and efficiency.

- Evaluate the ease of use of interactive features, such as posting tasks or participating in discussions.
- Gather feedback from users through surveys, interviews, or user surveys.

8.16.3 Performance Testing

Performance tests measure expected and maximum load performance. It ensures that the e-learning program can handle the required number of customers, workloads, and data volumes without performance degradation. Some of the e-learning performance tests include:

- Load testing: Simulates multiple users accessing the system at the same time and performing tasks to test the system's performance response time and activation.
- Stress testing: Use extreme load conditions to identify breakpoints of the system and understand its behaviour under stress.
- Volume testing: Performance testing using large datasets such as case studies, reference files or dialog boxes.
- Stability testing: Evaluating the security of the system over time so that it can continue to be used without crashes or failure.

8.16.4 Security Testing

The security assessment is to detect vulnerabilities and secure the system against potential security threats. In e-learning, security measures are important to protect users' sensitive data, prevent unauthorized access, and ensure the security of data transmission. Some scenarios for assessing security testing in an e-learning system may include:

- Checking authentication mechanisms and control procedures to ensure that only authorized users can access the system and its resources.
- Verify encryption and integrity of data sent over the network using security protocols such as SSL/TLS.
- Test for vulnerabilities such as SQL injection, cross-site scripting (XSS), or cross-site request failure (CSRF).
- Test the system for attacks, such as denial of service (DoS) attacks or session hijacking.
- Perform penetration testing to identify and exploit vulnerabilities in embedded systems, network, or processes.

8.16.5 Compatibility Testing

Compatibility testing to ensures that the e-learning system works well on different platforms, browsers, and devices. It includes compatibility testing with multiple operating systems, web browsers, and display solutions to ensure a consistent user experience. Some scenarios for social testing in e-learning systems may include:

- Testing the system in different web browsers such as Chrome, Firefox, Safari, and Internet Explorer/Edge.

8.17 Performance Evaluation of the Secure E-Learning System

The performance evaluation of a secure e-learning security is important to ensure that the business is meeting its needs without major impact. Below are some important factors and metrics to consider when evaluating the effectiveness of e-learning security.

8.17.1 Response Time

Response time is measured as quickly as how the system responds to user requests. It is very important to ensure that logging in, accessing class materials, submitting assignments, and interacting with the system are completed on time. Response times should be within the acceptable range to ensure a good customer experience.

8.17.2 Scalability

Scalability measures the ability of a system to manage increased users or underperforming resources. It is important for e-learning, which can have an immediate impact on user activity at peak times or when conducting tests.

8.17.3 Resource Utilization

Monitor resources usage such as CPU, memory, and network bandwidth to help identify potential bottlenecks or areas where resource allocation can be optimized to improve performance.

8.17.4 Load Testing

Load testing simulates real usage scenarios to evaluate system performance under expected load and peak load conditions. Load testing helps identify performance, issues, and areas for improvement by assigning the system to multiple virtual users.

8.17.5 Stress Testing

Stress testing involves pushing the body beyond its ability to determine its breaking point and how it recovers from a heavy load. These tests help identify potential operational weaknesses and measure system health.

8.17.6 Concurrency Testing

Concurrency test evaluates how the system's ability to manage many simultaneous or dissimilar users. Helps identify any contention issues that may arise when multiple user's access shared resources at the same time.

8.17.7 End-User Experience

Running user experience (UX) tests and collecting feedback from real users can provide valuable insights into functionality and experience insights. Understanding user expectations and trouble spots can guide improvements.

8.17.8 Security Impact

Performance evaluation should consider the impact of safety measures on physical activity. Encryption, authentication, and access control mechanisms, while important to security, can introduce some overhead that needs to be evaluated.

It is important to conduct performance evaluation during various stages of e-learning system's development, such as during initial testing, after adding updates or functionality, and before production deployment.

8.18 Results and Discussion

Data mining techniques have proven to be very useful in many fields and applications. These technologies allow organizations to extract insights, models, patterns, and information from big and complex data, resulting in improved decision-making, problem solving, and forecasting. The following are some advantages of data mining techniques.

8.18.1 Knowledge Discovery

Data mining techniques facilitate the research process. It seeks knowledge from data. By analysing and discovering large volumes of structured and unstructured data, these techniques can reveal hidden patterns, relationships, and patterns that may not be visible through traditional analysis. This information search allows organizations to gain a deep understanding of their data, identify valuable insights, and make informed decisions.

8.18.2 Predictive Analytics

Data mining techniques are commonly used in predictive analytics to enable organizations to make accurate predictions and forecasts based on historical data models. Using algorithms such as regression, classification, and clustering, data mining identifies patterns that help predict future outcomes, trends, or behaviour. This predictive capability helps businesses improve operations, identify business patterns, and make informed decisions.

8.18.3 Pattern Recognition

Data mining techniques are effective in pattern recognition. By analysing big data, this system can identify recurring patterns, anomalies and vulnerabilities that may be important to businesses. Pattern recognition helps in detecting fraud, identifying bad behaviour, improving processes, and making recommendations based on past patterns.

8.18.4 Risk Assessment and Fraud Detection

Data mining techniques are useful in risk assessment and fraud detection. By analysing historical data and patterns, these technologies can identify risks and fraud in areas such as finance, insurance, and cyber security. Data mining helps organizations identify suspicious patterns, flag potential fraud cases, and take appropriate action to mitigate risk.

8.18.5 Process Optimization and Efficiency

Data mining techniques can identify inefficiencies in business processes, allowing organizations to improve performance and improve overall results. By analysing process data, organizations can identify areas for improvement, streamline operations, and eliminate unnecessary or time-consuming tasks, saving costs, increasing productivity, and increasing customer satisfaction.

8.18.6 Decision Support Systems

Data mining techniques provide insights and support decision making. The technologies can help identify patterns, evaluate different situations, and suggest effective solutions by analysing historical data-mining-based decision support systems help organizations make decisions, reduce risk, and increase efficiency.

However, it is important to note that the effectiveness of data mining techniques depends on many factors such as data quality, choice of quality, choice of algorithm, model validity and intelligence of data analysts. Correct application and interpretation of data mining techniques requires knowledge of the domain and a good understanding of the specific problem at hand. In Addition, privacy and ethical value should be considered when processing sensitive or personal information.

8.19 Comparison with Existing E-Learning Systems

When comparing a secure e-learning system with existing e-learning systems, several factors need to be considered. Here are some key points to compare and evaluate:

8.19.1 Security Features

A secure e-learning system should have good security features to protect user data, ensure secure security, and prevent unauthorized access or data leakage. Comparing the security features such as encryption, access controls authentication methods, and secure transport method will help evaluate the level of security provided by each system.

8.19.2 Data Privacy and Confidentiality

Data privacy is extremely important in e-learning, especially when it comes to user's sensitive information. Comparing how each system handles data privacy, anonymity processes, data retention policies, and privacy compliance will provide insight into their commitment to protecting user privacy.

8.19.3 Access Control and User Management

The effectiveness of access control mechanisms and user management capabilities can significantly impact the security and usability of an e-learning system. Comparing how user roles and permissions are managed, authentication mechanisms employed (such as multi-factor authentication), and the granularity of access controls will help evaluate the system's ability to enforce appropriate access restrictions.

8.19.4 Performance and Scalability

Efficiency is important in e-learning system because it directly affects user experience and performance. Comparing performance metrics, scalability, and the ability to manage multiple users and high-end devices will help gauge the ability of each system to meet customer needs and standards used.

8.19.5 Usability and User Experience

The usability and user experience of an e-learning system are important to its success and engagement. Comparing the usability of features such as user interface, ease of navigation, availability of features such as search, classroom organization, content presentation, and discussion content will help evaluate the system's usability and user-friendliness of the system.

8.19.6 Course Content Management

Competing in the ability to create, manage and deliver course content is essential. Features such as content collection tools, support for multimedia content support, proficiency assessment and learner performance monitoring will affect the effectiveness of the e-learning training.

By comparing these factors, organizations can make informed decisions about choosing an e-learning security program that meets their specific security requirements, performance, user need, and overall business goals.

8.20 Conclusion

It is essential to design and implement e-learning security to ensure the protection of user's sensitive information, prevent unauthorized access, and provide a safe learning environment. By combining data mining techniques, access control methods, authentication methods, encryption methods and secure transmission methods, the system can solve various security problems in e-learning platforms.

Data mining techniques play an important role in extracting insights, models, and information from big data, enabling organizations to make data-driven decisions, enhanced forecasting, and improved processes. These technologies help improve decision making and problem solving with great benefits for knowledge discovery, predictive analytics, pattern recognition and customer segmentation.

References

Bhadani S, Shekhawat D (2015) A survey on secure e-learning. Procedia Comput Sci 48:547–553

Fung BC, Wang K, Yu PS (2010) Privacy-preserving data publishing: a survey of recent developments. ACM Comput Surv (CSUR) 42(4):1–53

Hameed A, Hassan S, Mahmood A (2016) Secure access control and authentication for e-learning system using biometric approach. Comput Electr Eng 50:208–222

Han J, Pei J, Kamber M (2011) Data mining: concepts and techniques. Morgan Kaufmann

Kantardzic M (2011) Data mining: concepts, models, methods, and algorithms. John Wiley & Sons

Tan PN, Steinbach M, Kumar V (2013) Introduction to data mining. Pearson

Ullah I, Yousaf MH, Kim DS (2016) Secure and efficient user authentication framework for e-learning systems. Comput Hum Behav 55:580–589

Witten IH, Frank E, Hall MA (2016) Data mining: practical machine learning tools and techniques. Morgan Kaufmann

Yordanova K, Georgiev V (2019) Security issues in e-learning systems. In: International conference on information technologies & information society, pp 187–195

Chapter 9
An Overview of the Internet of Things Within the Framework of Retail 4.0

Dipti Kiran, Rohit Kumar, Naveen Kumar, and Ashok Kumar Mishra

Abstract The adoption of the Internet of Things (IoT) as part of the Retail 4.0 framework is causing a radical transformation in the retail industry. This integration combines digitalization, innovation, and a customer-centric approach to reinvent traditional retail operations. This article provides a thorough examination of the function and consequences of IoT in Retail 4.0, highlighting its uses in real-time inventory management, individualized marketing, enhanced consumer experiences, and increased operational effectiveness. Data security, interoperability, scalability, and ethical issues are among the problems that are evaluated alongside the possible advantages, which include enhanced decision-making through data analytics and reduced procedures. The significance of edge computing, blockchain technology, predictive analytics, and sustainable practices are highlighted as future directions in IoT for retail are explored. In order to overcome these obstacles and completely realize the promise of IoT in retail, the study emphasizes the necessity of cooperative efforts, legislative frameworks, and creative solutions. By doing so, IoT can drive significant digital transformation, optimize retail operations, and deliver superior customer experiences, ultimately shaping the future of the retail industry in a connected and data-driven world.

Keywords Internet of things · Retail business · Digital revolution · E-commerce · Supply chain · Artificial intelligent

---

D. Kiran
Department of Commerce, Guru Ghasidas University, Bilaspur, C.G, India

R. Kumar (✉) · N. Kumar
University School of Management, Gautam Buddha University, Greater Noida, UP, India
e-mail: rohitguraru96@gmail.com

N. Kumar
e-mail: Naveen.kumar@gbu.ac.in

A. K. Mishra
Dean Department of Commerce, Guru Ghasidas Vishwavidyalaya, Bilaspur, C.G, India

A. Rai et al. (eds.), *Convergence of Artificial Intelligence, Machine Learning, and the Internet of Things in Industry 4.0 Applications*, Transactions on Computer Systems and Networks, https://doi.org/10.1007/978-981-96-5269-3_9

9.1 Introduction

Over the ages, the retail industry has experienced enormous adjustments due to shifts in customer behavior, technology, and economic situations. The development of retail, from antiquated marketplaces to modern multichannel experiences, is a reflection of greater social changes and progress. This overview traces the key stages in the development of retail, highlighting the major milestones and innovations that have shaped the industry (Yeap et al. 2022).

Marketplaces where the hub of economic activity in ancient civilizations, which is where retail got its start. In Mesopotamia, around 3,000 BCE, merchants traded goods in bazaars, a practice that spread to ancient Egypt, Greece, and Rome. These early markets were characterized by direct transactions between producers and consumers. During the medieval period, retail evolved with the emergence of guilds and trade fairs. Guilds regulated quality and prices, ensuring fair trade practices. Trade fairs, such as the Champagne Fairs in France, became important venues for international commerce, attracting merchants from across Europe.

9.1.1 The Rise of Shops and Department Stores

The seventeenth and eighteenth centuries saw the rise of permanent shops and the beginnings of modern retailing. In England, for example, the Royal Exchange, established in 1565, became a hub for retail activity. As a result of the rise of the middle class and the creation of urban canters during the eighteenth century Industrial Revolution, retail underwent additional transformation. The nineteenth century marked the advent of department stores, which revolutionized shopping by offering a wide range of products under one roof. Stores like Harrods in London (founded in 1834) and Macy's in New York (founded in 1858) provided a new shopping experience with fixed prices and return policies, a departure from the haggling typical of earlier retail practices. These stores also introduced window displays and in-store promotions, laying the groundwork for modern marketing techniques (Chaney 1983).

9.1.2 The Birth of Chain Stores and Supermarkets

The late 19th and early twentieth centuries saw the proliferation of chain stores, which standardized the shopping experience across multiple locations. Pioneers like A&P (The Great Atlantic & Pacific Tea Company) in the United States popularized the concept of chain grocery stores, offering consistent pricing and product availability. The 1930s and 1940s witnessed the rise of supermarkets, a significant shift in food retailing. Supermarkets like King Kullen, which opened in 1930 in New York, introduced the self-service model, allowing customers to browse aisles and select their goods. This innovation reduced labor costs and increased efficiency, leading to the widespread adoption of the supermarket format (Davison 2016).

9.1.3 The Era of Shopping Malls and Big-Box Retailers

The post-World War II economic boom in the mid-twentieth century spurred the development of suburban shopping malls. These enclosed spaces offered a convenient and climate-controlled shopping environment, attracting consumers to a diverse array of stores and entertainment options under one roof. The first modern shopping mall, Southdale Centre, opened in Minnesota in 1956, setting a template for future developments. The twentieth century witnessed the rise of big-box stores such as Walmart and Target. These stores offered a vast selection of goods at low prices, leveraging economies of scale and efficient supply chains (Lee et al. 2006). Walmart, founded in 1962, became the world's largest retailer, influencing retail practices globally with its focus on cost leadership and logistical efficiency (Scanlon 2011).

9.1.4 The Digital Revolution and E-commerce

The advent of the internet in the 1990s marked a new era in retail. Online retailers like Amazon, founded in 1994, disrupted traditional retail by offering a vast selection of products with the convenience of home delivery. E-commerce grew rapidly, with advances in technology enabling secure online transactions and efficient fulfillment systems (Gerlich 2001). The twenty-first century has seen the rise of omnichannel retailing, where businesses integrate online and offline channels to provide a seamless customer experience (Mahammadh 2015). Innovations such as click-and-collect services, mobile shopping apps, and personalized marketing have become standard practices. The COVID-19 epidemic hastened the transition to online shopping even more, prompting some shops to make significant investments in digital infrastructure in response to evolving consumer needs. The evolution of retail reflects broader societal changes and technological advancements. From ancient marketplaces to the digital revolution, each stage of retail development has brought new innovations and challenges. As the industry continues to adapt to emerging trends and technologies, the future of retail promises to be dynamic and consumer-centric, with an increasing focus on personalization, sustainability, and immersive experiences (Tian and Stewart 2006).

9.1.5 Understanding Retail 4.0

Retail 4.0 represents the latest transformation in the retail industry, driven by the integration of advanced technologies and the digitalization of retail operations. By utilizing big data, blockchain, artificial intelligence (AI), the Internet of Things (IoT), and other cutting-edge technologies, Retail 4.0 expands on previous iterations of the retail evolution and enhances the shopping experience while streamlining operations and generating more individualized and productive customer interactions. This overview provides a comprehensive understanding of Retail 4.0, its key components, benefits, and challenges (Malar 2019).

9.1.6 The Evolution to Retail 4.0

Retail 4.0 is a natural progression from earlier stages of retail evolution (Har et al. 2022) (Image 9.1).

Key Components of Retail 4.0

- Internet of Things (IoT)-IoT devices, such as smart shelves, beacons, and connected point-of-sale (POS) systems, collect and transmit data in real-time. These devices help in inventory management, personalized marketing,

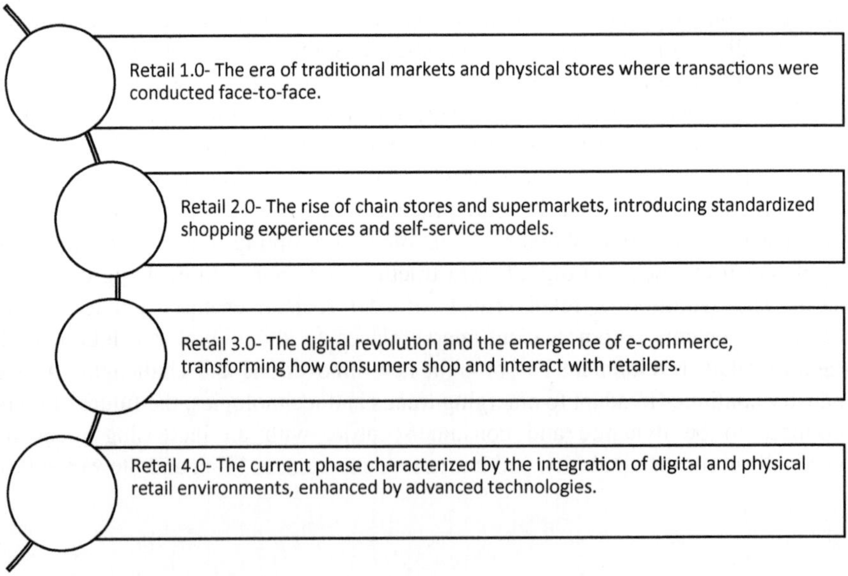

Image 9.1 Evolution of retail. *Source* Developed by author

and improving customer experiences by providing relevant information and recommendations.

- Artificial Intelligence (AI) and machine learning (ML)-These two fields use algorithms to examine enormous volumes of data in order to forecast customer behavior, improve pricing plans, and target advertising campaigns. AI-powered chatbots and virtual assistants enhance customer service by providing timely responses and recommendations.
- Big Data and Analytics-Retailers may learn about consumer preferences, buying habits, and market trends by processing and analyzing massive databases. Data analytics helps in making informed decisions, optimizing supply chains, and improving inventory management.
- Blockchain-Blockchain technology lowers the risk of fraud and counterfeiting by ensuring supply chain transparency and traceability. It enhances trust between retailers and consumers by providing verifiable information about product origins and authenticity.
- Virtual reality (VR) and augmented reality (AR): Retail spaces that are immersive are created by these technologies. To increase engagement and lower return rates, customers may virtually try on goods, see how furniture will look in their homes, and browse virtual stores.
- Omnichannel Integration-Retail 4.0 emphasizes seamless integration across multiple channels, including online, mobile, and physical stores. This integration ensures a consistent and personalized experience, allowing customers to shop anytime, anywhere, and on any device (Zhichao n.d.).

9.1.7 Benefits of Retail 4.0

Retail 4.0 represents the integration of advanced technologies such as the Internet of Things (IoT), Artificial Intelligence (AI), big data analytics, and blockchain into the retail sector. These technologies transform traditional retail operations and customer interactions, offering numerous benefits that drive the evolution of the industry. Here, we explore some of the key benefits of Retail 4.0 (Bakharev et al. 2020).

Enhanced Customer Experience

One of the most significant advantages of Retail 4.0 is the enhancement of customer experience. Personalized marketing and real-time recommendations allow retailers to cater to individual preferences, creating a more engaging shopping experience. By analyzing data from various customer touchpoints, retailers can offer personalized offers, product suggestions, and targeted promotions, which increase customer satisfaction and loyalty. Immersion technologies like as virtual reality (VR) and augmented reality (AR) allow shoppers to view products in their proper contexts before making a purchase, which makes shopping more pleasurable and engaging. Retailers stand out in a competitive market thanks to their ability to anticipate and successfully address the demands of their customers.

Improved Operational Efficiency

Automation and data-driven decision-making are at the core of Retail 4.0, significantly improving operational efficiency. Technologies like IoT enable real-time tracking and management of inventory, reducing instances of overstocking or stockouts. AI-driven analytics give merchants insights into market trends and consumer behavior, enabling them to streamline their supply chains and cut waste. These efficiencies lead to cost savings and more streamlined operations, which are crucial for maintaining profitability in the fast-paced retail environment. For example, predictive maintenance of equipment and facilities can prevent downtime and ensure smooth operations, enhancing the overall productivity of the retail business.

Increased Sales and Revenue

Retail 4.0 drives sales and revenue growth through personalized offers and targeted promotions. Retailers may customize their marketing tactics to draw in and keep customers by using big data analytics to find trends and opportunities. Predictive analytics helps retailers forecast demand and plan inventory accordingly, reducing lost sales due to stockouts and excess inventory costs. Additionally, seamless omnichannel experiences—where customers can interact with brands across various online and offline platforms—encourage higher spending and repeat purchases. By providing a consistent and convenient shopping experience, retailers can boost customer engagement and drive sales.

Transparency and Trust

Blockchain technology is a cornerstone of Retail 4.0, providing transparency in supply chains and enhancing trust between retailers and consumers. Blockchain technology makes it possible to track and authenticate each transaction and product movement, guaranteeing the legitimacy and moral sourcing of goods. This transparency is increasingly important in today's market, where consumers are more conscious of the origins and ethical implications of their purchases. Customers can verify the authenticity of products and trust that they meet specific standards, leading to increased brand loyalty and a stronger reputation for retailers committed to transparency.

Flexibility and Agility

Retail 4.0 gives merchants the adaptability and nimbleness they need to adjust to shifting consumer tastes and market situations. Retailers can provide a variety of shopping and delivery choices, including curb side pickup, same-day delivery, and buy online, pick up in-store (BOPIS). This is made possible by omnichannel strategy. Retailers can accommodate a wide range of customer wants and react quickly to market movements because to this flexibility. With the help of advanced analytics, merchants can stay ahead of the competition by adjusting their strategy in response to fluctuations in the market and client preferences. Agile supply chain management facilitated by real-time data and automation allows retailers to pivot quickly in response to disruptions, ensuring continuous availability of products and maintaining customer satisfaction (Bruni and Piccarozzi 2022).

9.1.8 Challenges of Retail 4.0

Retail 4.0 offers numerous benefits, but it also presents several challenges that retailers must navigate to fully leverage its potential. These challenges encompass data privacy and security, technology integration, skill gaps, customer adaptation, and the cost of implementation. A comprehensive understanding of these obstacles is crucial for retailers aiming to transition smoothly into the era of Retail 4.0.

Data Privacy and Security

The proliferation of data in Retail 4.0 raises significant concerns about privacy and security. Customer data is being used more and more to tailor experiences and streamline processes, therefore protecting this data is crucial. Retailers are subject to strict restrictions, such as the California Consumer Privacy Act (CCPA) in the US and the General Data Protection Regulation (GDPR) in Europe. Strong data protection safeguards are required by this legislation, granting customers control over their personal data. Refusal to comply may incur heavy fines and harm a retailer's standing. Protecting against data breaches and upholding consumer trust requires the implementation of robust cybersecurity measures, such as encryption, safe access controls, and frequent security audits.

Technology Integration

Integrating new technologies into existing systems is a complex and costly endeavor. Retailers often operate on legacy systems that are not designed to work with advanced technologies like IoT, AI, and blockchain. The challenge lies in creating a seamless integration where different technologies can communicate and function together effectively. This requires significant investment in infrastructure and technology upgrades. Retailers must also ensure minimal disruption to their operations during the transition period. Developing a clear integration strategy, choosing scalable solutions, and working with experienced technology partners can help mitigate these challenges and ensure a smoother implementation process.

Skill Gaps

The adoption of advanced technologies in Retail 4.0 necessitates a workforce with specialized skills. Managing and analyzing large datasets, developing and maintaining AI algorithms, and overseeing IoT devices require expertise that many retailers may lack. There is a pressing need for skilled personnel who can navigate these technologies and extract actionable insights. Retailers must invest in training and development programs to bridge these skill gaps. This includes upskilling current employees and attracting new talent with the necessary technical expertise. Collaborating with educational institutions and offering internships or apprenticeships can also help build a pipeline of skilled professionals ready to support Retail 4.0 initiatives.

Customer Adaptation

Not all customers are quick to embrace new technologies, presenting a significant challenge for retailers. While some consumers readily adopt innovations like mobile payments, AR/VR shopping experiences, and smart shelves, others prefer traditional shopping methods. Retailers must strike a balance between implementing advanced technologies and ensuring a user-friendly experience for all customers. It is essential to provide clear instructions and support for new technology use, such as tutorials and customer service assistance. Additionally, offering a choice between traditional and tech-enabled shopping options can help cater to diverse customer preferences and ensure that no segment of the customer base feels alienated.

Cost of Implementation

Implementing Retail 4.0 technology can be prohibitively expensive initially, especially for small and medium-sized businesses (SMEs). The investment required for purchasing and integrating new technologies, training staff, and maintaining advanced systems is substantial. Retailers need to carefully evaluate the potential return on investment (ROI) and develop a strategic plan to justify these expenses. Finding funding sources like grants, loans, or collaborations might help reduce the financial load. Starting with pilot projects and gradually scaling up successful implementations can help manage costs and demonstrate the value of Retail 4.0 initiatives (Kumar et al. 2020).

9.1.9 What is the Internet of Things (IoT)?

The Internet of Things (IoT) is an innovative concept that represents the next major step in the digital transformation of our environment. The Internet of Things, or IoT, is essentially a network of everyday objects, ranging from household appliances to industrial machinery, that are connected and equipped with sensors, software, and other integrated technologies that enable them to communicate, collect, and analyze data. This seamless connectivity allows for unprecedented automation, efficiency, and real-time response across various domains (Fleisch 2010; Greengard 2021).

Understanding IoT Components

IoT is built on three fundamental components: sensors/devices, connectivity, and data processing. The actual things with embedded technologies that collect data from their surroundings are called sensors or gadgets. These may consist of GPS devices, temperature sensors, motion detectors, and more. Connectivity makes ensuring that information gathered by these sensors is shared with other gadgets or centralized systems via networks like Bluetooth, Wi-Fi, cellular, or satellite communications. Data processing involves analyzing and acting upon the collected data, often utilizing advanced algorithms and artificial intelligence (AI) to make informed decisions or trigger specific actions.

Applications of IoT

IoT's versatility is evident in its wide range of applications across various sectors. IoT devices are used in healthcare to remotely monitor patient vitals, improving patient care and lowering hospital admission rates. Wearable technology tracks vital signs such as heart rate and sleep habits, enabling people to make well-informed decisions about their health. IoT sensors are used in agriculture to offer real-time data on crop health, weather, and soil conditions. This allows for more productive and sustainable farming practices. By streamlining traffic, cutting energy use, and boosting public safety, smart cities use IoT to make living in cities better. For example, real-time traffic data is used by smart traffic signals to modify their timings, which lowers emissions and congestion. In retail, IoT enhances the shopping experience through smart shelves that track inventory levels and beacons that send personalized offers to shoppers' smartphones based on their location within the store (Ramson et al. 2020).

9.1.10 Benefits of IoT

The primary advantage of IoT lies in its ability to turn data into actionable insights. By automating routine tasks and enabling real-time monitoring, IoT improves operational efficiency and reduces costs. For businesses, this translates to optimized supply chains, better resource management, and enhanced customer experiences. In industrial settings, predictive maintenance saves time and money by addressing possible problems before they lead to equipment breakdown, minimizing downtime. IoT offers more personalization and convenience for customers. The ability to remotely control smart home appliances like security cameras, lighting controls, and thermostats brings comfort and peace of mind. Wearable fitness trackers encourage healthy living and provide individualized health insights. IoT also facilitates the creation of new business models, such as subscription-based services and usage-based pricing, driving innovation and growth (Taherdoost 2023).

9.1.11 Challenges and Considerations

Despite its transformative potential, IoT faces significant challenges. Since there are several access points for cyberattacks due to the extensive network of linked devices, data security and privacy are serious problems. It is essential to guarantee strong security procedures and adherence to laws such as GDPR. In addition, efficient data management and storage solutions are needed due to the enormous amount of data created by IoT devices. Interoperability among diverse devices and standards remains a hurdle, necessitating collaborative efforts to establish universal protocols (Taherdoost 2023).

9.1.12 IOT in Retail 4.0

The Internet of Things (IoT) is a cornerstone of Retail 4.0, transforming the retail industry by enabling smarter operations, enhancing customer experiences, and driving efficiency. The internet of things (IoT) is a network of networked objects that exchange data and communicate with one another. These gadgets are used in retail settings and include sensors, beacons, smart shelves, linked point-of-sale (POS) systems, and more. This succinct synopsis delves into the important part that IoT plays in the evolution of retail. IoT devices provide real-time stock level data, revolutionizing inventory management. When inventory is low or goods are lost, store managers can receive automatic inventory monitoring alerts via smart shelves that are fitted with RFID tags or weight sensors. In addition to optimizing inventory levels and guaranteeing that popular items are always available, this lowers the danger of stockouts and overstocking. Real-time inventory tracking also allows for better demand forecasting and supply chain management, minimizing waste and improving efficiency (Singh et al. 2020). IoT technology enables highly personalized shopping experiences. Beacons, which are small wireless devices, can communicate with customers' smartphones as they enter a store. By analyzing data from previous purchases and browsing behavior, retailers can send personalized offers, discounts, and product recommendations to customers in real-time. This targeted approach not only enhances the shopping experience but also increases the likelihood of purchases, boosting sales and customer loyalty. IoT devices streamline store operations through automation and enhanced monitoring. Connected POS systems can track sales data in real-time, providing insights into peak shopping times and popular products. This information helps retailers optimize staffing levels and product placements. Additionally, smart HVAC systems can adjust lighting and temperature based on foot traffic and occupancy, creating a comfortable shopping environment while reducing energy costs. The integration of IoT in supply chains enhances transparency and traceability. Sensors and GPS devices on shipments provide real-time tracking information, allowing retailers to monitor the location and condition of goods throughout the supply chain. This level of visibility helps in identifying and addressing issues such as delays or damage promptly. Blockchain technology, when combined with IoT, can further ensure the authenticity and ethical sourcing of products, building trust with consumers. Retailers can use IoT to manage assets more efficiently. For example, RFID tags on shopping carts can help track their usage and location within the store, ensuring they are readily available for customers. Similarly, maintenance of equipment such as refrigeration units and POS systems can be scheduled proactively based on data from IoT sensors, reducing downtime and repair costs. IoT technology also plays a crucial role in enhancing security within retail environments. Surveillance cameras equipped with IoT capabilities can monitor store activity in real-time, detecting unusual patterns and alerting security personnel to potential threats. IoT-enabled alarm systems and smart locks can further secure store premises, providing peace of mind to both retailers and customers. IoT is a transformative force in Retail 4.0, driving enhanced inventory management, personalized customer experiences, smart store operations, improved supply chain transparency, efficient asset management, and heightened security. By leveraging the capabilities of IoT, retailers can create more efficient, responsive, and customer-centric operations. However, to fully realize the potential of IoT, retailers must address challenges related to data security, privacy, and technology integration, ensuring a seamless and secure shopping experience for all.

9.1.13 The Role of Buy Now, Pay Later (BNPL) in Retail 4.0

Buy Now, Pay Later (BNPL) services have emerged as a significant innovation within Retail 4.0, transforming the shopping experience by offering consumers flexible payment options. BNPL allows customers to make purchases and pay for them over time through interest-free installments, enhancing affordability and convenience. As retail continues to evolve with the integration of advanced technologies, BNPL plays a crucial role in driving sales, improving customer satisfaction, and shaping modern retail strategies. BNPL services improve the customer experience by providing greater financial flexibility. Consumers, especially younger demographics, often seek ways to manage their finances without incurring high-interest debt. BNPL meets this demand by allowing shoppers to spread the cost of their purchases over several months without interest. This option makes high-ticket items more accessible and reduces the psychological barrier to making significant purchases. Moreover, BNPL integrates seamlessly with both online and in-store shopping experiences. At checkout, customers can select BNPL as a payment option, simplifying the purchasing process and enhancing convenience. The availability of multiple payment plans caters to different financial situations, further personalizing the shopping experience. One of the primary benefits of BNPL for retailers is the potential to drive sales and improve conversion rates. By offering flexible payment options, retailers can attract customers who might otherwise be hesitant to make a purchase due to budget constraints. This increased affordability can lead to higher average order values (AOV), as customers are more likely to add additional items to their cart when they can pay in installments. BNPL services also reduce cart abandonment rates. Many online shoppers abandon their carts during the checkout process due to the total cost of their purchase. BNPL addresses this issue by breaking down the payment into manageable chunks, encouraging customers to complete their transactions. BNPL can enhance customer loyalty and retention by providing a positive shopping experience. Customers who find value in flexible payment options are more likely to return to retailers that offer BNPL services. Additionally, BNPL providers often have loyalty programs and incentives that encourage repeat purchases, further strengthening the relationship between retailers and consumers. Retailers can also use BNPL data to gain insights into customer behavior and preferences. This information can be leveraged to create targeted marketing campaigns and personalized offers, increasing customer engagement and loyalty. Retail 4.0 emphasizes seamless integration across multiple shopping channels, and BNPL is a key component of this strategy. BNPL services are available both online and in physical stores, providing a consistent payment experience regardless of the shopping channel. This integration supports the omnichannel approach, ensuring that customers can enjoy the same flexibility and convenience whether they are shopping on a website, a mobile app, or in a brick-and-mortar store. BNPL can be integrated with other digital tools and technologies used in Retail 4.0, such as mobile wallets, digital receipts, and personalized marketing platforms. This holistic integration enhances the overall shopping experience and aligns with the goals of Retail 4.0 to create a connected and customer-centric retail environment. BNPL services play a significant role in promoting financial inclusion. Traditional credit options are often inaccessible to individuals with limited credit histories or lower credit scores. BNPL providers typically use alternative methods to assess creditworthiness, allowing a broader range of consumers to access financing options. This inclusivity enables more people to participate in the retail economy and enjoy the benefits of flexible payment terms. BNPL offers numerous advantages, it also presents challenges that retailers must consider. The ease of access to BNPL can lead to overconsumption and potential financial difficulties for some consumers if not managed responsibly. Retailers must partner with reputable BNPL providers that prioritize responsible lending practices and offer clear, transparent terms to customers.

The implementation of BNPL services requires investment in technology and integration with existing payment systems. Retailers need to ensure a smooth and secure checkout process, maintaining customer trust and protecting sensitive financial information. BNPL is a transformative element of Retail 4.0, providing significant benefits to both consumers and retailers. By enhancing affordability, driving sales, and integrating with omnichannel strategies, BNPL services contribute to a more flexible, inclusive, and customer-centric retail environment. As the retail industry continues to evolve, BNPL will play an increasingly important role in shaping the future of shopping experiences, driving innovation, and meeting the demands of modern consumers (Yeap et al. 2022).

9.1.14 The Impact of IoT on Supply Chain

The Internet of Things (IoT) has revolutionized various industries, and its impact on the supply chain is particularly profound. By enabling real-time data collection, analysis, and communication among connected devices, IoT enhances visibility, efficiency, and responsiveness throughout the supply chain. This overview explores how IoT transforms supply chain operations, from inventory management and logistics to customer satisfaction and sustainability (Image 9.2).

1. **Enhanced Visibility and Transparency**

   - Real-Time Tracking-IoT devices, such as GPS-enabled trackers and RFID tags, provide real-time visibility into the location and status of shipments. This capability allows businesses to monitor the movement of goods from

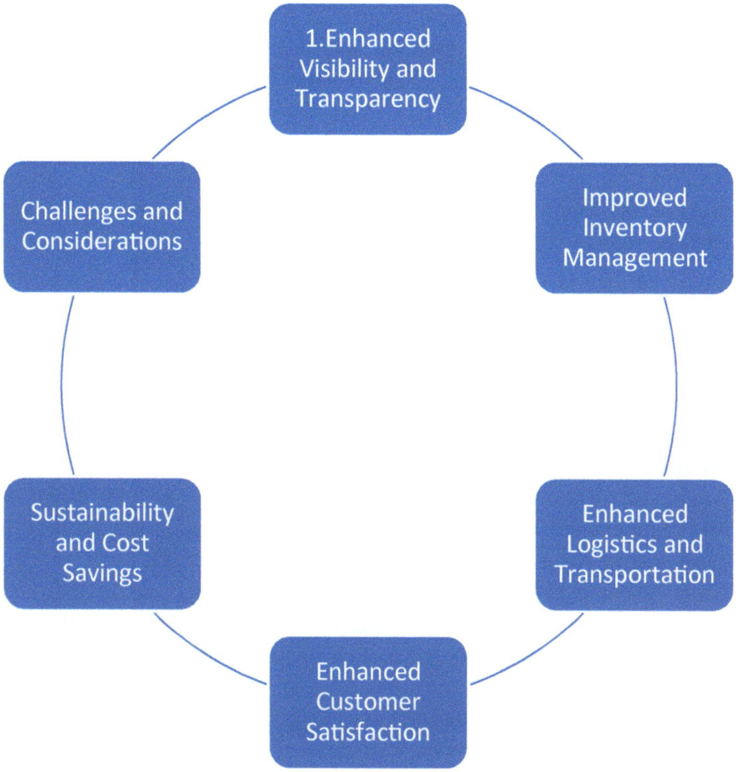

Image 9.2 Depicting circular process. *Source* Developed by author

production to delivery, ensuring that they are aware of any delays or disruptions immediately.

- Condition Monitoring-IoT sensors can monitor the condition of products, particularly those that are sensitive to environmental conditions like temperature, humidity, or light. For example, in the pharmaceutical and food industries, sensors can ensure that products are stored and transported under optimal conditions, reducing the risk of spoilage or damage.
- Supply Chain Transparency-IoT enhances transparency by providing a detailed view of the entire supply chain. Businesses can track the origin of raw materials, monitor the production process, and verify the authenticity of products. This transparency is particularly important for industries like pharmaceuticals, where counterfeiting is a major concern.

2. **Improved Inventory Management**

- Automated Inventory Tracking-IoT-enabled smart shelves and RFID tags provide real-time data on inventory levels. This automation helps in reducing manual counting errors and ensures that inventory data is always up-to-date.

- Demand Forecasting-By analyzing data from IoT devices, businesses can better predict demand patterns and adjust inventory levels accordingly. This capability helps in maintaining optimal stock levels, reducing both stockouts and excess inventory.
- Warehouse Efficiency-IoT devices can optimize warehouse operations by tracking the movement of goods and equipment. Automated guided vehicles (AGVs) and drones, guided by IoT, can streamline picking, packing, and shipping processes, improving overall efficiency.

3. **Enhanced Logistics and Transportation**

- Route Optimization-IoT-enabled fleet management systems can analyze real-time traffic and weather data to optimize delivery routes. This capability reduces fuel consumption, lowers transportation costs, and ensures timely deliveries.
- Predictive Maintenance-IoT sensors on transportation vehicles can monitor engine performance, tire pressure, and other critical parameters. Predictive maintenance helps in identifying potential issues before they lead to breakdowns, reducing downtime and maintenance costs.
- Asset Tracking-Businesses can track the location and condition of assets such as containers, pallets, and equipment in real-time. This visibility helps in optimizing asset utilization and reducing losses due to theft or misplacement.

4. **Enhanced Customer Satisfaction**

- Accurate Delivery Estimates-Real-time tracking and route optimization ensure more accurate delivery estimates. Customers can receive real-time updates on their orders, improving their overall shopping experience.
- Quality Assurance-IoT devices that monitor product conditions help ensure that customers receive products in optimal condition. For instance, IoT sensors can alert businesses to any temperature fluctuations that might affect perishable goods, allowing them to take corrective actions before delivery.
- Personalized Services-Businesses can learn more about the preferences and behavior of their customers by analyzing data from IoT devices. By providing individualized services and promotions, this data can increase consumer pleasure and loyalty.

5. **Sustainability and Cost Savings**

- Energy Efficiency-IoT devices can monitor and optimize energy usage in warehouses and transportation fleets. This capability reduces energy consumption and lowers operational costs, contributing to more sustainable supply chain practices.
- Waste Reduction-Real-time monitoring and predictive analytics help in minimizing waste throughout the supply chain. Businesses can reduce excess inventory, optimize production processes, and ensure that products are transported and stored under ideal conditions.

- Resource Optimization-IoT-enabled asset tracking and utilization help in optimizing the use of resources such as vehicles, equipment, and storage space. This optimization reduces costs and enhances operational efficiency.

6. **Challenges and Considerations**

- Data Security and Privacy-The extensive data generated by IoT devices poses significant security and privacy challenges. Businesses must ensure robust cybersecurity measures to protect sensitive information and comply with data protection regulations.
- Integration with Existing Systems-It can be difficult and expensive to integrate IoT technology with current supply chain systems. Businesses need to ensure seamless integration to fully realize the benefits of IoT.
- Initial Investment and ROI-The initial cost of implementing IoT technology can be high. Businesses must carefully evaluate the return on investment (ROI) to justify the expenditure and ensure long-term benefits (Kothari et al. 2018).

9.1.15 Personalized Marketing and Customer Engagement

In today's competitive business landscape, personalized marketing and customer engagement have become essential strategies for brands aiming to stand out and build lasting relationships with their customers. By leveraging data and technology, businesses can tailor their marketing efforts to meet individual customer preferences, enhancing their overall experience and fostering loyalty. This overview explores the concepts of personalized marketing and customer engagement, their importance, and effective strategies to implement them. Personalized marketing involves creating marketing messages and experiences that are tailored to the individual needs and preferences of each customer. This approach contrasts with traditional mass marketing, where a single message is broadcast to a broad audience. Personalized marketing is enabled by data collection and analysis, which provide insights into customer behavior, preferences, and purchase history (Bleier et al. 2018).

9.1.16 Key Components of Personalized Marketing

Data Collection-The foundation of personalized marketing is data. Businesses collect data through various channels, including website analytics, social media interactions, purchase history, and customer surveys. This data is used to build detailed customer profiles.

Segmentation-Based on shared traits including behavior, tastes, and demography, customers are divided into various categories. Businesses can target particular groups with pertinent communications by using segmentation.

Targeted Messaging-Personalized marketing involves crafting specific messages tailored to meet the distinct needs and preferences of different customer groups. This can include personalized emails, product recommendations, and tailored advertisements.

Dynamic Content-Dynamic content refers to content that adapts according to the viewer's profile or behavior. For instance, a website might show various product recommendations or offers based on the visitor's previous browsing activities.

Automation-Marketing automation tools enable businesses to deliver personalized messages at scale. Automated workflows can trigger personalized emails, SMS messages, or social media posts based on customer actions (Image 9.3).

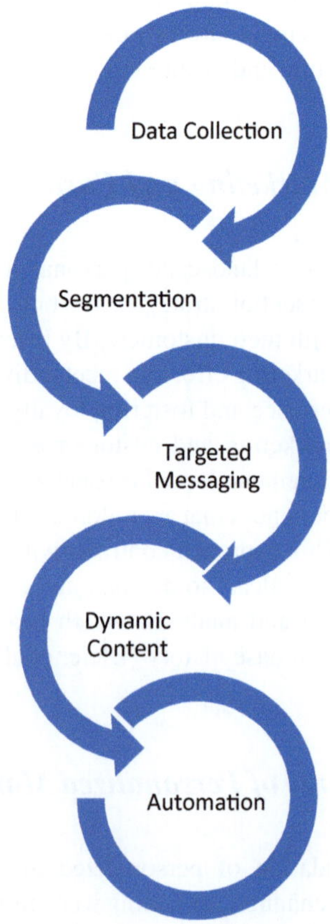

Image 9.3 Flowchart of automation. *Source* Developed by author

9.1.17 Importance of Personalized Marketing and Customer Engagement

Personalized marketing and engagement strategies are crucial for retaining customers. Retained customers are more valuable over time, contributing to consistent revenue streams and reducing the cost of acquiring new customers. In a crowded market, personalized experiences set brands apart from competitors. A unique and personalized customer journey enhances brand perception and loyalty. Personalized marketing and engagement enable brands to swiftly respond to evolving customer preferences and market trends. By staying in tune with customer needs, businesses can maintain relevance and competitiveness. Continuous interaction provides insightful information about the behavior and preferences of customers, which can inform marketing plans, product development, and other company choices. For modern businesses, personalized marketing and customer engagement are crucial for fostering strong, enduring relationships with customers. With the use of technology and data, brands are able to create customized experiences that connect with specific consumers, increasing engagement, loyalty, and conversion rates. Implementing effective strategies such as multichannel communication, interactive content, loyalty programs, and personalized customer service ensures that businesses can meet and exceed customer expectations, driving sustained growth and success.

The future of IoT in Retail 4.0 is poised to bring about transformative changes, shaping the way retailers operate, interact with customers, and manage their supply chains. Here's a description into what the future holds for IoT in the retail industry.

- Enhanced Customer Experience
  IoT will enable retailers to create hyper-personalized experiences for customers, offering tailored product recommendations, pricing, and promotions based on real-time data analysis of customer preferences, behavior, and context. Retailers will leverage IoT devices like smart mirrors, interactive displays, and RFID tags to enhance in-store experiences, allowing customers to try virtual outfits, receive personalized recommendations, and navigate stores efficiently.
- Seamless Omnichannel Integration
  IoT will enable the smooth integration of online and offline channels, offering customers a cohesive shopping experience. Shoppers can begin their journey on one platform and effortlessly switch to another without any interruptions. Advanced inventory management systems will offer real-time visibility into stock levels across all channels, ensuring that customers can obtain products whenever and wherever they need them.
- Smarter Supply Chains
  IoT sensors and AI algorithms will enable predictive analytics in supply chains, forecasting demand, optimizing inventory levels, and anticipating disruptions. This proactive approach will improve efficiency and reduce costs. IoT and blockchain technologies will work together to enhance transparency and traceability in supply chains, ensuring the authenticity and ethical sourcing of products, which is crucial for consumer trust.

- Enhanced Security and Sustainability
  With the proliferation of IoT devices, security will be a top priority. Retailers will invest in robust cybersecurity measures to protect customer data, prevent breaches, and ensure the integrity of IoT ecosystems. IoT will support sustainability initiatives by enabling energy-efficient operations, reducing waste through optimized processes, and providing insights for eco-friendly practices in packaging, logistics, and product lifecycle management.
- AI-Powered Insights and Automation
  IoT-generated data, combined with AI and machine learning, will provide actionable insights for retailers. These insights will inform decision-making, improve operational efficiency, and drive innovation in marketing strategies. IoT-enabled automation will streamline routine tasks, such as inventory management, replenishment, and customer service. This automation will free up human resources for more strategic and customer-focused activities.
- Immersive Technologies
  IoT will enable immersive augmented reality (AR) and virtual reality (VR) experiences for customers, allowing them to visualize products in real-world settings, virtually try before buying, and engage with interactive product demonstrations. Retailers will leverage IoT data to create targeted and interactive marketing campaigns, delivering personalized content through AR/VR experiences, IoT-connected devices, and location-based promotions (Rasool et al. 2020).

9.2 Literature Review

Retail 4.0 signifies a paradigm shift in the retail industry, leveraging advanced technologies like the Internet of Things (IoT) to redefine customer experiences, optimize operations, and drive innovation. This literature review explores the key aspects, benefits, challenges, and future directions of IoT integration within the framework of Retail 4.0. The concept of Retail 4.0 emerged from Industry 4.0, emphasizing the digital transformation of retail operations. IoT plays a pivotal role in this transformation by connecting physical devices and systems to the internet, enabling data collection, analysis, and automation in real-time. IoT integration in retail offers multifaceted benefits (Yeap et al. 2022). Real-time inventory management is one of the key advantages, allowing retailers to monitor stock levels, track product movements, and implement dynamic pricing strategies based on demand fluctuations. IoT enables personalized marketing strategies through data-driven insights into customer behavior and preferences. By analyzing customer interactions with IoT-enabled devices, retailers can tailor marketing campaigns, recommend products, and enhance overall customer satisfaction. Improved operational efficiency is another significant outcome of IoT integration (Davison 2016). Automated processes, predictive maintenance of equipment, and optimized supply chain management contribute to cost reductions, increased productivity, and streamlined operations (Lee et al. 2006).

Despite its transformative potential, IoT integration in retail also presents challenges. Data security and privacy concerns are paramount, considering the vast amount of sensitive customer information collected and transmitted by IoT devices Interoperability among diverse IoT devices and systems is another challenge. Ensuring seamless communication and compatibility across platforms and protocols is crucial for the effective functioning of IoT ecosystems in retail environments (Gerlich 2001). The complexity of managing IoT systems and the need for skilled personnel proficient in IoT technologies pose additional challenges. Training programs and workforce development initiatives are essential to address this skills gap and maximize the potential of IoT in retail. IoT integration enhances consumer engagement and experience in retail (Scanlon 2011). Connected devices such as smart mirrors, interactive displays, and personalized recommendations create immersive shopping experiences, driving customer loyalty and brand advocacy (Bruni and Piccarozzi 2022). The future of IoT in Retail 4.0 is marked by ongoing innovation and research. Predictive analytics using IoT-generated data for demand forecasting, blockchain technology for secure data management, and the integration of artificial intelligence (AI) algorithms for automated decision-making processes are key areas of focus (Kumar et al. 2020). The integration of IoT within the framework of Retail 4.0 presents a transformative opportunity for the retail industry (Fleisch 2010). While challenges exist, the benefits of enhanced customer experiences, operational efficiency, and data-driven insights underscore the significance of IoT in shaping the future of retail (Ramson et al. 2020).

9.3 Retail 4.0

Retail 4.0 represents the next phase in the evolution of retail, characterized by the integration of advanced technologies that revolutionize the way businesses operate, interact with customers, and manage their supply chains. At the core of Retail 4.0 are several key technologies that collectively enhance efficiency, customer experience, and overall competitiveness. Let's delve into these technologies and their implications for the retail industry (Har et al. 2022).

- Internet of Things (IoT)-IoT plays a pivotal role in Retail 4.0 by connecting physical devices and systems to the internet, enabling data collection, analysis, and automation in real-time. In retail, IoT devices include smart shelves that monitor inventory levels, RFID tags for tracking products throughout the supply chain, and customer-facing devices like smart mirrors for personalized shopping experiences. The data generated by IoT devices allows retailers to optimize inventory management, improve operational efficiency, and deliver personalized services to customers.
- Artificial Intelligence (AI) and Machine Learning (ML)-AI and ML technologies empower retailers to analyze vast amounts of data, predict consumer trends, and automate various processes. AI-driven chatbots provide personalized customer

support, recommendation engines offer tailored product suggestions, and predictive analytics help optimize pricing strategies and inventory management. ML algorithms also enable retailers to detect patterns in consumer behavior, anticipate demand fluctuations, and make data-driven decisions to enhance business outcomes.

- Big Data Analytics-Big Data analytics leverages extensive datasets to derive meaningful insights and guide strategic decisions. In Retail 4.0, retailers gather data from various sources, such as sales transactions, customer interactions, social media, and IoT devices. Analyzing this data allows retailers to better understand customer preferences, market trends, and operational efficiency. Big Data analytics also helps retailers segment their customer base, customize marketing efforts, and enhance customer engagement and loyalty.
- Cloud Computing-Cloud computing provides retailers with scalable and cost-effective infrastructure for storing, processing, and accessing data and applications. Retailers leverage cloud platforms to deploy AI and ML algorithms, host e-commerce websites, and streamline collaboration and communication across geographically dispersed teams. Cloud-based solutions also facilitate seamless integration with third-party services, enhance data security and privacy, and enable retailers to adapt quickly to changing market dynamics.
- Augmented Reality (AR) and Virtual Reality (VR)-AR and VR technologies provide immersive experiences that boost customer engagement and increase sales. In retail, AR applications let customers visualize products in their own space before buying, use virtual fitting rooms for clothing, and browse interactive product catalogues. Meanwhile, VR experiences create virtual store environments, simulate real-world scenarios for employee training, and offer virtual tours to showcase products and services.
- Blockchain Technology-Blockchain technology ensures transparency, security, and traceability in retail operations, particularly in supply chain management and payment processing. Blockchain-based solutions enable retailers to track product provenance, authenticate luxury goods, and prevent counterfeit products. Smart contracts powered by blockchain automate contract execution, facilitate secure transactions, and reduce transaction costs, making blockchain a valuable tool for enhancing trust and efficiency in retail transactions.
- Robotics and Automation-Robotics and automation technologies streamline repetitive tasks, improve operational efficiency, and enhance the customer experience in retail. Automated fulfillment centers use robotics for order picking, packing, and shipping, reducing fulfillment times and costs. In-store robots assist with inventory management, floor cleaning, and customer assistance, freeing up human employees to focus on more complex tasks and providing shoppers with enhanced service and convenience (Shankar et al. 2021).

9.3.1 *Implications for Retailers*

See (Image 9.4).

9.3.2 *Open Research Issues and Challenges of Internet of Things (IoT) in Retail 4.0*

The integration of Internet of Things (IoT) technologies in the retail sector, commonly known as Retail 4.0, has revolutionized business operations and customer engagement. However, alongside these benefits, several research issues and challenges must be addressed to fully leverage IoT's potential in retail. This article delves into these challenges and proposes potential research directions. A significant obstacle is ensuring data security and privacy. There is a greater chance of data breaches, cyberattacks, and unauthorized access due to the extensive use of connected devices that gather massive volumes of user data. Strong security protocols, encryption schemes, and data anonymization strategies must be developed through research in order to protect sensitive data and win over customers. Another issue that merchants face is the heterogeneous ecosystem of IoT devices and platforms lacks standardization and interoperability, which makes it challenging to integrate and operate disparate systems with ease. Developing industry-wide standards, protocols, and frameworks

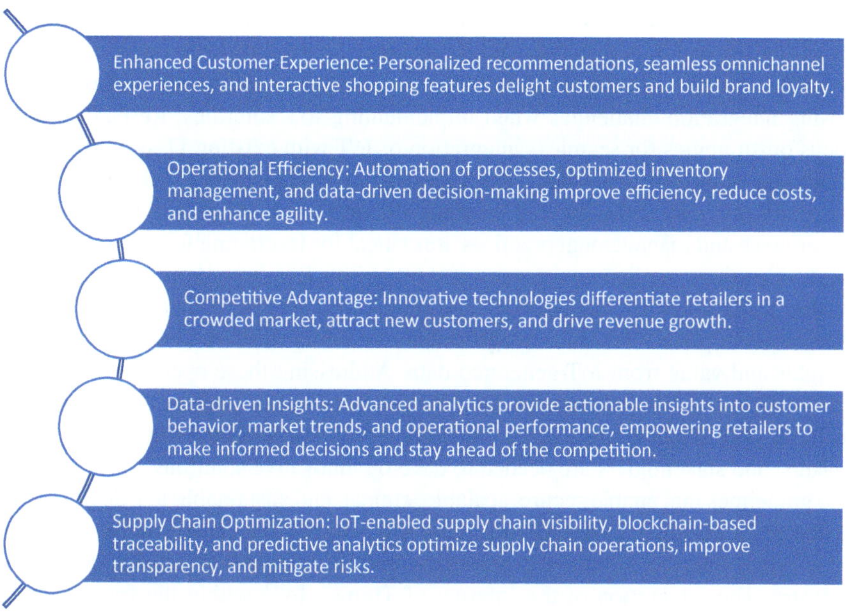

Image 9.4 Implications stages of retailers. *Source* Developed by author

to support data sharing, cooperation, and interoperability across diverse IoT platforms and devices should be the main goal of research. Large-scale IoT ecosystem management becoming more difficult as IoT installations in retail grow. It is necessary to address issues including firmware updates, lifecycle management, device provisioning, and network scalability. The goal of research should be to provide management platforms, automation tools, and scalable architectures that can effectively handle the increasing number of linked devices. IoT devices in retail generate enormous amounts of data, which calls for real-time processing and analytics capabilities. Processing data closer to the source, or edge computing, can assist lower latency, speed up response times, and facilitate real-time decision-making. The main areas of research should include developing edge analytics techniques, optimizing edge computing infrastructures, and tackling issues with data consistency and synchronization. IoT technology use in retail also brings up moral and legal issues with data protection, permission, responsibility, and transparency. Research should aim to develop ethical guidelines, regulatory frameworks, and best practices for IoT deployment in retail, ensuring compliance with data protection laws and industry standards while maintaining customer trust and confidence. IoT devices consume significant amounts of energy, leading to concerns about environmental sustainability and carbon footprint. Research efforts should focus on designing energy-efficient IoT devices, optimizing power consumption, implementing renewable energy solutions, and promoting sustainable practices in IoT deployment and management. Despite the potential benefits, customer acceptance and adoption of IoT-enabled services in retail can be challenging (De vass et al. 2021). Factors such as usability, accessibility, affordability, and perceived value influence customer decision-making. Research is needed to understand customer preferences, behaviors, and motivations toward IoT technologies, and to design user-centric solutions that address their needs and preferences effectively. Many retailers still rely on legacy systems and infrastructure, posing integration challenges when implementing IoT solutions. Research should focus on strategies for seamless integration of IoT with existing IT systems, legacy databases, point-of-sale systems, and supply chain management platforms, ensuring compatibility, data integrity, and minimal disruption to operations. Effective data governance and management practices are critical for leveraging IoT data effectively while ensuring compliance, accuracy, and reliability. Research should address issues related to data quality, data lifecycle management, data governance frameworks, data ownership, and data sharing agreements, enabling retailers to derive actionable insights and value from IoT-generated data. Addressing these open research issues and challenges is crucial for unlocking the full potential of Internet of Things (IoT) in Retail 4.0. Collaborative research efforts involving academia, industry, policymakers, and stakeholders are needed to develop innovative solutions, best practices, and guidelines that enable secure, scalable, ethical, and sustainable IoT deployments in the retail sector. By overcoming these challenges, IoT can revolutionize retail operations, enhance customer experiences, and drive digital transformation in the industry. The integration of the Internet of Things (IoT) within the framework of Retail 4.0 represents a transformative shift in the retail industry, ushering in a new era of digitalization, innovation, and customer-centricity. This overview has highlighted the key aspects, benefits, challenges, and future directions of IoT in Retail 4.0, underscoring its potential to revolutionize retail operations, enhance customer experiences, and drive business growth (Čolaković and Hadžialić 2018).

9.4 Key Aspects of IoT in Retail 4.0

IoT in Retail 4.0 encompasses a wide range of applications and technologies, including connected devices, data analytics, AI, cloud computing, and automation. These technologies converge to create a seamless and interconnected retail ecosystem where physical and digital experiences merge, enabling retailers to gain real-time insights, optimize operations, and deliver personalized services to customers.

9.4.1 Benefits of IoT Integration

The integration of IoT brings numerous benefits to retailers. Real-time inventory management, personalized marketing strategies, enhanced customer experiences, and improved operational efficiency are among the key advantages. IoT-enabled solutions empower retailers to make data-driven decisions, streamline processes, and stay competitive in a rapidly evolving market landscape.

9.4.2 Challenges and Considerations

IoT integration in retail has potential, but it also comes with drawbacks, including issues with data security and privacy, interoperability, scalability of IoT ecosystems, and ethical considerations. It will take teamwork, creative thinking, and established legal frameworks to overcome these obstacles and guarantee the ethical and long-term implementation of IoT technology in retail.

9.5 Conclusion

The integration of the Internet of Things (IoT) within the Retail 4.0 framework marks a revolutionary transformation in the retail industry, blending digital innovation with customer-centric strategies. This convergence redefines traditional retail operations, leveraging IoT's capabilities to offer personalized marketing, real-time inventory management, improved customer experiences, and enhanced operational efficiency. However, the journey toward fully realizing IoT's potential in retail is not without challenges, including data security, interoperability, scalability, and ethical concerns. IoT's applications in Retail 4.0 have already begun to reshape how retailers interact with consumers and manage their operations. Personalized marketing driven by IoT allows retailers to deliver tailored promotions and recommendations, enhancing customer satisfaction and loyalty. Real-time inventory management enabled by IoT sensors and devices ensures optimal stock levels, reducing losses due to overstocking or stockouts and streamlining supply chain operations. These improvements contribute to a more efficient and responsive retail environment, capable of meeting the dynamic needs of today's consumers. Customer experience, a critical factor in retail success, is significantly enhanced through IoT. Smart devices and interconnected systems create immersive and engaging shopping experiences, whether in physical stores or online. IoT facilitates seamless omnichannel retailing, providing consumers with a consistent and convenient shopping journey across various platforms. The ability to gather and analyze vast amounts of data in real time allows retailers to anticipate customer needs and preferences, delivering personalized interactions that drive engagement and retention. Operational efficiency is another area where IoT demonstrates substantial impact. Automation and data-driven decision-making streamline processes, reduce costs, and optimize resource utilization. Predictive maintenance of equipment and facilities, enabled by IoT, minimizes downtime and extends the lifespan of assets (Čolaković and Hadžialić 2018). These efficiencies contribute to a leaner, more agile retail operation capable of adapting swiftly to market changes and consumer demands. Despite these advantages, the deployment of IoT in retail is fraught with challenges. Data security is critical since there is a greater chance of cyberattacks and data breaches due to the widespread use of connected devices. To preserve confidence and safeguard customer data, retailers need to put strong security measures in place and adhere to laws like the CCPA and GDPR. Standardized protocols and cooperative efforts are necessary to achieve seamless integration, as interoperability among various IoT devices and platforms continues to be a major obstacle. Another crucial concern is scalability, since the growth of IoT networks necessitates large investments in technology and infrastructure. Small and medium-sized retailers (SMEs) in particular would find it difficult to defend these expenditures in the absence of evident and quick returns. Concerns about data privacy and the correct use of client information also give rise to ethical issues. While using IoT technologies, retailers need to balance these moral issues in order to gain and keep the trust of their customers. Addressing the challenges associated with IoT in retail requires a multifaceted approach. Collaborative efforts among retailers, technology providers, and regulatory bodies are essential to developing standardized protocols, ensuring data security, and fostering interoperability. Regulatory frameworks must evolve to keep pace with technological advancements, providing guidelines that balance innovation with consumer protection. Retailers must also invest in innovative solutions and continuous improvement to stay ahead of the curve and fully harness IoT's potential. Within the Retail 4.0 framework, the Internet of Things is a revolutionary force that can improve customer experiences and optimize retail operations. Retailers can significantly impact the digital landscape and shape the future of the retail sector in a connected and data-driven world by embracing the opportunities and addressing the problems posed by the Internet of Things. The successful integration of IoT will depend on the collective efforts of all stakeholders to create a secure, interoperable, and innovative retail environment that meets the needs of consumers and businesses alike.

9.5.1 Future of IOT in Retail 4.0

IoT in Retail 4.0 has a bright future because of a number of new trends and technologies that will likely spur further development. Edge computing provides faster and more effective data handling, lowering latency and enhancing real-time decision-making by processing data closer to the source rather than depending on centralized cloud servers. Blockchain technology provides transparency and security in supply chains, enhancing trust between retailers and consumers by verifying product authenticity and ethical sourcing. Predictive analytics, powered by AI and machine learning, enables retailers to foresee trends and opportunities, allowing them to make proactive and informed decisions. Sustainable practices, supported by IoT, promote environmental responsibility through efficient resource management and reduced waste. These technologies collectively contribute to a more innovative, resilient, and sustainable retail ecosystem.

References

Bakharev V, Mityashin G, Katrashova Y, Strelnikov A, Bugaenko A, Karachev V (2020) The impact of industry 4.0 technologies on retail development. In: Proceedings of the international scientific conference-digital transformation on manufacturing, infrastructure and service, pp 1–7

Bleier A, De Keyser A, Verleye K (2018) Customer engagement through personalization and customization. In: Customer engagement marketing, pp 75–94

Bruni R, Piccarozzi M (2022) Industry 4.0 enablers in retailing: a literature review. Int J Retail Distrib Manag 50(7):816–838

Chaney D (1983) The department store as a cultural form. Theory Cult Soc 1(3):22–31

Čolaković A, Hadžialić M (2018) Internet of things (IoT): a review of enabling technologies, challenges, and open research issues. Comput Netw 144:17–39

Davison B (2016) Super city: Los Angeles and the birth of the supermarket, 1914–1941. Calif Hist 93(3):9–27

De Vass T, Shee H, Miah S (2021) IoT in supply chain management: Opportunities and challenges for businesses in early industry 4.0 context. Oper Supply Chain Manag: Int J 14(2):148–161

Fleisch E (2010) What is the internet of things? An economic perspective. Econ Manag Financ Mark 5(2):125–157

Gerlich RN (2001) The digital revolution

Greengard S (2021) The internet of things. MIT press

Har LL, Rashid UK, Te Chuan L, Sen SC, Xia LY (2022) Revolution of retail industry: from perspective of retail 1.0 to 4.0. Procedia Comput Sci 200:1615–1625

Kothari SS, Jain SV, Venkteshwar A (2018) The impact of IOT in supply chain management. Int Res J Eng Technol 5(8):257–259

Kumar MS, Raut RD, Narwane VS, Narkhede BE (2020) Applications of industry 4.0 to overcome the COVID-19 operational challenges. Diabetes Metab Syndr: Clin Res Rev 14(5):1283–1289
Lee MY, Atkins KG, Kim YK, Park SH (2006) Competitive analyses between regional malls and big-box retailers: a correspondence analysis for segmentation and positioning. J Shopp Cent Res 13(1):81–98
Mahammadh VR (2015) E-commerce: a business revolution in twenty first century. Int J Mark Technol 5(9):202–213
Malar MS (2019) Industry 4.0 transforming global retail industry
Ramson SJ, Vishnu S, Shanmugam M (2020) Applications of internet of things (IoT)—An overview. In: 2020 5th international conference on devices, circuits and systems (ICDCS). IEEE, pp 92–95
Rasool A, Shah FA, Islam JU (2020) Customer engagement in the digital age: a review and research agenda. Curr Opin Psychol 36:96–100
Scanlon ER (2011) Big box, no more quick fixes: a historical account of consumption, retail and discount shopping typologies (Doctoral dissertation, Massachusetts Institute of Technology)
Shankar V, Kalyanam K, Setia P, Golmohammadi A, Tirunillai S, Douglass T et al (2021) How technology is changing retail. J Retail 97(1):13–27
Singh G, Srivastav S, Gupta A, Garg V (2020) Companies adoption of IoT for smart retailing in industry 4.0. In: 2020 international conference on intelligent engineering and management (ICIEM). IEEE, pp 487–492
Taherdoost H (2023) Security and internet of things: benefits, challenges, and future perspectives. Electronics 12(8):1901
Tian Y, Stewart C (2006) History of e-commerce. In: Encyclopedia of e-commerce, e-government, and mobile commerce. IGI Global, pp 559–564
Yeap JA, Ooi SK, Ramayah T, Md. Kassim N, Kunchamboo V (2022) Retail internet of things: a retailer-and consumer-oriented literature review. Glob Bus Manag Res 14
Zhichao Y (n.d.) Retail development 4.0

Chapter 10
Supervised Machine Learning Algorithms for Retinal Vessel Detection in Retinal Images

Asha Rani Mishra

Abstract Retinal pictures are essential for a number of uses, including the identification of people and the discovery of diseases. Detection of the blood vessels that are present in the retinal image is very helpful for an ophthalmologist to easily diagnose the fundus image. Also based on the status of the retinal blood vessels, early identification of diabetes can be achieved. This paper presents an empirical analysis of three supervised models—K Nearest Neighbour (K-NN), Support Vector Machine (SVM), and Decision Tree (DT) in detecting retinal blood vessels Grey Scaling, Gaussian Blur, and Gamma Correction methods are used to extract features. To measure the efficacy performance of the supervised models evaluation metrics—Precision, Recall and F1-Score are used. The highest accuracy is obtained in case using Decision Tree Algorithm as 95.75%.

Keywords Retinal image · K-NN · SVM · Decision tree · Blood vessels · Feature extraction

10.1 Introduction

Major health issues affecting the elderly and older citizens are eye disease and vision loss. The retina is the most crucial internal element of the eye. Retinal pictures are essential for a number of uses, including the identification of people or diseases. Detection of retinal blood vessels, can also contribute significantly to the early diagnosis of diabetes.

Internal to the eye is the retina. The centre of the retina is circular or oval shaped optical disc. The primary blood veins of the retina radiate outward from the optical

A. R. Mishra (✉)
Department of Artificial Intelligence and Machine Learning, Galgotias College of Engineering and Technology, Greater Noida, UP, India
e-mail: asha1.mishra@gmail.com

A. Rai et al. (eds.), *Convergence of Artificial Intelligence, Machine Learning, and the Internet of Things in Industry 4.0 Applications*, Transactions on Computer Systems and Networks, https://doi.org/10.1007/978-981-96-5269-3_10

nerve's core. The interconnection of blood vessels in the retina acts as an anatomical feature which aids significantly in the finding of various diseases. Retinal blood vessels come in two different varieties: arteries and veins. However, manually identifying blood vessels is a difficult task because of the intricacy and poor contrast of the blood vessels in the retinal image. An ophthalmoscope is used by ophthalmologists to examine the retina. The retinal images is frequently utilized in the judgement and treatment of many different illnesses such as glaucoma and diabetic retinopathy (Wankhade et al. 2016).

The identification of retinal blood vessels during illness diagnosis helps in understanding of retinal blood vessels and disease (Joshi et al. 2012) which is more efficient for illness diagnosis over conventional technologies include great anti-counterfeiting strength, tiny imaging devices, low cost, simple image collecting with contactless operating universality, and liveliness (Sharma and Hempriya 2013). Additionally, because blood vessels are internal to the live organism, the environment outside the skin has less of an impact on the illness identification system (Zabihi et al. 2012). As a result, using retinal blood vessels to diagnose diseases is thought to be one of the most potential future treatments.

For an ophthalmologist, it's not easy to detect every eye disease by just looking the scans of the eyes, especially problems related with retina (Kaur et al. 2012). The key challenge was that prior updates in this retinal detection problem were unable to forecast all the vessels located in the eye, and the accuracy was poor (Karegowda et al. 2011). This study aims to assist ophthalmologists in the diagnosis of eye diseases such arteriosclerosis, diabetes, and hypertension. Autonomous blood vessel extraction algorithms can save medical expenditures, improve patients' vision, and save time.

10.2 Literature Review

For the extraction of blood vessels from retinal pictures, a number of methods have been suggested. Chaudhuri et al. (1989) employed the 2D Gaussian filtering approach for segmenting retinal blood vessels. The best approximate representation of the blood vessels is obtained using a 2D Gaussian-based matching template. All potential orientations are covered by rotating the x and y parameters. For feature extraction, one other method is also performed namely K-means. The K value indicates how many regions there are in the image. After an initialization phase, each instance is assigned to the nearest cluster in the K-means method. Then, until the system achieves stability, the centres of each cluster are recalculated. To segment blood arteries, the K-means algorithm is used for upgrading with vascular enhancement filters like Gabor, Gaussian, and Frangi and top-hat transforms. Vessel enhancement filters output value is then subjected to the K-means algorithm. K (region count) is configured to be 3. The image has been divided into three sections, as indicated by this. These are the instances where vesselness is most likely. The zones with a high and medium likelihood of vesselness are joined in the final stage, yielding a binary vessel map. And in conclusion of proposed sample paper, Gaussian Filter with K-mean gave the accuracy of 95.50%.

Diego Marin et al. (2011) employed neural networks for categorization of pixels. Blood vessel enhancement and Grey-level homogenization in the original fundus image are the first steps in the blood vessel detection process. Next, features are extracted for pixel numerical representation, and finally, a classifier is used to identify whether a pixel is a blood vessel or not. To make monochrome input images, the green band was used from the original RGB retinal images. As a result, arteries and other blood-containing components of the retinal layer are better visible and have stronger contrast in the green channel. 'Feature Vector' are representation of a pixel in terms of certain quantifiable characteristics obtained from feature representation process to characterise pixels. It is used in the classification step to determine whether or not a particular set of pixels corresponds to an actual blood artery. Colour fundus images usually have noise, poor contrast, and noticeable lighting changes. A preprocessing consisting of the following operations is carried out to fix these defects and produce images that are better suited for gathering the pixel information required in the classification stage. It seems like a good idea to choose characteristics that describe grey-level variation in candidate pixels' surroundings.

In order to extract the retinal blood artery network, Yavuz and Köse (2017) developed a unique method that includes preprocessing to get the dataset ready for segmentation. Gabor, Frangi, and Gauss filters for image enhancement and top-hat transform are used for preprocessing. K-means and Fuzzy C-means (FCM) are applied to get a binary vessel map using hard and soft clustering, Last step is post processing to remove isolated regions that were incorrectly segmented. The Vessel Enhancement Filtering method is presented by Gao et al. (2020) which is utilizes Eigen values of the Hessian matrix.

A novel technique is presented by Budai et al. (2013), in which the Hessian matrix for each picture pixel is first created, followed by the generation of a Gaussian pyramid. The vessels and background are well contrasted thanks to the Gaussian resolution hierarchy (Gaussian pyramid) that is constructed using the green channel of the fundus picture. There are three levels in the hierarchy (0–2). The original image (level 0) has the maximum resolution, but each subsequent level has halved width and height. Because of the downscaling, this hierarchy improves segmentation time and enables for the use of the same 3×3 neighbourhood analysis to extract vessels of various diameters. Each level of the pyramid is subjected to a neighbourhood study. Three greyscale images are produced as a result, and each of them has pixels that indicate the chance of mapping a pixel to a vessel class. The Hessian matrix of the 3×3 neighbourhood for each pixel of the picture is then calculated in neighbourhood analysis. The two eigenvalues are similar in homogenous regions, resulting in values near to one. If a vessel is present, the enhanced photos will have a greater pixel value than the homogenous regions. Each level of the pyramid is subjected to this scrutiny. The method can achieve accuracy of 94.4% approx.

According to Rehman et al. (2022), there are two upgrading phases that is done after preprocessing and picture quality enhancement phase. It rely on filtering and comparison histogram that are suggested for segmenting retinal vessels. Enhancement of classified image is done by using morphological operators for more precise segmentation. The suggested method is evaluated using the dice coefficient.

Sadeghzadeh et al. (2010) suggested employing wavelet transforms. This study presents a novel technique for retinogram vessel detection. A detailed, multi-scale description of local structure is provided by the Dual Tree Complex Wavelet Transform (DT-CWT).The pixels are classified as vessel or non-vessel based on their DT-CWT coefficients using a random forest classifier.

Kaur et al. (2016) suggested that retinal haemorrhages be detected automatically. 50 retinal fundus photos were taken from the collection, and 4546 blobs were analysed as part of the study. The sensitivity and specificity of the suggested research algorithm were 90.42% and 93.53%, respectively. The algorithm suggested in the study will help ophthalmologists with automatic bleeding diagnosis and could be useful in medical imaging.

A radial projection technique built on SVM has reportedly been used for the categorization of retinal vessels, according to You et al. (2011). The benefit of using the centrelines of vessels to identify low contrast narrow vessels is a feature of the suggested SVM technique. The accuracy using the DRIVE and STARE datasets was 94.34% and 94.97%, respectively.

Tuba et al. (2017) used SVM classification using overlapping block-based technique described by DCT coefficients and chromaticity was employed in this method to segment retinal blood vessels. On the DRIVE database, which is accessible to the general public, an overlapping block-based technique was tested. The results yielded scores of 97.73% (sensitivity), 67.49% (specificity), and 95.48% (accuracy), respectively. The advantage of this algorithm is that it can categorize thicker arteries more accurately than larger retinal vessels (Table 10.1).

10.3 Methodology

In this paper, three supervised algorithms—K-NN, SVM, and Decision Tree are used for detection of retinal blood vessels using DRIVE and STARE dataset. Sequence of steps followed for classification and performance evaluation is shown in Fig. 10.1.

Step 1: Image Preprocessing

To limit the noise in the raw images, first raw images which is in RGB format is converted into single-channel greyscale images and then a Gaussian filter is applied. The mathematical equation for conversion of RGB to grey scale is shown in Eq. (10.1).

$$I_{\text{gray}} = I_{\text{red}} \times 0.33 + I_{\text{green}} \times 0.33 + I_{\text{blue}} \times 0.33 \tag{10.1}$$

where I_x is X channel intensity of a particular pixel. Gaussian filter which is a linear filter reduces the noise level by blurring the image. The mathematical equation for

Table 10.1 Existing work on machine learning-based retinal detection

Authors and year	Algorithm used	Feature extraction method	Performance metrics achieved
Chaudhuri et al. (1989)	K means	Detect piecewise linear segments of blood vessels using Gaussian Filter	Accuracy of 95.50%
Marín et al. (2011)	Neural networks	Pixel classification using grey level	Accuracy of 94.4%
Yavuz and Köse (2017)	K-means and fuzzy C-means (FCM)	Retinal blood vessel network detection using Gabor, Frangi and Gauss filters	Accuracy of 95.94% with K-means and 95.71% with fuzzy C-means
Gao et al. (2020)	Hessian matrix as multi scale filter along with divergence value of the normalized gradient	Vessel enhancement filtering method	Sensitivity of 75.81%, specificity of 95.50% and accuracy of 94.01%
Budai et al. (2013)	Gaussian pyramid and Hessan matrix as feature extractor	Multi resolution method for segmenting blood vessels on DRIVE and STARE dataset	Accuracy of 94.4%
Rehman et al. (2022)	K nearest neighbour (K-NN) and support vector machine (SVM)	Segmentation of retinal vessels using Dice coefficient	Accuracy of 92%
Sadeghzadeh et al. (2010)	Random forest classifier	Dual tree complex wavelet transform with DT-CWT coefficient	Sensitivity of 90.42%, specificity of 93.53%
Kaur et al. (2016)	Random forest based classifier	Morphological operations	Sensitivity of 90.42%, specificity of 93.53%
You et al. (2011)	Support vector machine (SVM)	Radial projection method using complex wavelet on DRIVE and 94.97% on STARE	Average accuracy of 94.34%
Tuba et al. (2017)	Support vector machine (SVM)	DCT coefficients	Sensitivity of 97.73% specificity of 67.49% and accuracy of 95.48%

Gaussian filter is shown in Eq. (10.2).

$$G(x, y) = 1/2\pi\sigma 2 \exp\left(-x^2 + y^2/2\pi\sigma^2\right) \tag{10.2}$$

Step 2: Feature Vector Formation

Feature vector is a numeric or symbolic representation of features of an object (Images). Feature vector are widely used in machine learning because numerical

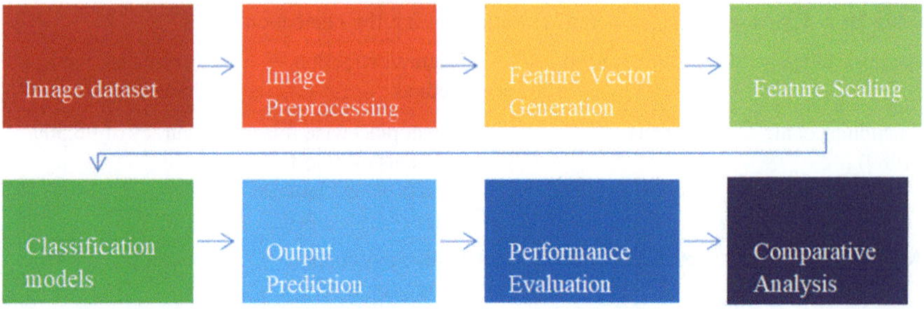

Fig. 10.1 Workflow in retinal vessel detection

representation of an object helps us in applying some mathematical operation for find the relationship among the objects. he retinal scan photos were utilized to produce the three attributes shown below for each pixel—pre-processed image greyscale intensity, gradient magnitude, and maximal Eigen value of hessian matrix. These features were chosen to optimize the models' efficacy while keeping the overall number of features to a minimum in order to increase calculation speed.

Gradient Magnitude

The norm of a vector comprising the image's gradient in the x and y directions is used to determine this value. Its purpose is to detect image edges. The Eq. (10.3) explains how to calculate the gradient magnitude of an image.

$$\|\nabla F\| = \sqrt{}(F_x)^2 + (F_y)^2 \tag{10.3}$$

Maximum Eigen Value of Hessian Matrix

The Hessian matrix provides an explanation for the 2nd order local changes in image intensity close to the chosen pixel. Previous studies have shown that the resultant Hessian matrix may be used to construct an orthonormal coordinate system that is aligned with the second order structure of an image, and this technique is known as Eigen vector decomposition represented by Eq. (10.4).

$$\lambda_+ = 1/2(F_{xx} + F_{yy}) + \sqrt{}(F_{xx} + F_{yy})^2 + (2F_{xy})^2 \tag{10.4}$$

Step 3: Feature Scaling

Feature Scaling is essential for machine learning algorithms because many algorithms such as K-NN, SVM, etc., use distance between the data points in model building and in predicting the class label. So, it is important to scale our dataset otherwise the feature which contain higher value range starts dominating in distance calculation and our model become biased towards that feature. So that in this research, min max scaling on our dataset. Min max scaling formula is as follows

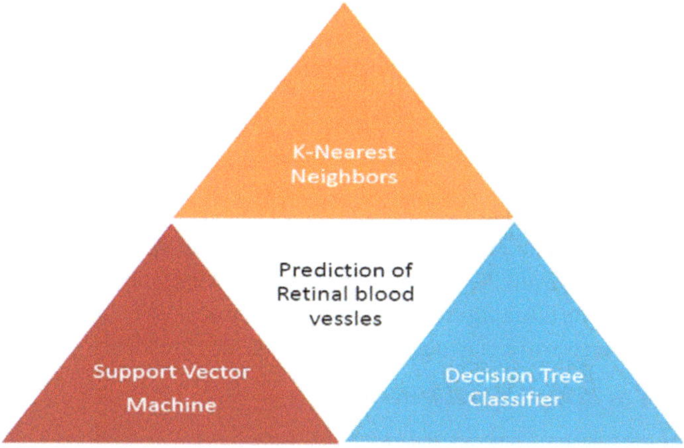

Fig. 10.2 Supervised models for prediction of retinal blood vessels

$$X_{\text{new}} = (X_i - X_{\min})/(X_{\max} - X_{\min}) \tag{10.5}$$

Step 4: Supervised Models for Classification of Retinal Vessels

Supervised algorithms used for retinal detection are K-NN, SVM, and Decision Tree as shown in Fig. 10.2.

K-NN Algorithm in Retinal Vessel Detection

By randomly initializing a selection of pixels and their extremely close neighbours, the k-NN method was changed to be unsupervised. Then, these clustered points were subjected to the standard k-NN approach, which classifies each pixel based on its k nearest neighbours (Tewatia and Mishra 2017). This procedure was repeated till convergence is achieved. To calculate distance Eq. (10.6) is used:

$$D(x_i, x_j) = \left\|x_i - x_j\right\|^2 \tag{10.6}$$

The x's values represent the feature vectors corresponding to the pixels. The k-NN technique is used to group similar kind of pixels that have the same features as judged by this measure. A random sample of 1,000 pixels, together with their ten closest neighbours is considered. This equates to 10,000 pixels out of a total of 423,500 (2%). The k-NN technique might then be implemented regularly once these variables are set (Mishra 2019; Mishra et al. 2021). The total number of neighbours to consider while classifying pixels has a significant impact on the algorithm's accuracy and runtime. When examining the trade-off between performance and speed, a value of 50 was obtained empirically. While this value has a somewhat longer runtime than existing SVM approaches, it provides significantly better accuracy.

A post-processing phase was conducted to ensure that the image formed matched the dataset training label images after the formation of two clusters—vessel and non-vessel. 0 values are mapped with 'non-vessel' cluster while the 'vessel' cluster is represented with a value of 255. Because most retinal images include more non-vessel pixels than vessel pixels, white colour shows presence of the blood vessel forecasts the dark surrounding area.

SVM in Retinal Vessel Detection

SVM analysis was utilized to better understand the present approaches for detecting retinal blood vessels. The 3D feature vectors and data set specified in the preceding sections were used to train the model. Because there were only a few samples available, testing was done utilizing cross validation. Because SVMs are meant to optimize the margin between positive and negative examples, they are ideal for this application.

Decision Tree in Retinal Vessel Detection

A decision tree is a hierarchical structure in which each leaf node represents a class label. Each internal node represents a 'test' on an attribute and each branch reflects the test's outcome or decision. The paths from root to leaf serve as a representation of the classification rules. The training set consists of labelled features and their respective class labels. The training set's consistency is maintained by the manner the tree is constructed. A training set's decision tree makes it possible to explain the many examples connected to each feature in the training set in an easy-to-understand manner, which is very beneficial (knowledge gain process). More specifically, it can be said that the Decision Tree Algorithm performs tests at each node starting from the Root Node, and based on the results, it creates branches out of the nodes. It then moves on to the newly generated nodes one at a time, and repeats the testing process until all of the nodes become Leaf Nodes. Axis-Parallel test, Linear combination of features test, and Non-linear combination of features test are the three main types of tests that direct Decision Tree Algorithm to create new nodes.

10.4 Results and Discussion

The obtained retinal scan images included a total of 423,500 pixels and were 605 × 700 pixel images. These pixels each had a corresponding maximum eigenvalue, gradient magnitude, and greyscale intensity. Therefore, 423,500 distinct 31 feature vectors were used to represent each of the 20 retinal scans. The retinal images for the training set are shown in Fig. 10.3, and the training labels for the images are shown in Fig. 10.4 (Table 10.2).

The performance is evaluated on the DRIVE database with 20 images in which 15 images are used as training set and rest 5 images are used as test set. Three supervised learning algorithms were applied on the dataset to locate retinal vessels. Predicted images by the supervised models are shown in Fig. 10.5.The K-NN model were able to achieve the accuracy of 94.4 and 78.5% vessels (pixels) are correctly labelled out of total vessels present in the test set. The SVM model is able to achieve the accuracy of 95.7 and 74.8% vessels are correctly labelled out of total vessels present in the

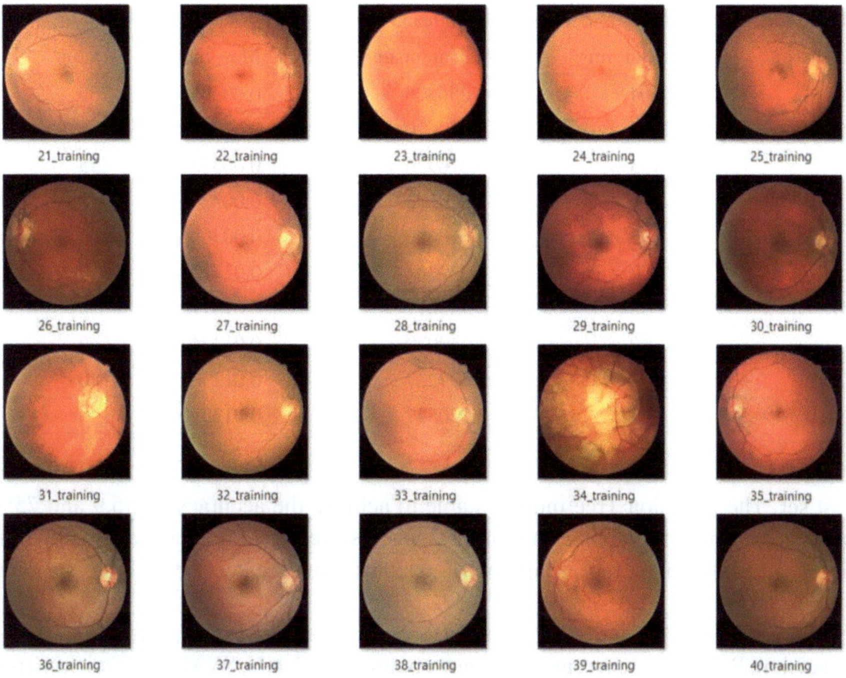

Fig. 10.3 Retinal images

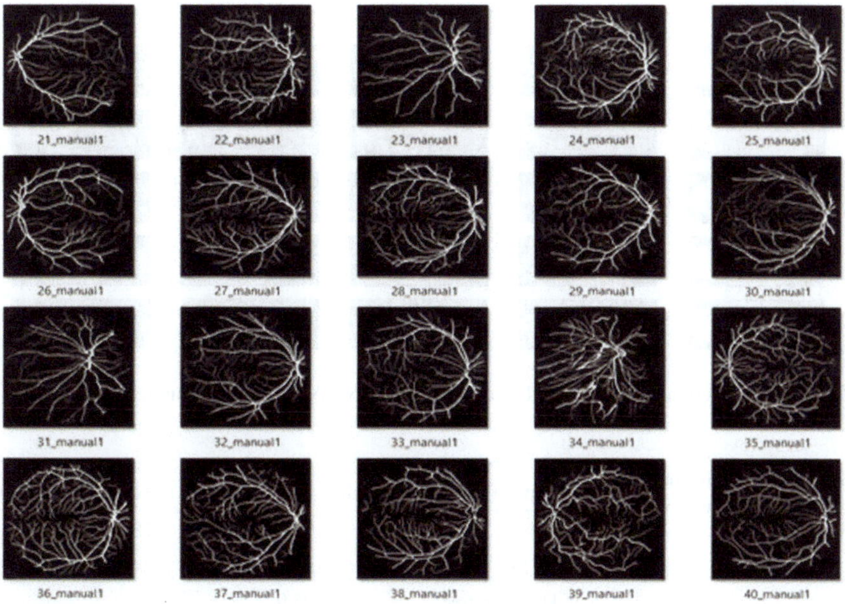

Fig. 10.4 Training label

Table 10.2 Output corresponding to different features vectors

S. No	Blur_img	Gradient_magnitude	Mev_hm	Output
0	104	193.256306	0.003030	255
1	124	117.957620	0.003606	255
2	133	45.276926	0.003723	255
3	96	171.819673	0.003202	255
4	93	179.610690	0.003029	255
5	57	212.339351	0.017217	0
6	58	216.333077	0.015775	0
7	60	218.449079	0.014247	0
8	62	218.229237	0.012909	0
9	63	219.020547	0.012050	0

test set. The Decision Tree model is able to achieve the accuracy of 95.9 and 75% vessels are correctly labelled out of total vessels present in the test set as shown in Table 10.3. Graphical representation of performance of supervised model is shown in Fig. 10.6.

Tables 10.4, 10.5, and 10.6 show performance metrics on the dataset in terms of Precision, Recall, F1 score, and accuracy obtained by K-NN, SVM and Decision Tree model, respectively. Figure 10.7 represents classification report for retinal image '36' using K-NN supervised model.

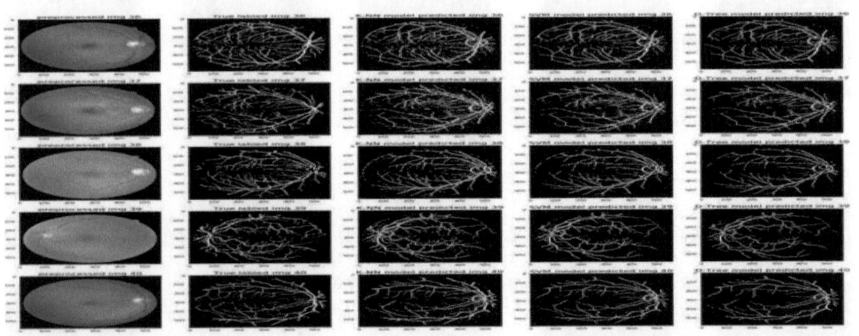

Fig. 10.5 Prediction of retinal blood vessels using K-NN, SVM, and decision tree classifier

Table 10.3 Performance comparison of supervised models

Type of supervised model	Accuracy (in %)	Sensitivity (in %)
K-NN	94.4	78.5
SVM	95.7	74.8
Decision tree	95.9	75

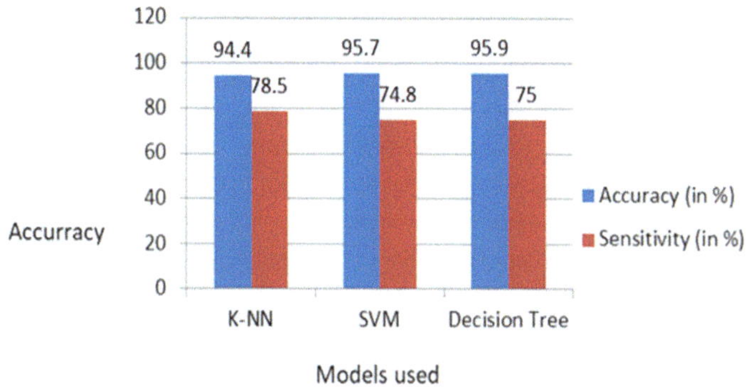

Fig. 10.6 Accuracy of supervised models

Table 10.4 Performance measures of K-NN model

S. No	Sample image	Cluster type: non-vessel			Cluster type: non-vessel			Accuracy
		Precision	Recall	F1-score	Precision	Recall	F1-score	
0	36	0.970912	0.958977	0.964907	0.694575	0.764547	0.727883	0.937832
1	37	0.980421	0.939927	0.959747	0.561776	0.804029	0.661418	0.928049
2	38	0.978353	0.965000	0.971631	0.676240	0.773958	0.721806	0.948512
3	39	0.978618	0.968877	0.973723	0.700574	0.774764	0.735804	0.952200
4	40	0.984135	0.966618	0.975298	0.665451	0.809928	0.730615	0.954746
5	Average	0.978488	0.959880	0.969061	0.659723	0.785445	0.715505	0.944268

Table 10.5 Performance measures of SVM model

S. No	Sample image	Cluster type: non-vessel			Cluster ytype: vessel			Accuracy
		Precision	Recall	F1-score	Precision	Recall	F1-score	
0	36	0.966744	0.979509	0.973084	0.811693	0.723860	0.765265	0.951706
1	37	0.978681	0.962583	0.970565	0.666598	0.781076	0.719310	0.946718
2	38	0.975365	0.980470	0.977911	0.781107	0.737824	0.758849	0.959528
3	39	0.975663	0.983585	0.979608	0.808834	0.738959	0.772319	0.962568
4	40	0.980553	0.983926	0.982236	0.795341	0.761970	0.778298	0.967108
5	Average	0.975401	0.978014	0.976681	0.772715	0.748737	0.758808	0.957526

Average accuracy is shown in Fig. 10.8 while average values for Precision, Recall, and F1 score of used supervised methods are shown in Table 10.7.

Table 10.6 Performance measures of decision tree model

S. No	Sample image	Cluster type: non-vessel			Cluster type: vessel			Accuracy
		Precision	Recall	F1-score	Precision	Recall	F1-score	
0	36	0.960245	0.986031	0.972967	0.853218	0.665450	0.747726	0.951167
1	37	0.973966	0.976023	0.974993	0.744017	0.727610	0.735722	0.954310
2	38	0.968881	0.987273	0.977990	0.831370	0.664290	0.738498	0.954310
3	39	0.969421	0.989082	0.979153	0.861874	0.668054	0.748848	0.961501
4	40	0.978167	0.987274	0.982700	0.824880	0.731211	0.775226	0.967872
5	Average	0.970136	0.985136	0.977561	0.821072	0.691323	0.749204	0.958850

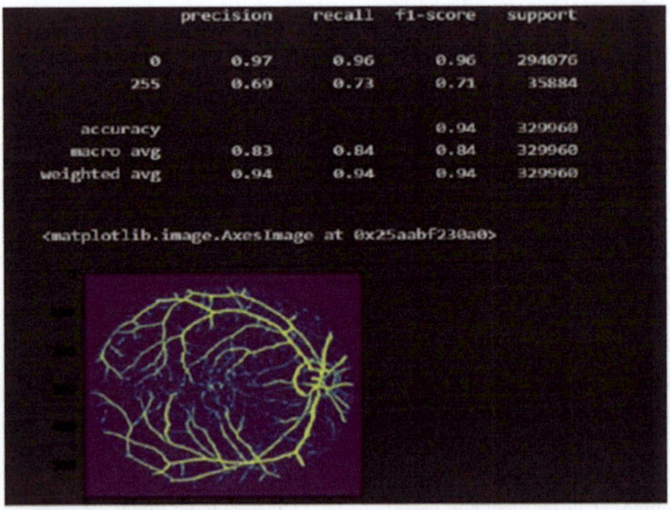

Fig. 10.7 Classification report for sample image 36 using K-NN model

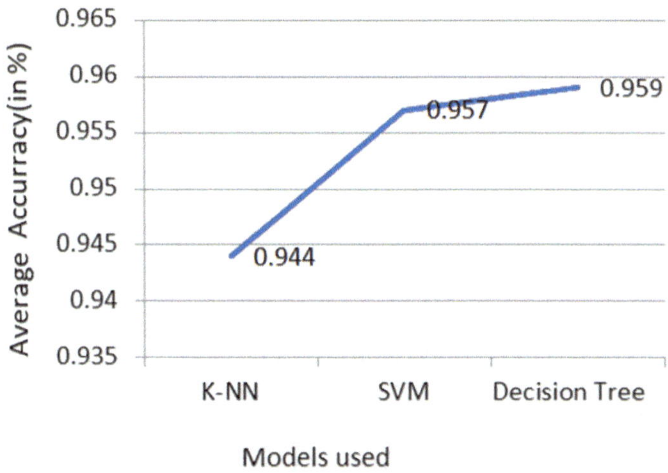

Fig. 10.8 Average accuracy of supervised models

Table 10.7 Average accuracy of the used supervised models

Supervised algorithm used	Average values for cluster type: non-vessel			Average values for cluster type: vessel			Average accuracy
	Precision	Recall	F1-score	Precision	Recall	F1-score	
K-NN	0.978	0.960	0.969	0.660	0.785	0.715	0.944
SVM	0.975	0.978	0.977	0.772	0.749	0.758	0.957
Decision tree	0.970	0.985	0.978	0.821	0.691	0.749	0.959

10.5 Conclusion

Using Grey Scale Intensity, Gradient magnitude, Max Eigenvalue for Hessian Matrix as feature generation methods supervised models achieved better accuracy as compare to existing works by detecting major and the miner blood vessels effectively. Despite the fact that the K-NN algorithm is slightly less accurate overall, it is interesting investigating because it does not require a training set. The K-NN algorithm's drawback is that it takes a while to process each image, which may not be practical in the field. Use of Support Vector Machine and Decision Tree algorithms, produced more superior results. Although another type of unsupervised method might be used to improve speed and accuracy in future. The suggested system's average vessel segmentation accuracy in the DRIVE dataset was 94.42% when using the k-NN model, 95.75% when using the SVM model, and 95.88% when using the Decision Tree model.

References

Budai A, Bock R, Maier A, Hornegger J, Michelson G (2013) Robust vessel segmentation in fundus images. Int J Biomed Imaging

Chaudhuri S, Chatterjee S, Katz N, Nelson M, Goldbaum M (1989) Detection of blood vessels in retinal images using two-dimensional matched filters. IEEE Trans Med Imaging 8(3):263–269. https://doi.org/10.1109/42.34715

Gao J, Chen G, Lin W (2020) An effective retinal blood vessel segmentation by using automatic random walks based on centerline extraction. BioMed Res Int

Joshi S, Karule PT (2012) Retinal blood vessel segmentation. IJEIT 1(3)

Karegowda AG, Nasiha A, Jayaram MA, Manjunath AS (2011) Exudates detection in retinal images using back propagation neural network. 25(3)

Kaur J, Dr. Sinha HP (2012) An effective blood vessel detection algorithm for retinal images using local entropy thresholding. IJERT 1(4)

Kaur N, Chatterjee S, Acharyya M, Kaur J, Kapoor N, Gupta S (2016) A supervised approach for automated detection of hemorrhages in retinal fundus images. In: 2016 5th international conference on wireless networks and embedded systems (WECON). IEEE, pp 1–5

Marín D, Aquino A, Gegundez-Arias ME, Bravo JM (2011) A new supervised method for blood vessel segmentation in retinal images by using gray-level and moment invariants-based features. IEEE Trans Med Imaging 30(1):146–158. https://doi.org/10.1109/TMI.2010.2064333

Mishra AR (2019) Impact of feature representation on supervised classifiers—A comparative analysis. Global Sci-Tech 11(2):69–74

Mishra AR, Pippal SK, Kumar A, Singh D, Singh A (2021) Clear vision—Obstacle detection using bat algorithm optimization technique. In: 2021 9th international conference on reliability, Infocom technologies and optimization (Trends and future directions) (ICRITO), Noida, India, pp 1–5. https://doi.org/10.1109/ICRITO51393.2021.9596467

Rehman A, Harouni M, Karimi M, Saba T, Bahaj SA, Awan MJ (2022) Microscopic retinal blood vessels detection and segmentation using support vector machine and K-nearest neighbors. Microsc Res Tech 85(5):1899–1914

Sadeghzadeh R, Berks M, Astley S, Taylor C (2010) Detection of retinal blood vessels using complex wavelet transforms and random forest classification. In: Proceedings of medical image understanding and analysis (MIUA), pp 127–131

Sharma HA (2013) Detection of blood vessels and diseases in human retinal images. Int J Comput Sci Commun Eng IJCSCE Special issue on "Emerging Trends in Engineering & Management", IECTE 2013

Tewatia R, Mishra A (2017) A hybrid approach for intrusion detection based on pattern matching and clustering using neural network. J Netw Commun Emerg Technol 7(3):17–21

Tuba E, Mrkela L, Tuba M (2017) Retinal blood vessel segmentation by support vector machine classification. In: 27th international conference. IEEE, pp 1–6

Wankhade MB, Gurjar AA (2016) Analysis of disease using retinal blood vessels detection. IJECS 05(12)

Yavuz Z, Köse C (2017) Blood vessel extraction in color retinal fundus images with enhancement filtering and unsupervised classification. J Healthc Eng

You X, Peng Q, Yuan Y, Cheung Y-M, Lei J (2011) Segmentation of retinal blood vessels using the radial projection and semi-supervised approach. Pattern Recogn 44(10–11):2314–2324

Zabihi SM, Pourreza HR, Banaee T (2012) Vessel extraction of conjuctival images using LBPs and ANFIS. Int Schol Res Netw. ISRN Machine vision, 2012, Article ID 424671, 6 pages

Chapter 11
Signal Processing Techniques and Data Collection Through Sensors in Industry 4.0 Application

Amrita Rai, Karabi Kalita, and Om Prakash

Abstract Modern systems use signal processing in anything from smaller, more power-efficient Internet of Things components to sophisticated sensor monitoring in automated cars and smart automated machines. This chapter focuses an overview of signal processing techniques and data collection related to sensors and sensors based applications. Effective analysis and handling of signals produced by sensors inside a system constitute a critical and crucial component of a well operating measurement or automatic systems. Depending on the applications, the raw signal obtained from the sensors is typically passes through few hardware circuits. The raw signals are processed through preamplifiers circuits for removing unwanted noise and providing strength of signals. In this chapter, various methods of data handling and signal processing methods will be described and discussed. Also includes the sensor systems in many fields of applications such as navigation, medical science, wearable devices for sports and health monitoring, and various IoT-based applications.

Keywords Signal processing · Data collection · Sensors · Industry 4.0 · Internet of things (IoT) · Wireless sensor networks (WSNs) · Data aggregation · Smart manufacturing · Machine learning · Edge computing · Real-time monitoring · Industrial automation

---

A. Rai · K. Kalita (✉)
Department of Electronics & Communication Engineering, LLOYD Institute of Engineering and Technology, Greater Noida, India
e-mail: karabi.kalita@lloydcollege.in

A. Rai
e-mail: amrita.rai@liet.in

O. Prakash
Department of Electronics & Communication Engineering, SVCET, Chittoor, India
e-mail: om4096@gmail.com

A. Rai et al. (eds.), *Convergence of Artificial Intelligence, Machine Learning, and the Internet of Things in Industry 4.0 Applications*, Transactions on Computer Systems and Networks, https://doi.org/10.1007/978-981-96-5269-3_11

11.1 Introduction

Sensors are advanced devices designed to observe parameters like temperature changes, seismic activity, or spot people and moving vehicles. They come in various types like seismic, passive infrared intrusion (PIR), acoustic microphone arrays, magnetic sensors, and still video cameras, each serving specific purposes. Sensors, whether stationary or mobile, communicate through a processing unit known as a sink, allowing them to share information and work together, forming sensor networks used in military, healthcare, security, and surveillance (Akyildiz 2002).

A sensor network serves as a communication infrastructure with specialized transducers, monitoring, recording, and responding to various phenomena (Rahman 2010). The communication infrastructure and protocols of sensor networks differ notably from recent Internet-based systems due to their distinct requirements (Biradar et al. n.d.). Technological advancements focus on implementing sensor networks using small, low-cost, low-power, multifunctional, and distributed sensors (Rahman 2010).

Sensor nodes are dispersed around a sensor field in the wireless sensor network architecture shown in Fig. 11.1. These dispersed sensor nodes can gather information and relay it back to the sink. In a wireless design, data routing back to the sink is made easier by a multi-hop system. In turn, the sink uses satellite channels or the Internet to interact with the task manager node (Yong-Min et al. 2009).

A sensing unit, a processing unit, a transceiver unit, and a power unit are the four fundamental parts of a sensor node, as shown in Fig. 11.2. In addition to these, there may be application-specific elements like a location-finding system, a power generator, and a mobilizer. The sensing unit is divided into sensors and analog-to-digital converters (ADCs).

Sensors generate analog signals when they detect something, and the ADCs transform those signals into digital signals that are supplied to the processing unit (Hariharan and Sreelekshmi 2017). The processing unit controls the processes that allow the sensor node to work with other sensors to complete designated sensing tasks. It

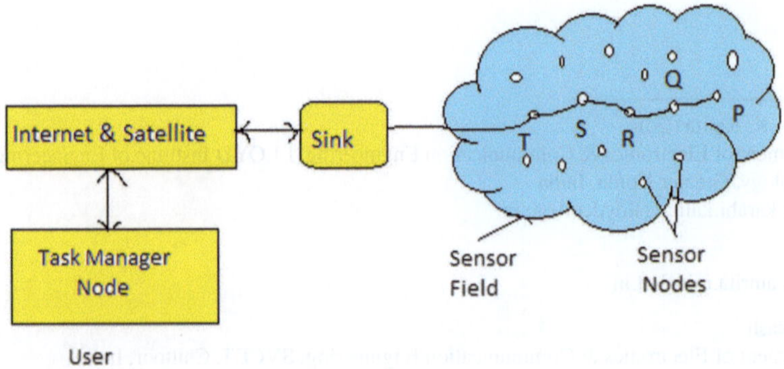

Fig. 11.1 Typical wireless sensor network

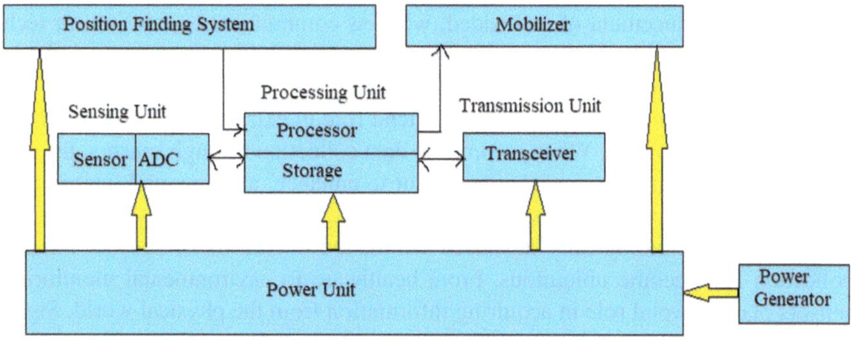

Fig. 11.2 Components of a sensor node

is frequently connected to a tiny storage unit. Moreover, a transceiver unit makes it easier for the node to connect to the network (Akyildiz 2002).

The protocol stack that the sink and sensor network employ is comparable to the computer network (Akyildiz 2002). Several of these protocols may be utilized in next-generation wireless sensor networks (WSNs). For instance, the application layer may be supplied with available routes from the network layer, channel availability from the link layer, and residual energy information from the physical layer through upward information flow as shown in Fig. 11.3. Moreover, between the application layer and the physical layer, downward information flow or back-and-forth information flow may be employed. For instance, the physical layer may receive transmission characteristics from the application layer, such as the operating frequency and transmit power to be used during transmission.

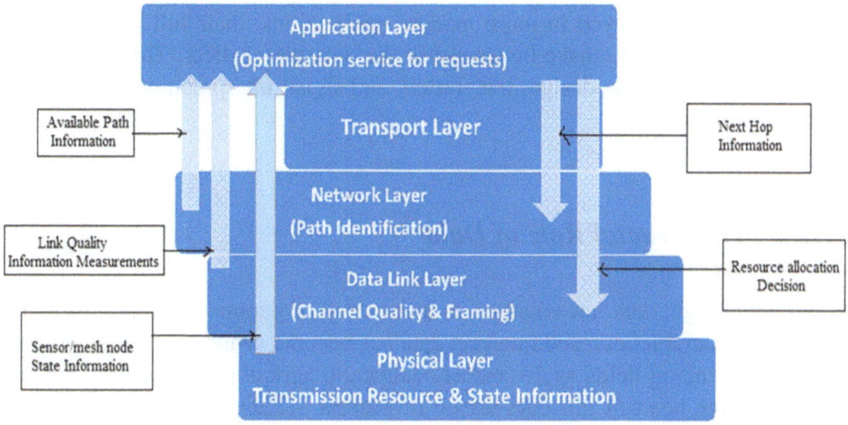

Fig. 11.3 Protocol stack for sensor network

With the advancement of embedded, wireless communication, and sensor technologies in recent years, wireless sensor networks (WSNs) have drawn a lot of attention (Hariharan and Sreelekshmi 2017).

Signal processing techniques play a crucial role in extracting meaningful information from sensor data. When it comes to data collection through sensors, the goal is not just to capture raw measurements but to enhance, analyze, and interpret the signals for various applications.

In today's technologically advanced world, the utilization of sensors for data collection has become ubiquitous. From healthcare to environmental monitoring, sensors play a pivotal role in acquiring information from the physical world. Signal processing techniques are employed to extract meaningful insights from the raw data generated by these sensors.

11.2 Contextualizing the Technological Landscape

11.2.1 Pervasiveness of Sensors

Sensors have permeated nearly every aspect of our lives, from the ubiquitous presence of smartphones equipped with an array of sensors to specialized sensors in industries like healthcare, environmental monitoring, and manufacturing. These sensors act as sentinels, capturing a myriad of physical phenomena and translating them into data streams that form the basis for informed decision-making (Laerhoven 2005).

For a second, set aside considerations about computer size, memory, and processing power, to achieve ubiquitous and pervasive computing, sensors and the data they provide are just as crucial as any of these components. Although sensors are now widely employed in more modern applications, their full potential must be realized if we are to make better use of the data they collect. This necessitates better approaches to the design, integration, and signal interpretation of sensors in computer systems.

11.2.2 The Crucial Role of Data

In the digital age, data has emerged as a currency of paramount importance. The ability to collect, process, and interpret data has become synonymous with empowerment, influencing fields as diverse as healthcare, agriculture, transportation, and beyond. At the heart of this data-driven revolution lies the symbiotic relationship between sensors and signal processing.

In summary, signal processing techniques enhance the quality, accuracy, and relevance of data collected through sensors. These techniques enable the extraction of valuable information, facilitate real-time decision-making, and contribute to the success of various applications in fields such as healthcare, environmental monitoring, industrial automation, and more.

11.3 A Brief Review of Signal Processing Techniques for Sensors

Signal processing techniques play a crucial role in extracting meaningful information from sensor data. Various sensors capture signals from the environment, and processing these signals enhances their utility and facilitates interpretation. Here's a brief review of some common signal processing techniques for sensors:

- **Filtering**: Low-pass, High-pass, and Band-pass Filters: These filters are used to remove or pass specific frequency components of a signal. For instance, low-pass filters can eliminate high-frequency noise from sensor signals.
- **Noise Reduction**: Noise reduction in sensor data can be achieved through a variety of techniques. Simple averaging or moving average methods provide a straightforward approach to reducing random noise by computing the mean of neighboring data points. This process effectively diminishes fluctuations within the data. On the other hand, adaptive filtering methods such as the Kalman filter offer a more advanced solution by dynamically modifying filter parameters to optimize noise reduction according to fluctuations in environmental conditions or sensor characteristics. These strategies allow for more precise noise reduction tailored to particular scenarios, thereby improving the dependability and precision of sensor data analysis across a range of applications.
- **Signal Smoothing**: Signal smoothing plays a vital role in improving data quality, especially when dealing with noisy datasets. One highly effective technique for achieving this is Savitzky-Golay Filtering, well-known for its ability to smooth signals while preserving important features. This method proves particularly beneficial in scenarios characterized by noisy sensor data, where preserving the integrity of critical information is paramount. By employing Savitzky-Golay Filtering, fluctuations and irregularities in the signal can be reduced, resulting in a cleaner and more easily interpretable dataset without compromising the significance of underlying features. Widely applied across diverse domains requiring accurate signal analysis, this approach serves as a valuable tool for enhancing data quality and facilitating more informed decision-making processes.
- **Feature Extraction**: Feature extraction is a crucial step in data analysis, enabling the identification of meaningful patterns and information from raw data. One powerful technique used for this purpose is the wavelet transform, which decomposes signals into different frequency components. This decomposition allows for the detection and characterization of specific patterns or events present within the data, making it particularly useful in applications such as signal processing and image analysis. Another commonly employed method for feature extraction

is principal component analysis (PCA), which effectively reduces the dimensionality of the data while retaining essential information. By identifying the principal components that contribute the most to the variance in the dataset, PCA aids in extracting key features and patterns, facilitating tasks such as pattern recognition and classification. These techniques, including wavelet transform and PCA, play pivotal roles in various fields where accurate feature extraction is essential for data interpretation and decision-making.

- **Time-Frequency Analysis**: Time-frequency analysis plays a crucial role in understanding signal dynamics evolving over time. One widely used method for this purpose is the short-time Fourier transform (STFT), which provides insights into how a signal's frequency content changes across different time intervals. Particularly useful for detecting transient events and temporal variations within signals, the STFT aids in capturing crucial signal features. Additionally, the wavelet transform serves as another versatile tool, not limited to feature extraction but also employed for time-frequency analysis. By decomposing signals into various scales, the wavelet transform reveals intricate variations in signal characteristics across both time and frequency domains. These techniques, encompassing the STFT and wavelet transform, find applications in diverse fields such as signal processing, biomedical engineering, and communication systems. They empower researchers and practitioners to delve deeper into time-varying phenomena, facilitating informed decision-making based on comprehensive analyses.
- **Signal Enhancement**: Signal enhancement techniques are vital for improving signal quality across diverse applications. Deconvolution is utilized to boost signal resolution by counteracting the impacts of system distortion, thereby restoring the original signal for more accurate analysis. Meanwhile, homomorphic filtering offers an effective solution for enhancing signals in nonlinear systems, frequently applied in audio and image processing tasks. Its adeptness at addressing nonlinear distortions makes it particularly valuable, contributing to the removal of unwanted noise and the enhancement of desired features within the signal. These techniques, including deconvolution and homomorphic filtering, serve as powerful tools in signal processing, facilitating clearer data interpretation and informed decision-making processes
- **Pattern Recognition**: Pattern recognition plays a pivotal role in various fields, including sensor data analysis, where the identification of patterns or anomalies is essential for decision-making processes. Machine learning algorithms, such as neural networks, support vector machines, and decision trees, offer sophisticated techniques for pattern recognition tasks. These algorithms are adept at learning from data patterns and extracting meaningful features to classify or detect anomalies within sensor data. By leveraging these machine learning techniques, analysts can automate the process of recognizing complex patterns in large datasets, thereby enhancing efficiency and accuracy in data analysis and decision-making processes.
- **Signal Denoising**: Signal denoising is essential for improving the quality and reliability of sensor data, particularly in applications where precise signal amplitude is crucial. Thresholding techniques offer a simple yet effective approach to

noise reduction in sensor signals. By setting thresholds based on signal characteristics, these methods can efficiently remove noise while preserving important signal features. Thresholding techniques are particularly valuable in scenarios where signal amplitude plays a significant role, as they enable the extraction of meaningful information from noisy data. Leveraging threshold-based denoising methods enhances the accuracy of data analysis and decision-making processes in various fields, ranging from biomedical engineering to environmental monitoring.
- **Adaptive Signal Processing**: Adaptive signal processing, particularly through adaptive filtering techniques, provides a dynamic approach to signal analysis and enhancement. These algorithms possess the capability to continuously adjust their parameters in response to the evolving characteristics of the input signal, rendering them highly suitable for dynamic and changing environments. By adapting to variations in the signal, adaptive filtering methods can effectively mitigate noise, enhance signal clarity, and extract relevant information even in challenging conditions. This adaptability makes them invaluable in numerous applications such as telecommunications, audio processing, and biomedical signal analysis, where signals may exhibit non-stationary behavior or encounter unpredictable environmental factors. Adaptive signal processing techniques contribute significantly to improving the robustness and performance of systems operating in real-world scenarios, enabling more accurate data interpretation and informed decision-making.
- **Digital Signal Processing (DSP)**: Digital signal processing (DSP) is a cornerstone of modern technology, offering powerful tools for analyzing and manipulating signals in various domains. Among these tools, the fast Fourier transform (FFT) stands out as a key technique for transforming signals from the time domain to the frequency domain. By decomposing signals into their constituent frequency components, the FFT unveils crucial information about the signal's spectral characteristics and frequency distribution. This insight is invaluable for tasks such as identifying dominant frequencies, detecting periodic phenomena, and analyzing signal harmonics. Widely employed across diverse fields including telecommunications, audio processing, and image analysis, the FFT plays a pivotal role in signal processing applications, facilitating efficient and accurate analysis of complex signals and enabling innovative advancements in technology and research.

In summary, the selection of signal processing techniques depends on the nature of the sensor data, the specific application, and the desired outcome. Combining multiple techniques in a well-designed signal processing pipeline often yields optimal results for extracting valuable information from sensor signals (Fig. 11.4).

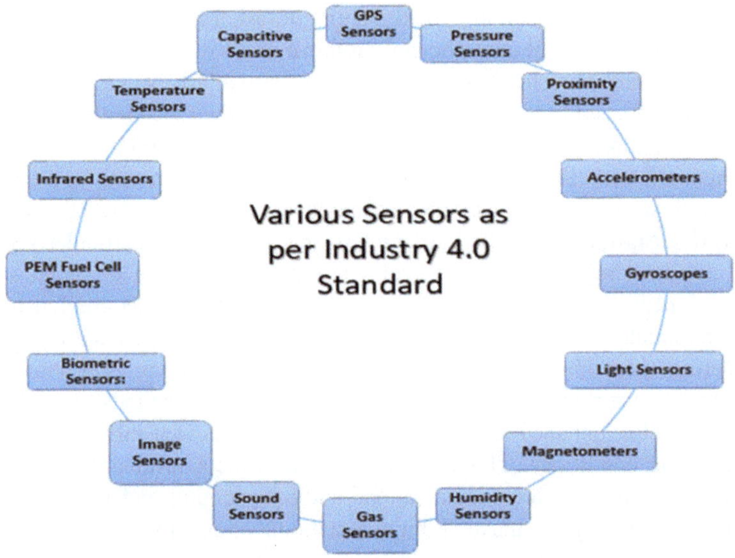

Fig. 11.4 Various sensors used in industry as per 4.0 standards

11.4 Different Types of Sensor and Their Signal and Data

Sensors are devices that detect and measure physical properties or changes in the environment and convert them into electrical signals. The type of sensor and the nature of the signal/data it produces depend on the physical quantity it is designed to measure. Here are some common types of sensors and the signals/data they generate: (Keshari et al. 2019; Kim et al. 2013; Lee et al. 2009; Rajesh Kannan 2022; Yick et al. 2008)

- **Temperature Sensors**: Temperature sensors are essential devices that provide valuable data by generating analog signals directly proportional to the temperature they measure. These signals typically manifest as voltage or current outputs, facilitating straightforward integration into various systems. The data they offer is presented in familiar units such as degrees Celsius or Fahrenheit, ensuring compatibility with standard measurement conventions. By accurately translating thermal changes into actionable data, temperature sensors play a crucial role in diverse applications ranging from industrial processes to environmental monitoring.
- **Pressure Sensors**: Pressure sensors serve as vital components in a wide array of industries, offering insights into the physical forces exerted within systems. These sensors are adept at translating pressure variations into meaningful signals, which can take the form of analog voltage or current outputs, as well as digital signals, depending on the specific requirements of the application. Regardless of the signal type, pressure sensors deliver precise data reflecting the pressure levels encountered, often expressed in units such as Pascals (Pa) or other pressure units.

This data empowers engineers and operators to monitor and adjust systems with accuracy, ensuring optimal performance and safety across various environments, from manufacturing plants to automotive applications.

- **Light Sensors (Photodetectors)**: Light sensors, commonly known as photo detectors, are indispensable tools for detecting and measuring light levels in a variety of applications. These sensors typically produce analog signals, either voltage or current, which are directly proportional to the intensity of light they detect. The data provided by light sensors is expressed in luminance values, typically measured in lux, or radiant flux, often measured in watts. This information enables precise monitoring and control of lighting conditions in environments ranging from indoor spaces to outdoor settings, facilitating tasks such as energy management, safety monitoring, and ambient light adjustments in electronic devices.
- **Proximity Sensors**: Proximity sensors, integral in numerous industries, emit digital signals to detect object presence, providing binary data crucial for automation and safety systems. This information enables swift and accurate responses to object presence or absence, enhancing efficiency and safety in various applications.
- **Accelerometers**: Accelerometers, pivotal in motion sensing, capture acceleration forces experienced by objects, producing analog or digital signals indicative of acceleration in units like m/s^2 or g, enabling precise motion tracking and enhancing safety across devices from smartphones to automotive systems.
- **Gyroscopes**: Gyroscopes, vital for measuring orientation and rotation, provide angular velocity data either through analog or digital signals, enabling precise navigation and orientation tracking. By accurately capturing rotational movement, gyroscopes enhance performance across industries, from aerospace systems to consumer electronics, improving functionality and stability.
- **Magnetometers**: Magnetometers are pivotal in detecting and measuring magnetic fields, offering data in analog or digital formats, crucial for navigation, geophysical exploration, and device functionality. By providing magnetic field strength values in units like Tesla or Gauss, they facilitate precise mapping of magnetic fields, benefiting navigation systems, geological surveys, and magnetically sensitive equipment calibration, thereby advancing technology and scientific research.
- **Humidity Sensors**: Humidity sensors are indispensable in monitoring and regulating humidity levels, offering precise data on relative humidity through analog signals, aiding in climate control, agriculture, and industrial processes. By accurately measuring humidity levels and presenting them as percentage values, these sensors play a crucial role in maintaining optimal conditions for human comfort, product quality, and equipment performance in diverse environments.
- **Gas Sensors**: Gas sensors are pivotal instruments for detecting and quantifying specific gases present in various environments. They typically produce analog signals in the form of voltage or current outputs, enabling accurate measurement of gas concentrations. This data is conveyed in units such as parts per million (ppm) or other relevant metrics, aiding in environmental monitoring, safety control, and industrial processes where gas detection is paramount.

- **Sound Sensors (Microphones)**: Sound sensors, commonly known as microphones, are indispensable tools for capturing and converting sound waves into electrical signals. These sensors typically generate analog signals, varying in voltage or current, in direct proportion to the detected sound pressure levels. The data they provide manifests as audio signals, usually in waveform format or processed audio data, facilitating applications such as speech recognition, audio recording, and noise monitoring across various industries and consumer devices.
- **Image Sensors (Cameras)**: Image sensors, integral components of cameras, convert light into digital signals representing pixel values. These signals capture the nuances of the scene, enabling the creation of detailed image data. This data is commonly stored in formats such as JPEG, PNG, or RAW, facilitating diverse applications in photography, videography, surveillance, and medical imaging, among others. Through their ability to faithfully capture and reproduce visual information, image sensors play a crucial role in modern imaging technology.
- **Biometric Sensors**: Biometric sensors capture unique physiological or behavioral traits, generating diverse signals like electrical for ECG or pressure for fingerprint recognition. The collected data spans fingerprint patterns, facial features, iris patterns, and heart rate, among others, crucial for identity verification, access control, and personal health monitoring. These sensors play a pivotal role in modern security and healthcare systems due to their versatility and ability to accurately analyze biometric information.
- **Proton Exchange Membrane (PEM) Fuel Cell Sensors**: Proton exchange membrane (PEM) fuel cell sensors generate electrical current as a signal, reflecting the operation and efficiency of the fuel cell. The data derived from these sensors typically represents the electrical power output in watts, providing insights into the performance of the fuel cell system. Monitoring this power output is crucial for optimizing energy production and ensuring the effective operation of PEM fuel cell systems in various applications.
- **Infrared Sensors (IR)**: Infrared sensors, commonly abbreviated as IR sensors, detect levels of infrared radiation emitted by objects in their vicinity. The data retrieved from these sensors often pertains to temperature or motion-related information, enabling applications such as thermal imaging or motion detection systems. By measuring infrared radiation levels and analyzing the resulting data, IR sensors facilitate diverse functions ranging from temperature monitoring in industrial processes to motion sensing in security systems.
- **Global Positioning System (GPS) Sensors**: GPS sensors receive signals from satellites to determine precise geographic coordinates, providing data on latitude, longitude, altitude, and time. This information is instrumental in navigation systems, tracking devices, and location-based services, enabling accurate positioning and route planning. By harnessing satellite signals, GPS sensors offer invaluable data for a wide range of applications, enhancing efficiency and convenience in various industries and everyday life.
- **Capacitive Sensors**: Capacitive sensors detect changes in capacitance, reflecting alterations in proximity or touch-related interactions with objects. The data yielded by capacitive sensors provides information regarding proximity or touch

events, offering versatile applications in touchscreens, proximity switches, and object detection systems.

These sensors are widely used in various applications, including consumer electronics, industrial automation, healthcare, environmental monitoring, and more. The signals and data generated by sensors are crucial for monitoring, control, and decision-making in diverse fields.

11.5 Data Handling Techniques of Sensors

Handling data from sensors involves various techniques to ensure accurate, reliable, and meaningful information extraction. Below are some key data handling techniques commonly used with sensors:

- **Signal Conditioning**: Signal conditioning is a crucial process in sensor technology aimed at enhancing the quality and usability of sensor signals. Amplification is employed to boost weak sensor signals, ensuring they reach a level suitable for further processing and analysis. Filtering techniques, such as low-pass, high-pass, or band-pass filters, are applied to eliminate noise or unwanted frequency components, thereby improving signal accuracy. Additionally, normalization is utilized to scale sensor outputs to a standardized range, promoting consistency and compatibility across different systems and applications
- **Data Sampling**: Data sampling involves setting the sampling rate, which determines how often sensor data is recorded to capture dynamic changes accurately. Adhering to the Nyquist Theorem is essential, ensuring that the sampling rate is at least double the highest frequency in the signal to prevent aliasing and accurately reconstruct the original signal from the sampled data. Proper data sampling is crucial for maintaining signal integrity and fidelity in various sensing applications.
- **Data Acquisition Systems (DAS)**: Data acquisition systems (DAS) utilize specialized hardware to sample, process, and digitize analog sensor signals, facilitating efficient data capture and analysis. These systems incorporate analog-to-digital converters (ADCs) to convert continuous sensor signals into digital format, enabling further processing and storage of the acquired data. DAS play a critical role in various applications, providing accurate and reliable data acquisition for scientific research, industrial monitoring, and automation tasks.
- **Calibration**: Calibration involves adjusting sensor readings to correct inaccuracies or drift, ensuring data accuracy. This is achieved by employing calibration curves or mathematical models to convert raw sensor data into precise physical measurements, enhancing measurement reliability. By implementing calibration techniques, sensors can provide accurate and consistent data for various applications, bolstering confidence in the acquired measurements.
- **Data Compression**: Data compression aims to decrease the volume of sensor data to optimize storage and transmission efficiency. This involves employing techniques such as run-length encoding, delta encoding, and various compression

algorithms to condense the data. By reducing redundancy and encoding information more compactly, data compression enables faster transmission and more economical use of storage resources without sacrificing information integrity. Efficient data compression plays a vital role in managing large datasets generated by sensor networks in various applications.

- **Noise Reduction**: Noise reduction techniques utilize filtering methods like moving averages or Kalman filters to diminish random noise in sensor signals. Additionally, systematic noise sources are identified and eliminated through strategic sensor placement or shielding measures. By implementing these strategies, noise reduction enhances the accuracy and reliability of sensor data, improving the quality of information for analysis and decision-making purposes.
- **Data Fusion**: Data fusion integrates data from multiple sensors, like Kalman filtering, to enhance accuracy and reliability, providing a comprehensive view of the environment. This approach improves situational awareness and decision-making across various domains such as robotics, navigation, and environmental monitoring.
- **Data Pre-processing**: Data pre-processing involves the removal of outliers or erroneous data points and smoothing data using techniques like Savitzky-Golay Filtering to eliminate high-frequency noise. These steps ensure that the data is clean, consistent, and suitable for further analysis or modeling purposes.
- **Time Synchronization**: Time synchronization aligns timestamps of data from different sensors to enable accurate temporal correlation, which is vital for applications requiring precise timing relationships between sensor data. This ensures that events recorded by multiple sensors are properly synchronized, enhancing the reliability and effectiveness of data analysis in various domains.
- **Interpolation and Extrapolation**: Estimating missing data points involves interpolation between sampled values or extrapolation based on historical data to predict future values, enhancing the completeness and continuity of datasets. These techniques are essential for maintaining data integrity and facilitating seamless analysis in various applications across industries.
- **Data Storage**: For data storage, selecting suitable formats like databases, binary files, or JSON depends on the volume and structure of sensor data, ensuring efficient storage and retrieval. Implementing proper indexing and organization enhances retrieval efficiency, facilitating quick access to relevant data for analysis and decision-making processes.
- **Data Transmission Protocols**: Data transmission protocols are optimized to efficiently transmit sensor data over networks, minimizing latency and prioritizing critical data for real-time applications. This ensures the timely and reliable delivery of sensor information, crucial for various industries and applications requiring rapid response and decision-making.
- **Security Measures**: To safeguard sensor data, encryption and authentication mechanisms are implemented during transmission and storage, ensuring confidentiality and integrity. Additionally, measures are taken to prevent tampering and unauthorized access, bolstering overall security and protecting sensitive information from compromise.

- **Power Management**: Power management strategies in wireless sensor networks include implementing techniques like duty cycling and sleep modes to minimize energy consumption, ensuring prolonged operational lifespans and efficient resource utilization. These measures optimize power usage, enabling sensors to operate effectively while conserving energy resources.
- **Real-Time Processing**: Real-time processing of sensor data facilitates rapid decision-making, crucial for time-sensitive applications. Edge computing or fog computing is employed for on-site data processing, reducing latency and enhancing responsiveness in real-time analysis and decision-making processes.

Effective data handling techniques are essential for maximizing the utility of sensor data in various applications, ranging from IoT devices and industrial automation to healthcare and environmental monitoring.

11.6 Smart Sensors Tendencies and Perspectives

As of my last knowledge update in January 2022, the field of smart sensors has been rapidly evolving, driven by advances in technology, miniaturization, connectivity, and the Internet of Things (IoT). The Internet of Things (IoT) facilitates the connection of both animate and inanimate objects through diverse network media, aiming to instigate revolutionary changes by enhancing dynamics and convenience. This technology has significantly expanded the realm of smart devices, empowering various objects with intelligent capabilities through the integration of sensors, RFID, and other forms of embedded computing, thereby fostering pervasive connectivity, smart services, and a plethora of business prospects (Sehrawat and Gill 2019). Here are some tendencies and perspectives in the realm of smart sensors:

- **Miniaturization and Integration**: Smart sensors continue to become smaller, more compact, and integrated with advanced functionalities. This trend allows for deployment in various environments, including wearable devices and IoT applications.
- **Wireless Connectivity**: Integration of wireless communication technologies (e.g., Bluetooth, Wi-Fi, Zigbee) enables seamless connectivity, facilitating real-time data transmission and remote monitoring.
- **IoT Integration**: Smart sensors play a crucial role in IoT ecosystems, contributing to the creation of smart cities, industrial automation, and connected devices. The ability to collect, process, and transmit data is integral to IoT functionality.
- **Energy Efficiency**: Power consumption is a critical consideration for smart sensors, especially in remote or battery-powered applications. Energy-efficient designs, low-power modes, and energy-harvesting technologies are increasingly important.
- **Edge Computing**: Smart sensors are evolving to perform more processing at the edge of the network, reducing the need for centralized data processing. This is particularly beneficial for applications requiring real-time decision-making.

- **Advanced Sensing Technologies**: Integration of advanced sensing technologies, such as hyperspectral imaging, terahertz sensing, and quantum sensors, allows for more sophisticated and precise data capture.
- **Machine Learning and AI Integration**: Smart sensors are incorporating machine learning and artificial intelligence algorithms to analyze data locally, enabling quicker and more context-aware decision-making. This also reduces the need for extensive data transmission.
- **Security and Privacy**: With the increasing volume of data generated by smart sensors, there is a growing focus on ensuring the security and privacy of the collected information. Encryption, secure communication protocols, and privacy-by-design principles are becoming more prevalent.
- **Environmental and Biomedical Applications**: Smart sensors are finding increased use in environmental monitoring, pollution detection, and healthcare applications. These sensors provide valuable data for improving public health and addressing environmental challenges.
- **Flexibility and Adaptability**: Smart sensors are designed to be adaptable to various environments and conditions. This adaptability is crucial for applications in dynamic and changing scenarios.
- **Sensor Fusion**: Combining data from multiple sensors (sensor fusion) enhances accuracy and provides a more comprehensive understanding of the monitored environment.
- **Cost Reduction**: Advances in manufacturing and materials contribute to cost reductions, making smart sensors more accessible for widespread deployment.
- **Standardization and Interoperability**: Efforts toward standardization ensure interoperability between different smart sensor devices and platforms, promoting a more cohesive and interconnected IoT ecosystem.
- **Human–Machine Interaction**: Smart sensors are increasingly being used for human–machine interaction, such as touch less interfaces, gesture recognition, and emotion detection.
- **Rapid Prototyping and Development**: The availability of development platforms and rapid prototyping tools accelerates the creation of innovative applications and solutions using smart sensors.

It is important to note that these tendencies are subject to continuous evolution as technology advances. New developments and trends may have emerged since my last update.

11.7 Signal Processing and Data Acquisition Methods for Multichannel Sensor

Signal processing and data acquisition methods for multichannel sensors are evolving to accommodate the increasing complexity and volume of data generated. One approach involves parallel processing techniques, where data from multiple channels is processed simultaneously to reduce latency and improve efficiency. Additionally, advanced algorithms such as wavelet transforms or principal component analysis (PCA) are employed to extract relevant information from multichannel sensor data, enhancing signal clarity and accuracy. Furthermore, the integration of real-time data fusion algorithms enables seamless integration and analysis of data from diverse sensor channels, providing a comprehensive understanding of the observed phenomena. These methods leverage advancements in computing power and algorithmic sophistication to unlock the full potential of multichannel sensor systems in various applications, including biomedical monitoring, environmental sensing, and industrial automation.

The multi-channel data acquisition system encompasses numerous subsystems dedicated to the collection, analysis, and control of data characteristics, facilitating the implementation of innovative concepts. This systematic approach involves both hardware and software components for measuring and managing data acquisition. The hardware system comprises signal analysis modules and acquisition devices, which play a crucial role in capturing and processing data. Central to this system is the multiplexer, responsible for efficiently managing inputs and receiving signals from various data acquisition networks. Automated data reading and control by computer systems further enhance the functionality of the system. Each signal within the data acquisition setup serves to capture specific elements for parallel analysis, with parameters adjusted based on the underlying principles of applied ideas. Through the integration of digital systems, signal conditioning processes are carried out routinely, ensuring optimal data quality and system performance daily (Bonney et al. 2023).

The individual analog signals are directed to the multiplexer either directly or after undergoing amplification and/or signal conditioning, as needed as shown in Fig. 11.5. Subsequently, these signals are sequentially converted into digital format using analog to digital converters. To optimize time efficiency, the multiplexer is designed to anticipate the next channel to be converted while the previous data is being processed and stored in the sample/hold.

Once the conversion process is completed, the status line from the converter prompts the sample/hold to revert to the sample mode and capture the signal from the next channel. Upon the completion of signal acquisition, either immediately or upon command, the sample/hold switches to the hold mode, triggering a new conversion process while the multiplexer selects the subsequent channel. While this method may be relatively slower compared to systems where sample/hold or A/D converter outputs are multiplexed, it offers the advantage of cost-effectiveness through the shared utilization of most subsystems.

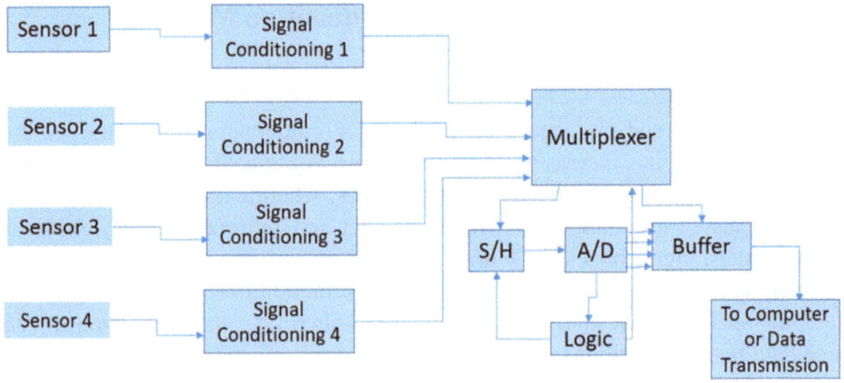

Fig. 11.5 Block diagram of signal processing and data acquisition methods for multichannel sensors

In cases where signal variations are exceptionally slow, it is possible to achieve sufficient measurement accuracy even without utilizing the sample/hold mechanism.

In engineering research, a plethora of electronic devices and software have been employed to record data derived from natural or synthetic sources. As an illustration, in numerous instruments as shown in Fig. 11.5, including seismic, resistivity, ground penetrating radar, magnetic, gravity, microgravity, and sonar devices, are used in geophysical research to measure various subsurface qualities and parameters. The development of low-cost alternative prototypes is crucial for academic and educational reasons, as the current gadgets are relatively costly (Kafadar and Sertcelik 2016). Data acquisition systems for a range of applications, including multichannel brain recording and stimulation (Kursu et al. 2015), early breast cancer detection, EKG analysis, temperature recording, and soil moisture monitoring (Riley et al. 2006), have been created by numerous scientists for this reason.

11.8 Efficient Data Collection Method in Static Sensor Network

Efficient data collection in a static sensor network is crucial for information retrieval for optimizing resource utilization, minimizing energy consumption, and ensuring timely and accurate. A key function of WSNs is data collection, where sensor nodes gather information and transmit it to central sink nodes through multi-hop wireless communication. Although the Shannon sampling theorem traditionally governs data collection in WSNs, it incurs high energy consumption and data redundancy due to the need for raw data collection and transmission. Compressed sensing, as a novel approach, mitigates these issues by employing low-speed sampling, compressing data, and employing nonlinear reconstruction algorithms, thereby reducing energy consumption and data redundancy, thus emerging as a promising research direction in the field (Cao et al. 2020).

Many physical or environmental conditions can be sensed by sensor nodes in wireless sensor networks, which are widely employed in commercial, military, medical, and scientific applications. Completing data gathering is a crucial role of wireless sensor networks. Through multi-hop wireless communication, sensor nodes gather sensor data and send it to sink nodes. Many wireless sensor network applications, including habit monitoring, battlefield monitoring, infrastructure monitoring, and environmental monitoring, depend on efficient data collecting.

The Shannon sampling theorem is currently the primary technique for gathering data on wireless sensor network signals in research and development. However, it requires a lot of raw data collection and transmission, which results in node energy consumption and data redundancy. There are numerous technologies used in compressed sensing projects that compress raw data by low-speed sampling before reconstructing the compressed signal using a nonlinear reconstruction method. Compressed sensing directs research away from the path of comprehensive data collection by limiting the amount of raw data that is collected. The traditional compressed sensing method in wireless sensor network data collection compresses and gathers data using the dense random matrix (Luo et al. 2010) as the observation matrix.

Here are several methods to achieve efficient data collection in a static sensor network:

- **Data Aggregation**: Data aggregation in sensor networks involves locally combining data from neighboring nodes before transmission to the sink node, reducing the amount of data transmitted and conserving energy. This process utilizes aggregation functions such as averaging or summation to condense the information collected by individual sensors, improving network efficiency and reducing communication overhead. By aggregating data at the source, sensor networks can enhance scalability, prolong network lifetime, and optimize resource utilization. Data aggregation, typically involving statistical extraction of values like maximum, minimum, and average, reduces communication and power consumption in dense sensor networks but sacrifices some original data structure (Campobello et al. 2016).
- **Clustering**: Clustering in sensor networks involves organizing nodes into clusters, each led by a cluster head, streamlining data transmission. By routing data from cluster members to the cluster head before forwarding aggregated information to the sink node, clustering minimizes long-distance transmissions, conserving energy and optimizing network efficiency. Various data collection methods exist in wireless sensor networks, with digitalization playing a significant role across industries. Cluster-based routing protocols have emerged as a promising solution to address network lifetime issues, offering benefits such as energy efficiency, enhanced connectivity, and efficient topology management, thus attracting considerable research attention in recent years (Kurian et al. 2022).

- **Duty Cycling**: Duty cycling involves periodically activating and deactivating sensor nodes, conserving energy by allowing nodes to sleep during idle periods and wake up for data collection and transmission. This technique effectively reduces energy consumption while maintaining the functionality of the sensor network.
- **Localized Data Processing**: Localized data processing entails analyzing and processing data within the sensor network before transmitting aggregated or summarized results to the sink node, effectively reducing the volume of raw data transmitted and conserving energy and bandwidth resources. This approach optimizes efficiency by minimizing the need for extensive data transmission and facilitating more streamlined communication within the network.
- **Data Prediction**: Utilize predictive modeling on historical sensor data to forecast future readings. Transmit only predicted values or the variance between predicted and actual readings, minimizing transmitted data volume while ensuring accuracy. This approach optimizes data transmission efficiency without sacrificing predictive quality.
- **Hierarchical Routing**: Hierarchical routing organizes sensor nodes into multiple levels, where each level aggregates and relays data upward. This architecture minimizes the necessity for long-distance transmissions, conserving energy consumption and optimizing network efficiency. By reducing the burden on individual nodes and facilitating efficient data transmission, hierarchical routing extends the lifespan of sensor nodes, enhancing the overall reliability and scalability of the sensor network.
- **Quality-Aware Data Collection**: Quality-aware data collection involves assigning varying levels of importance or quality to sensor data. This strategy prioritizes the transmission of critical or high-quality data, ensuring timely and accurate delivery for crucial applications or analyses. Meanwhile, less important data may undergo aggregation or lower sampling rates, optimizing bandwidth and resources without compromising essential data integrity. By tailoring data transmission based on quality considerations, this approach enhances overall system efficiency and effectiveness in diverse operational contexts.
- **Event-Driven Data Collection**: Event-driven data collection initiates data collection and transmission solely upon meeting predefined events or thresholds. This approach minimizes unnecessary transmissions during sensor inactivity, conserving energy and network resources. By focusing data transmission on relevant events, it optimizes system efficiency and responsiveness in dynamic environments.
- **Spatial Correlation**: Leverage spatial correlation between neighboring sensor nodes to mitigate redundancy in data transmission. By exploiting similarities in data collected by adjacent nodes, unique or significant information observed by each node needs to be transmitted. This strategy optimizes bandwidth usage and conserves energy, enhancing the efficiency of the sensor network.
- **Cross-Layer Optimization**: Cross-layer optimization involves jointly optimizing communication protocols, network layer, and application layer to enhance data collection efficiency. By considering interactions between layers, overhead can be reduced, leading to improved overall system performance. This approach fosters

synergy among different layers, resulting in a more streamlined and effective data collection process.
- **Energy-Efficient Routing Protocols**: Energy-efficient routing protocols like Low-Energy Adaptive Clustering Hierarchy (LEACH) or Threshold-sensitive Energy-Efficient sensor Network protocol (TEEN) prioritize energy conservation in sensor networks. These protocols dynamically adjust routing strategies to minimize energy consumption, prolonging the lifespan of battery-powered sensor nodes. By efficiently managing energy resources, such protocols enhance the sustainability and reliability of wireless sensor networks.
- **Time Synchronization**: Time synchronization of sensor nodes enables efficient time-based scheduling and coordination within the network. By ensuring that all nodes operate on a synchronized time scale, collision probability is reduced, leading to improved overall network efficiency. This synchronization facilitates precise data collection, transmission, and coordination of tasks, enhancing the reliability and performance of the sensor network.
- **Multi-hop Communication**: Multi-hop communication utilizes intermediate nodes to relay data instead of transmitting directly to the sink node, enabling more efficient and extended network coverage. By distributing data transmission across multiple hops, energy consumption is balanced more evenly throughout the network, prolonging the lifespan of individual nodes and enhancing overall network resilience. This approach optimizes resource utilization and improves the scalability of the sensor network architecture
- **Load Balancing**: Load balancing involves distributing data collection tasks evenly across sensor nodes to ensure balanced energy consumption throughout the network. By preventing some nodes from becoming depleted faster than others, load balancing enhances network longevity and stability. This approach optimizes resource utilization and promotes efficient operation of the sensor network.

Implementing a combination of these methods, based on the specific characteristics and requirements of the sensor network, can significantly enhance the efficiency of data collection in static sensor networks.

11.9 Conclusion

The ability to derive meaningful information from unprocessed sensor data is made possible by signal processing techniques, which are essential for comprehending and interpreting other physical events. The accuracy, dependability, and effectiveness of sensor-based data collection systems can all be improved by this procedure.

The capacity to remove noise and extract pertinent characteristics from complex signals is one of the main benefits of signal processing techniques. With this, we can find trends, patterns, and abnormalities in the data that may not be visible in its raw form. Additionally, data compression is a benefit of signal processing techniques since it lowers the need for transmission and storage without compromising accuracy.

The utilization of signal processing techniques is crucial for optimizing data collection via sensors. These methods allow for the refinement of collected data, enhancing system performance and facilitating a deeper understanding of observed phenomena. With ongoing technological advancements, the evolution of signal processing techniques promises to yield more advanced sensor systems capable of addressing multifaceted challenges across diverse domains.

References

Akyildiz IF, Su W, Sankarasubramaniam Y, Cayirci E (2002) A survey on sensor networks. IEEE Commun Mag 102–114

Biradar RV, Patil VC, Sawant SR, Mudholkar RR (n.d.) Classification and COMPA networks, Special Issue on Ubiquitous Computing Security Systems

Bonney MS, de Angelis M, Dal Borgo M, Wagg DJ (2023) Contextualisation of information in digital twin processes. Mech Syst Signal Process 184:109657. ISSN 0888-3270. https://doi.org/10.1016/j.ymssp.2022.109657

Campobello G, Segreto A, Serrano S (2016) Data gathering techniques for wireless sensor networks: a comparison. Int J Distrib Sens Netw 12(3). https://doi.org/10.1155/2016/4156358

Cao K, Liu H, Liu Y, Meng G, Ji S, Li G (2020) Efficient data collection method in sensor networks. Hindawi Complexity 2020:17. Article ID 6467891 https://doi.org/10.1155/2020/6467891

Hariharan N, Sreelekshmi A (2017) Survey on wireless sensor networks. Int J Sci Eng Sci 1(11)

Kafadar O, Sertcelik I (2016) A computer-aided data acquisition system for multichannel seismic monitoring and recording. IEEE Sens J 16(18):6866–6873. https://doi.org/10.1109/JSEN.2016.2592960

Keshari AK, Ramakrishnan R, Rao JP, Jayaraman V, Rama Murthy AS (2019) Development of multichannel data acquisition system for metal oxide sensor array as an electronic nose. In: 2019 6th international conference on signal processing and integrated networks (SPIN), Noida, India, pp 574–578. https://doi.org/10.1109/SPIN.2019.8711631

Kim HS et al (2013) Circuits, systems, and technologies for detecting the onset of sudden cardiac death through EKG analysis. IEEE Circuits Syst Mag 13(4):10–25

Kurian NS, Sandhya S, Rajeshwari N, Rakshini M, Priya CB, Joan NSJ (2022) A review of clustering protocols in wireless sensor network for energy efficient data transfer. In: 2022 international conference on sustainable computing and data communication systems (ICSCDS), Erode, India, pp 1093–1098. https://doi.org/10.1109/ICSCDS53736.2022.9760801

Kursu O, Vähäsöyrinki M, Rahkonen T (2015) Integrated 16-channel neural recording and stimulation circuit. Anal Integr Circuits Signal Process 84(3):363–372

Lee S, Pattem S, Sathiamoorthy M, Krishnamachari B, Ortega A (2009) Spatially-localized compressed sensing and routing in multi-hop sensor networks. GeoSens Netw 9:11–20

Luo J, Xiang L, Rosenberg C (2010) Does compressed sensing improve the throughput of wireless sensor networks. In: Proceedings of the 2010 IEEE international conference on communications, Cape Town, South Africa, pp 1–6

Rahman KC (2010) A survey on sensor network. J Comput Inf Technol 1(1):76–87

Rajesh Kannan (2022) Multichannel data acquisition system. Int J Res Appl Sci Eng Technol (IJRASET) 10(V)

Riley TC, Endreny TA, Halfman JD (2006) Monitoring soil moisture and water table height with a low-cost data logger. Comput Geosci 32(1):135–140

Sehrawat D, Gill NS (2019) Smart sensors: analysis of different types of IoT sensors. In: 2019 3rd international conference on trends in electronics and informatics (ICOEI), Tirunelveli, India, pp 523–528. https://doi.org/10.1109/ICOEI.2019.8862778

Van Laerhoven K (2005) The pervasive sensor. In: Murakami H, Nakashima H, Tokuda H, Yasumura M (eds) Ubiquitous computing systems. UCS 2004. Lecture Notes in Computer Science, vol 3598. Springer, Berlin, Heidelberg. https://doi.org/10.1007/11526858_1

Yick J, Mukherjee B, Ghosal D (2008) Wireless sensor network survey. Comput Netw 52(12):2292–2330

Yong-Min L, Shu-Ci W, Xiao-Hong N (2009) The architecture and characteristics of wireless sensor network. In: IEEE international conference on computer technology and development, ICCTD'09, vol 1

Chapter 12
Design a Photonic Crystal Fiber-Based Biosensor Using a Thin Metals Layer for Industry 4.0

Shradha Gupta, Sanjeev Sharma, and Sri Krishna Singh

Abstract A photonic crystal fiber biosensor has been designed by using a thin layer of metals. The structure of biosensors has become more effective by adding tiny coatings of metals. The present work examines the materials used for these coatings and how they impact the sensor's abilities. Nano-coated photonic crystal fiber biosensors are like tiny detectors that can sense very small amounts of specific substances. By adding extremely small coatings to these detectors, scientists have made them even better at detecting things accurately and quickly. A finite element method has been used for sensing performance and for modelling or irregular-shaped geometry. There are various investigations have been demonstrated in this design by adjusting the air-hole diameter, pitch, and depth of the metals layer coating. In simple terms, it's like finding new and improved ways to build super-sensitive detectors that can be used in important areas like medicine and protecting the environment.

Keywords Biosensor · Photonic crystal fiber · Sensitivity · FDTD · FMW · Photonic band gaps · Quality factor

---

S. Gupta (✉) · S. K. Singh
Department of Physics, ABES Institute of Technology, Ghaziabad, Uttar Pradesh, India
e-mail: shradhamay@gmail.com

S. Sharma
Department of Applied Science & Humanities, IMS Engineering College, Ghaziabad, Uttar Pradesh, India

S. K. Singh
Faculty of Department of Physics, Government College Lamta Balaghat, Bhondwa, Madhya Pradesh, India

A. Rai et al. (eds.), *Convergence of Artificial Intelligence, Machine Learning, and the Internet of Things in Industry 4.0 Applications*, Transactions on Computer Systems and Networks, https://doi.org/10.1007/978-981-96-5269-3_12

12.1 Introduction

Generally, photonic crystals are periodically arrangement of dielectric or semiconductors materials that contain photonic bandgap in which electromagnetic wave propagate in it and give various properties such as ODR reflection, transmittance, flip flops, wavelength division multiplexers etc. (Yablonovy 1993; Sharma et al. 2013, 2014a, 2015, 2016a, b; Kumar et al. 2010; Vimal et al. 2022). A temperature dependent photonic crystals materials have been used to design ODR mirror and optical filters (Wu and Gao 2016; Sharma et al. 2016a, b, 2017, 2021a; Suthar et al. 2012). Also, nonlinear photonic crystals play an important role to design DWDM, optical filters and flip flops (Sharma 2021; Sharma et al. 2014b, 2020a, b; Kalhan et al. 2018). Photonic crystals also used in optical fiber communication to improve the channel efficiency (Sharma and Kumar 2022a; Pandey et al. 2022; Gupta et al. 2021; Rao et al. 2023). Currently, photonic crystals are used in medical applications for designing biosensors.

In the era of Industry 4.0, technological advancements are directing industrial processes toward unparallel heights of efficiency, automation, and connectivity. As sectors work to satisfy the needs of this era marked by profound changes, sensing technologies play a pivotal role in monitoring and controlling various parameters critical to manufacturing processes. Among these, biosensors have emerged as key components, facilitating the real-time detection of biomolecules with high sensitivity and specificity. This research paper delves into the design and development of a cutting-edge biosensor by leveraging the unique properties of Photonic Crystal Fiber (PCF) and incorporating a thin metal layer. This innovative approach aims to enhance the performance and applicability of biosensors in Industry 4.0, addressing the growing need for rapid and accurate sensing solutions in diverse industrial applications.

The integration of Photonic Crystal Fiber, with its intricate micro structured design capable of guiding and manipulating light in unprecedented ways, presents an exciting opportunity to revolutionize biosensing capabilities. PCFs offer advantages such as increased sensitivity, resolution, broader detection range, and compatibility with various sensing modalities (Hameed et al. 2016; Chu et al. 2019; Divya et al. 2018; Hasan et al. 2017; Liu et al. 2018). Moreover, the incorporation of a thin metal layer further enhances the biosensor's capabilities by providing a platform for surface plasmon resonance (SPR) phenomenon, enabling label-free detection of biomolecules (Chu et al. 2019; Dash and Jha 2014). This research thus explores the synergistic effects of combining PCFs and thin metal layers, harnessing their unique characteristics to create a biosensor that not only meets the stringent requirements of Industry 4.0 but also pushes the boundaries of sensing technologies.

As the fourth industrial revolution continues to unfold, the demand for intelligent and interconnected systems grows, necessitating sensors that can seamlessly integrate into the fabric of smart manufacturing processes. This paper emphasizes the potential impact of the proposed biosensor in advancing the capabilities of Industry 4.0, offering a glimpse into the future of real-time, high-throughput, and reliable biosensing technologies. By exploring the design intricacies, fabrication processes, and performance evaluations of the PCF-based biosensor with a thin metal layer, this research aims to contribute valuable insights to the field of biosensing, fostering the development of novel solutions for the evolving landscape of modern industries.

12.2 Compare and Contrast

A range of studies have explored to study and improvise design of biosensors using the photonic crystal fiber to improve sensitivity and practicality for Industry 4.0. Hameed (Hameed et al. 2016) proposed a highly sensitive multichannel biosensor with a silver layer coated by gold, offering high-refractive index sensitivity and self-calibration capabilities. Mescia et al. (2009) optimized a sensor for methadone detection in water, emphasizing the importance of fiber cross section design. He focused for methadone detection by exploring the design of a silica-based photonic crystal fiber sensor by optimizing the fiber cross section for enhanced interaction with the analyte. Rifat et al. (2016, 2017) presented biosensors with external sensing approaches, using copper-graphene and gold layers, respectively, to simplify the structure and enhance performance. These studies collectively highlight the potential for photonic crystal fiber-based biosensors in Industry 4.0 applications. They introduced a simpler configuration using a copper-graphene-based plasmonic biosensor, which demonstrated promising results in detecting biomolecules and organic chemicals. Dortu discussed the performance of a photonic chip-based polymer biosensor with fabrication, emphasizing its potential for remote diagnostics due to its scalability and low cost (Dortu et al. 2011). These studies collectively highlight the potential of photonic crystal fiber biosensors in various applications, including Industry 4.0. The fiber design parameters such as the air hole radius and pitch size of the PCFs can be changed to see SPR phenomena in them and hence the performance of a sensor can also be controlled (Azzam et al. 2016; Dash and Jha 2014). SPR sensors have been extensively examined with many plasmonic materials like gold, silver and copper. To produce this SPR effect, either metal coating should be deposited in the inner air holes or on the outer surface of PCF. This Silver material displays very sharp resonance peak, however it is chemically unstable and very susceptible to oxidization, which reduces the sensing performance efficiency (Azzam et al. 2016; Dash and Jha 2014). It is difficult to keep this additional layer thickness uniform thus increasing manufacturing cost for it unlike gold that does not suffer from oxidization as it is chemically stable and gives a high resonance peak (Dash and Jha 2014; Schriver et al. 2013).

The metal film-coated PCF biosensors have been reported by many researchers (Gao et al. 2014; Qin et al. 2014; Rifat et al. 2015). When analyzing photonic crystal fibers that contain microscale air holes selectively coated with plasmonic metals or infiltrated with liquids, some fabrication challenges arise. These "inside sensing" biosensors can offer high sensitivity, but developing the production process proves difficult. Depositing multiple metal layers with a precisely controlled coating inside the narrow air channels poses technical issues. Maintaining coating accuracy throughout the intricate holes to build up various metal films also presents manufacturing complexities. While the sensing capabilities may excel, incorporating such tailored designs demands innovative fabrication solutions. In order to address the limitations of these biosensors, plasmonic crystal fiber surface plasmon resonance (PCF SPR) sensors coated externally with metal have been proposed. Various studies have explored D-shaped PCF SPR sensors (Gangwar and Singh 2016; Luan et al. 2015; Xie et al. 2017; Tian et al. 2012; Rifat et al. 2017), showing the improved sensing performances.

Different approaches to the design of a photonic crystal fiber-based biosensor using a thin metal layer for Industry 4.0 with focus of biometrics have been studied in recent years. A biosensor that uses metal and defect multilayer photonic crystal lattice to detect anemia has been proposed (Sharma et al. 2023). Their sensor measures reduction in hemoglobin and causes reduction in lattice refractive index. By the creation of a resonator in the photonic crystals to detect the Dengue virus, making this biosensor a promising tool for diagnosing Dengue fever (Sharma and Kumar 2022b). S. Sharma et al. designed various biosensor for healthcare applications by using 2D photonic crystals (Sharma et al. 2021b, 2022; Sharma and Gupta 2021). The schematic diagram of 2D photonic crystal is shown in Fig. 12.1.

Researchers have proposed developing nanophotonic biosensors that utilize evanescent field-based sensing using plasmon resonances in metals and Mie resonances in dielectric materials. These proposed biosensors would harness the optical properties of nanoscale structures to detect bioparticles with high sensitivity and specificity (Altug et al. 2022). They focused on sensitivity with ease-of-use and miniaturization. Rifat et al. (2016, 2017) both focus on the use of a simple and chemically stable metal layer, with (Rifat et al. 2017) using gold and (Rifat et al. 2016) using copper and graphene (Rifat et al. 2017, 2016). Both papers achieve high sensitivities and resolutions, making them suitable for a range of applications. Hameed et al. (2016) takes a different approach by designing a multichannel biosensor with a bimetallic layer, achieving even higher sensitivities and resolutions (Hameed et al. 2016). Mescia et al. (2009), on the other hand, focuses on the design of the photonic crystal fiber itself, optimizing its structure for biosensing applications (Mescia et al. 2009). Gupta et al. (2019) proposed a PCF-based biosensor with bimetallic selectively filled metal layers, offering multimolecular sensing (Gupta et al. 2019). Yun et al. (2019) enhanced the fluorescence of a biosensor platform using silica-coated silver nanoparticles (Yun et al. 2019). Kwoun et al. (2001) introduced a new sensing interface for biosensors, using nanofiber thin films (Kwoun et al. 2001). These papers collectively demonstrate the potential of photonic crystal fiber-based biosensors for Industry 4.0, with each approach offering its own unique advantages.

As the fourth industrial revolution continues to unfold, the demand for intelligent and interconnected systems grows, necessitating sensors that can seamlessly integrate into the fabric of smart manufacturing processes. This paper emphasizes the potential impact of the proposed biosensor in advancing the capabilities of Industry 4.0, offering a glimpse into the future of real-time, high-throughput, and reliable biosensing technologies. By exploring the design intricacies, fabrication processes,

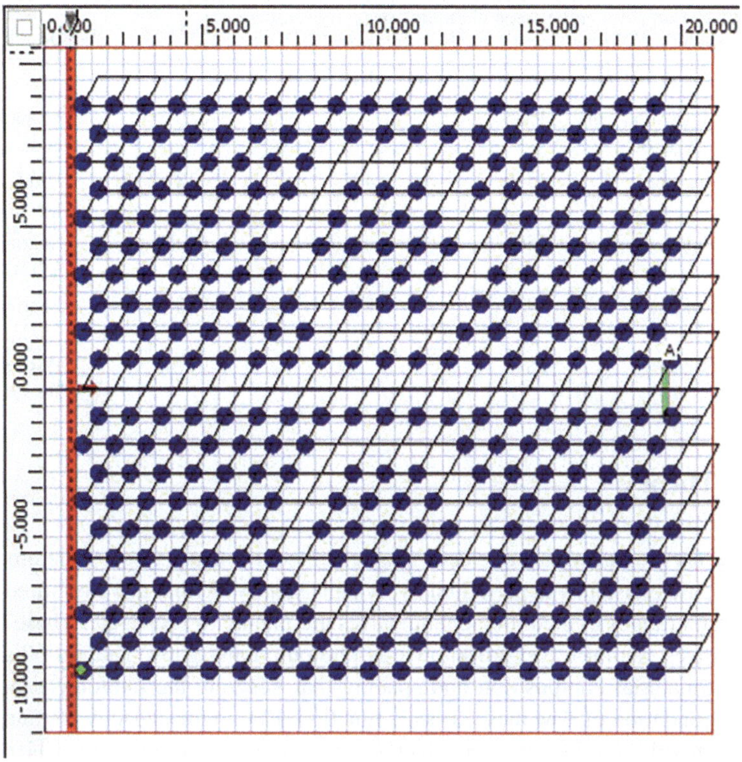

Fig. 12.1 Two-dimensional photonic crystal profile

and performance evaluations of the PCF-based biosensor with a thin metal layer, this research aims to contribute valuable insights to the field of biosensing, fostering the development of novel solutions for the evolving landscape of modern industries.

Table 12.1 represents a comparison between literatures that are summarized in this section. A Photonic crystal—based glucose in urine and blood provides a sensor of quality factor of 269 at operating wavelength 1585 nm (Mohammed et al. 2020). Also, a Photonic crystal-based biosensor provides a sensitivity of 427 nm/RIU with a quality factor of 7187.2 (Panda and Devi 2020). On the other hand, S. Sharma et al. proposed a biosensor for the detection of chikungunya provides a sensitivity of 3430 nm/RIU with a quality factor of 60.21 (Sharma et al. 2021b). A. Panda et al. proposed a biosensor for the detection of cancer provides a sensitivity of 2360 nm/ RIU with a quality factor of 99.76 (Dash and Jha 2017).

Table 12.1 Comparisons between photonic crystal biosensors

S. No	Detection target	Operating wavelength (nm)	Sensitivity (nm/RIU)	Quality factor	References
1	Glucose in urine and blood	1585	–	269	Mohammed et al. (2020)
2	Malaria	1550	427	7187.2	Panda and Devi (2020)
3	Chikungunya	1505	3430	60.21	Sharma et al. (2021b)
4	Anemia	930	892.85	990	Sharma et al. (2023)
5	Cancer	650	2360	99.76	Dash and Jha (2017)

12.3 Theory

Photonic crystals are a class of electromagnetic media with a periodic structure where the dielectric constant and periodicity vary consistently. Within this structured material, certain ranges of electromagnetic waves are unable to propagate. These ranges where waves cannot travel are known as photonic band gaps. The periodic alterations in material composition and spacing that generate the photonic band gaps are the foundation of how these crystals function. Due to the capacity for numerous applications like omnidirectional reflectors, photonic crystals have piqued the interest of many researchers seeking to explore their properties and potentials further. For instance, they may allow for the development of new optical devices or technologies through the precise manipulation of light. While investigation continues into harnessing their distinctive characteristics, these structured materials offer promising opportunities across different fields involving the use and control of electromagnetic waves. While investigating and studying the multi-layer graded structure, it has been found in differing the physical properties as those of homogeneous and conventional structure. Numerous scientists have delved into the fascinating realm of graded multilayered structures and their unique properties of transmission and reflection. In particular, graded photonic crystal (GPC) structures have caught the attention, with their alternating layers exhibiting changes in refractive indices while maintaining consistent thickness. As it is widely acknowledged, the band gap of a photonic crystal structure is greatly influenced by the contrast in refractive indices between these alternating layers and their relative thickness. This opens up exciting possibilities for experimenting with different combinations of parameters in order to unveil the band gaps and explore the reflective and transmissive capabilities of the photonic crystal structure.

Photonic crystal biosensors are a subject of extensive research, utilizing a variety of theories and modelling techniques. Among the prominent ones are the coupled mode theory and finite-difference time-domain (FDTD) simulation. The coupled mode theory, analyses the interaction between light and the periodic structure of the photonic crystal. This theory considers the propagation of light through the photonic crystal and how it couples with different modes within the structure. By understanding the coupling between light and the photonic crystal, researchers can predict the behaviors of the biosensor in response to changes in the surrounding environment, such as biomolecular binding events. Coupled mode theory provides insights into the design and optimization of photonic crystal biosensors for enhanced sensitivity and specificity in detecting biological analytes. Another theory commonly employed is Finite-Difference Time-Domain (FDTD) simulation, which utilizes numerical methods to solve Maxwell's equations in both space and time domains. FDTD simulations enable researchers to model the behaviors of light within the photonic crystal structure with high accuracy. By simulating various parameters such as geometry, refractive index, and wavelength, FDTD theory allows for the prediction of sensor performance and optimization of design parameters. This approach aids in understanding the complex light-matter interactions within the biosensor and facilitates the development of highly sensitive and selective devices for biological sensing applications.

Rigorous Coupled-Wave Analysis (RCWA) is a computational technique used to solve Maxwell's equations for the electromagnetic fields in periodic structures like photonic crystals. It provides a rigorous solution by decomposing the fields into their constituent plane waves and solving for the scattering coefficients. RCWA is particularly useful for analyzing the optical properties and performance of photonic crystal biosensors. Transfer Matrix Method (TMM) is another computational method used to analyze the propagation of light through multilayered structures, including photonic crystals. It calculates the transmission and reflection coefficients at each interface within the structure, allowing researchers to study how light interacts with the biosensor and how it is affected by changes in parameters such as refractive index or thickness.

The design of a photonic crystal structure that incorporates a linear gradient in the refractive index of alternating layers for a set number of layers has been proposed. In a standard photonic crystal, the suggested structure consists of a periodic arrangement of two different dielectric or semiconductor materials with low and high refractive index values (n_1 and n_2) and thicknesses a and b. In a ternary photonic crystal, the proposed configuration includes a repeating pattern of three dielectric or semiconductor materials with varying refractive indices (n_1, $n_2(x)$ and n_3) and thicknesses a, b, and c, respectively. To analyze the reflective properties, researchers have utilized the transfer matrix method (TMM). This computational technique calculates how light propagates through thin films and periodic multilayer structures by determining the reflection and transmission of electromagnetic waves at each boundary between layers. By applying TMM, scientists aim to better understand how modifying aspects like the refractive index gradient or number of alternating strata impact photonic bandgap formation within the crystal. Overall, investigating photonic structures with graded refractive indices may provide insights applicable to developing advanced optical devices with tailored properties. These theories and techniques, among others, provide valuable insights into the design, optimization, and performance of photonic crystal biosensors for various biological sensing applications.

12.4 Results and Discussion

For the study we used two methods to get results for PCF. We used the FDTD method. It also used the finite element method in COMSOL software. PCF-based SPR sensors work because light passes close to a metal and dielectric surface. This creates electrons on the surface between the metal and dielectric. This is called the interface. The close light causes surface plasmon waves or SPWs at the interface. The SPWs interact with the material in the sensing layer. A cross-section view of 2D photonic crystals fiber is shown in Fig. 12.2.

The SPR sensors' performance depends on the geometrical parameters of the PCF. These should be chosen such that there is easy interaction between evanescent field and metal surface. Sensitivity can be increased by making strong coupling between core guided mode and SPP mode. First we select lattice parameter for simulation. However, if core diameter(dc) is greater than the lattice parameter, it reduces effective refractive index (n_{eff}) of core and degrades guidance along the core (Hasan et al. 2017). On the other hand, a smaller value of dc would result in strong light confinement in the core that effectively reduces the generation of SPW (Wang et al. 2016).

Sensitivity is one way to measure how well a PCF-based SPR sensor performs. We used wavelength interrogation and amplitude interrogation to analyze the sensitivity of the suggested fiber. For the wavelength sensitivity in wavelength interrogation this formula can be used (Rifat et al. 2016):

$$S_{\lambda}(\text{nm/RIU}) = \Delta\lambda_{\text{peak}}/\Delta n_a,$$

where $\Delta\lambda_{\text{peak}}$ is the difference between peak wavelength shifts and Δn_a is the variation of analyte RI.

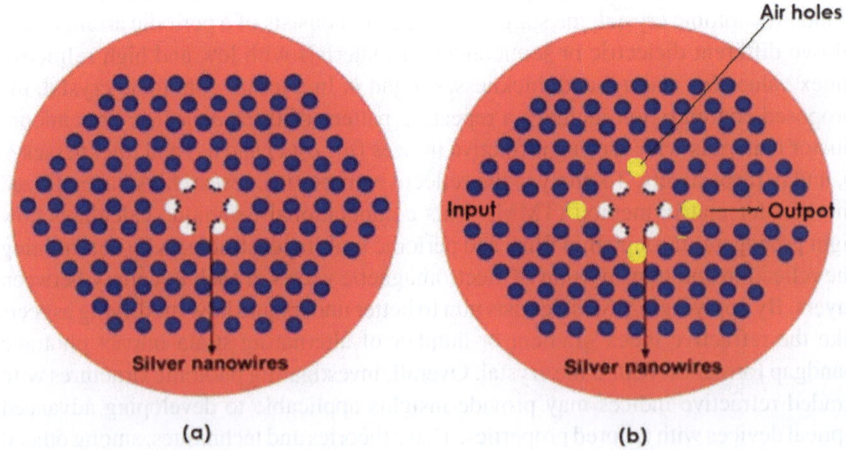

Fig. 12.2 Cross-section view of 2D photonic crystal fiber

Table 12.2 Comparisons between photonic crystal fiber biosensors

S. no	Detecting target	Operating frequency	Sensitivity %	Birefringence	Confinement loss	References
1	Skin cancer	2	96.74	0.0006	2.41×10^{-14} cm^{-1}	Kumar et al. (2021)
2	Blood components	1	87.68	5.1×10^{-5}	1.86×10^{-9} cm^{-1}	Bulbul et al. (2020)
3	Biosensor	1.3	74.55	0.0153	0.013×10^{-11} cm^{-1}	Habib et al. (2021)
4	Cancer	2.5	98	–	10^{-8} dB/m	Natesan et al. (2019)
5	Glucose detection	NA	23,267 nm/RIU	NA	NA	Hasan et al. (2017)
6	Analyte	NA	2200 nm/RIU	NA	NA	Foster et al. (2002)

Another important parameter of the sensor is the resolution of the sensor which represents how a small change of analyte RI can be detected by the sensor. The resolution of the proposed sensor can be obtained by the following formula (Eid et al. 2021):

$$R(\text{RIU}) = \Delta n_a \times \Delta\lambda_{\text{min}}/\Delta\lambda_{\text{peak}}$$

The Table 12.2 summarized a brief numerical analysis of the several photonic crystal fiber studied. It represents a comparison between photonic crystal fiber. Here, M. M. A. Eid et al. represents a Mono-rectangular core photonic crystal fiber (MRC-PCF) for skin and blood cancer detection using FEM and Comsol methods. The proposed biosensor provides a sensitivity of 96.74% and confinement loss of 2.41 $\times$ 10^{-14} cm^{-1} (Kumar et al. 2021). Also, A. Kumar et al. designed a Decagonal solid core PCF based refractive index sensor for blood cells detection that provides the sensitivity of 87.68% and confinement loss of 1.86×10^{-9} cm^{-1} (Bulbul et al. 2020). Similarly, A.M. Bulbul, Habib and Nateson proposed sensors that gives better sensitivity and confinement losses indicated in listed table (Habib et al. 2021; Natesan et al. 2019; Hasan et al. 2017; Foster et al. 2002).

12.5 Conclusion

To sum it up, the design of a Photonic Crystal Fiber (PCF) based biosensor that incorporates a thin metallic layer offers tremendous potential for Industry 4.0 applications. This novel bio-sensor can provide significant improvements in real time detection of biological analytes by exploiting the unique features of PCFs such as their extreme sensitivity to environmental changes and ability to confine light within a small cross section along with enhanced surface plasmon resonance effects facilitated through the use of metal layer.

The sensor sensitivity and specificity are improved when PCF is combined with a thin metal film, thus making it possible to detect low concentrations of biomolecules precisely. Additionally, this biosensor is compatible with Industry 4.0 technologies hence its seamless integration into automated manufacturing processes which enables fast and effective detection bio analytes across different industries.

Generally, with its thin metal layer, it is evident that this proposed PCF-based biosensor designed specifically for addressing Industry 4.0 issues has cutting-edge solutions that are performance oriented scalable and compatible to modern manufacturing processes. Its potentiality to change sensing capabilities across various sectors including healthcare and environmental monitoring indicates how significant.

As the fourth industrial revolution proceeds, the demand for intelligent and interconnected systems increases, requiring sensors that can seamlessly integrate with smart textiles is in the supply chain. By investigating the complex design, fabrication processes, and evaluation efficiencies of PCF-based biosensors with metal thin films, this review aims to contribute valuable insights biosensing industry, and promote the development of innovative solutions for today's evolving industry environment.

References

Altug H, Oh SH, Maier SA et al (2022) Advances and applications of nanophotonic biosensors. Nat Nanotechnol 17:5–16. https://doi.org/10.1038/s41565-021-01045-5

Azzam SI, Hameed MFO, Shehata REA et al (2016) Multichannel photonic crystal fiber surface plasmon resonance based sensor. Opt Quant Electron 48:142. https://doi.org/10.1007/s11082-016-0414-4J

Bulbul AA-M, Imam F, Awal MA, Mahmud MAP (2020) A novel ultra-low loss rectangle-based porous-core PCF for efficient THz waveguidance: design and numerical analysis. Sensors 20(22):6500

Chu S, Nakkeeran K, Abobaker AM, Aphale SS, Babu PR, Senthilnathan K (2019) Design and analysis of surface-plasmon-resonance-based photonic quasi-crystal fiber biosensor for high-refractive-index liquid analytes. IEEE J Sel Top Quantum Electron 25(2):1–9. https://doi.org/10.1109/JSTQE.2018.2873481

Dash JN, Jha R (2014) SPR biosensor based on polymer PCF coated with conducting metal oxide. IEEE Photonics Technol Lett 26(6):595–598. https://doi.org/10.1109/LPT.2014.2301153

Dash JN, Jha R (2017) Graphene-based birefringent photonic crystal fiber sensor using surface plasmon resonance. IEEE Photonics Technol Lett 26:1092–1095

Divya J, Selvendran S, Raja AS (2018) Photonic crystal-based optical biosensor: a brief investigation. Laser Phys 28(6). https://doi.org/10.1109/LPT.2014.2301153

Dortu F, Egger H, Kolari K, Haatainen T, Furjes P, Fekete Z, Bernier D, Sharp G, Lahiri B, Kurunczi S, Sanchez J-C, Turck N, Petrik P, Patko D, Horvath R, Eiden S, Aalto T, Watts S, Johnson NP, Rue RMDL, Giannone D (2011) Design and process development of a photonic crystal polymer biosensor for point-of-care diagnostics. In: Proceedings of the SPIE 8087, clinical and biomedical spectroscopy and imaging II, 80870D. https://doi.org/10.1117/12.889420

Eid MMA, Rashed ANZ, Bulbul AA-M, Podder E (2021) Mono-rectangular core photonic crystal ber (MRC-PCF) for skin and blood cancer detection. Plasmonics 16(3):717727

Foster I, Kesselman C, Nick J, Tuecke S (2002) The physiology of the grid: an open grid services architecture for distributed systems integration. Technical report, Global Grid Forum

Gangwar RK, Singh VK (2016) Highly sensitive surface plasmon resonance-based D-shaped photonic crystal fiber refractive index sensor. Plasmonics 1–6. https://doi.org/10.1007/s11468-016-0395-y

Gao D, Guan C, Wen Y, Zhong X, Yuan L (2014) Multi-hole fiber based surface plasmon resonance sensor operated at near-infrared wavelengths. Opt Commun 313:94–98. https://doi.org/10.1016/j.optcom.2013.10.015

Gupta A, Anand H, Singh A, Singh R, Tiwari A (2019) Photonic crystal fiber based surface plasmon resonance bio-sensors with bimetallic selectively filled metal layers. In: IEEE wireless and microwave technology conference (WAMICON), pp 1–3. https://doi.org/10.1109/WRAP47485.2019.9013679

Gupta S, Choudhary S, Babulal KS, Sharma S (2021) Optimizing and validating performance of 40 Gbps optical system. In: Modern electronics devices and communication systems: select proceedings of MEDCOM, pp 131–140

Habib A, Rashed ANZ, El-Hageen HM, Alatwi AM (2021) Extremely sensitive photonic crystal fiber based cancer cell detector in the terahertz regime. Plasmonics 16(4):12971306

Hameed MFO, Alrayk YKA, Shaalan AHA, El Deeb WS, Obayya SSA (2016) Design of highly sensitive multichannel bimetallic photonic crystal fiber biosensor. J Nanophotonics 10(4):046016. https://doi.org/10.1117/1.JNP.10.046016

Hasan MR, Akter S, Rifat AA, Rana S, Ali SA (2017) A Highly sensitive gold-coated photonic crystal fiber biosensor based on surface plasmon resonance. Photonics 4(18). https://doi.org/10.3390/photonics4010018

Kalhan A, Sharma S, Kumar A (2018) 16-Channel DWDM based on 1D defect mode nonlinear photonic crystal. AIP Conf Proc 1953:060046. https://doi.org/10.1063/1.5032777

Kumar A, Verma P, Jindal P (2021) Decagonal solid core PCF based refractive index sensor for blood cells detection in terahertz regime. Opt Quantum Electron 53(4):113

Kumar V, Singh KS, Ojha SP, Singh G, Anis M (2010) Enlargement of omni-directional reflection range by cascading photonic crystals. vol 4, pp 449–452

Kwoun SJ, Lec RM, Han B, Ko FK (2001) Polymer nanofiber thin films for biosensor applications. In: Proceedings of the IEEE 27th annual northeast bioengineering conference (Cat. No.01CH37201), Storrs, CT, USA, pp 9–10. https://doi.org/10.1109/NEBC.2001.924694; Robinson S, Dhanlaksmi N (2017) Photonic crystal-based biosensor for the detection of glucose concentration in urine. Photon Sens 7(1):1119

Liu M, Yang X, Shum P, Yuan H (2018) High-sensitivity birefringent and single-layer coating photonic crystal fiber biosensor based on surface plasmon resonance. Appl Opt 57(8):1883–1886. https://doi.org/10.1364/AO.57.001883

Luan N, Wang R, Lv W, Yao J (2015) Surface plasmon resonance sensor based on D-shaped microstructured optical fiber with hollow core. Opt Express 23(7):8576–8582. https://doi.org/10.1364/OE.23.008576

Mescia L, Prudenzano F, Allegretti L, Calò G, De Sario M, D'Orazio A, Maiorano L, Palmisano T, Petruzzelli V (2009) Design of silica-based photonic crystal fiber for biosensing applications. J Non-Cryst Solids 355(18–21):1163–1166. https://doi.org/10.1016/j.jnoncrysol.2009.01.047

Mohammed NA, Hamed MM, Khalaf AAM, El-Rabaie S (2020) Malaria biosensors with ultra-sensitivity and quality factor based on cavity photonic crystal designs. Eur Phys J Plus 135(11):933

Natesan A, Govindasamy KP, Gopal TR, Dhasarathan V, Aly AH (2019) Tricore photonic crystal bre based refractive index sensor for glucose detection. IET Optoelectron 13(3):118123

Nishiguchi K, Sueyoshi K, Hisamoto H, Endo T (2016) Fabrication of gold-deposited plasmonic crystal based on nanoimprint lithography for label-free biosensing application. Jpn J Appl Phys 55(8S3)

Panda A, Devi PP (2020) Photonic crystal biosensor for refractive index based cancerous cell detection. Opt Fiber Technol 54(10):2123

Pandey SK, Singh S, Maurya JB, Prajapati YK (2022) Design of a broadband dispersion compensated ultra-high nonlinear photonic crystal fiber. Opt Quant Electron 54(503)

Qin W, Li S, Yao Y, Xin X, Xue J (2014) Analyte-filled core self-calibration microstructured optical fiber based plasmonic sensor for detecting high refractive index aqueous analyte. Opt Lasers Eng 58:1–8. https://doi.org/10.1016/j.optlaseng.2014.01.003

Rao MPS, Singh V, Sharma S (2023) Design and analysis of suspended core photonic crystal fiber formed by semicircular air holes near the core. Int J Microw Opt Technol 18(1):86–95

Rifat AA, Mahdiraji GA, Chow DM, Shee YG, Ahmed R, Adikan FRM (2015) Photonic crystal fiber-based surface plasmon resonance sensor with selective analyte channels and graphene-silver deposited core. Sensors 15:11499–11510. https://doi.org/10.3390/s150511499

Rifat AA, Mahdiraji GA, Ahmed R, Chow DM, Sua YM, Shee YG, Adikan FMR (2016) Copper-graphene-based photonic crystal fiber plasmonic biosensor. IEEE Photonics J 8(1):1–8. https://doi.org/10.1109/JPHOT.2015.2510632

Rifat AA, Hasan MR, Ahmed R, Butt H (2017a) Photonic crystal fiber-based plasmonic biosensor with external sensing approach. J Nanophotonics 12(1):012503. https://doi.org/10.1117/1.JNP.12.012503

Rifat AA, Ahmed R, Mahdiraji GA, Adikan FRM (2017b) Highly sensitive D-shaped photonic crystal fiber based plasmonic biosensor in visible to near-IR. IEEE Sens J 17(9):2776–2783. https://doi.org/10.1109/JSEN.2017.2677473

Schriver M, Regan W, Gannett WJ, Zaniewski AM, Crommie MF, Zettl A (2013) Graphene as a long-term metal oxidation barrier: worse than nothing. ACS Nano 7(7):5763–5768. https://doi.org/10.1021/nn4014356

Sharma S, Kumar R, Singh KS, Kumar V, Jain D (2013) Design of a transmission TM mode filter using one-dimensional ternary photonic crystal. AIP Conf Proc 1536:705–706. https://doi.org/10.1063/1.4810422

Sharma S, Kumar R, Singh KS, Jain D (2014a) Design of an omnidirectional reflector using one dimensional ternary photonic crystal. Int J Eng Technol Res (special issue: STET-2014) 90–93

Sharma S, Kumar R, Singh KhS, Kumar V, Kumar A (2014b) Single channel tunable wavelength division demultiplexer using one dimensional defect mode nonlinear photonic crystal. Optik-Int J Light Electron Opt 125:4895–4897. https://doi.org/10.1016/j.ijleo.2014.04.034

Sharma S, Kumar R, Singh KS, Kumar V, Kumar A (2015) Omnidirectional reflector using linearly graded refractive index profile of 1D binary and ternary photonic crystal. Optik-Int J Light Electron Opt 126(11–12):1146–1149. https://doi.org/10.1016/j.ijleo.2015.03.029

Sharma S, Dwivedi D, Yadav A, Sengar A (2016a) Temperature dependence ZnS based one-dimensional photonic crystals. Int J Phys Res (IJPR) 6(1):1–6

Sharma S, Kumar R, Singh KS, Jain D, Kumar A (2016b) Enhanced ODR range using exponentially graded refractive index profile of 1D binary photonic crystal. AIP Conf Proc 1728

Sharma S, Kumar R, Singh KS, Kumar A (2017) Temperature dependence ODR reflection by using 1D binary and ternary photonic crystal. Optoelectron Adv Mater-Rapid Commun 19(5–6):319–324

Sharma S, Kumar A, Singh KS (2020a) Design of a tunable DWDM multiplexer using four defect layers of GaAs nonlinear photonic crystals. Optik-Int J Light Electron Opt 212:164652. https://doi.org/10.1016/j.ijleo.2020.164652

Sharma S, Gupta S, Suther B, Singh KhS (2020b) Design of a tunable transmission mode filter using 1D Ge based nonlinear photonic crystal. AIP Conf Proc 2220:050016. https://doi.org/10.1063/5.0001128

Sharma A (2021) Design of a tunable 8-channel DWDM covers all transmission windows using defect mode nonlinear photonic crystals. Opt Eng 60(1). https://doi.org/10.1117/1.OE.60.1.017105

Sharma S, Gupta S (2021) Detection of albumin and urea concentration in urine using 2d photonic crystals. J Phys: Conf Ser 2007(1):012015. https://doi.org/10.1088/1742-6596/2007/1/012015

Sharma S, Kumar V, Gupta S (2021a) Tunable transmittance using temperature dependence ZnS-based ID photonic crystals. In: Advances in smart communication and imaging systems. Springer, pp 325–330. https://doi.org/10.1007/978-981-15-9938-5_31

Sharma S, Kumar A, Singh KS, Tyagi H (2021b) 2D photonic crystal based biosensor for the detection of chikungunya virus. Optik 237:166575

Sharma S, Kumar A (2022) Analysis of silica based single-mode fiber doped with germanium at different transmission window. Silicon 14:1023–1028. https://doi.org/10.1007/s12633-020-00884-6

Sharma S, Kumar A, Agrawal R (2022a) Design of a biosensor for the detection of glucose concentration in urine using 2D photonic crystals. In: Computational intelligence in healthcare applications, pp 297–307

Sharma S, Kumar A (2022b) Design of a biosensor for the detection of dengue virus using 1D photonic crystals. Plasmonics 17:675–680. https://doi.org/10.1007/s11468-021-01555-x

Sharma S, Singh M, Tewari RK et al (2023) A biosensor for the detection of anemia using metal ag and defect multilayer photonic crystal. Plasmonics. https://doi.org/10.1007/s11468-023-02088-1

Suthar B, Kumar V, Kumar A, Singh KhS, Bhargava A (2012) Thermal expansion of photonic bandgap for 1-D photonic crystals. Progr Electromagn Res Lett 32:81–90

Tian M et al (2012) All-solid D-shaped photonic fiber sensor based on surface plasmon resonance. Opt Commun 285(6):1550–1554. https://doi.org/10.1016/j.optcom.2011.11.104

Vimal, Sharma S, Sharma AK, Tiwari R (2022) Tunable filter at second transmission window containing 1D Ternary superconductor/dielectric photonic crystals. In: Smart innovation, systems and technologies, vol 239. Springer, pp 189–195. https://doi.org/10.1007/978-981-16-2857-3_20

Wang G, Li S, An G, Wang X, Zhao Y, Zhang W, Chen H (2016) Highly sensitive D-shaped photonic crystal fiber biological sensors based on surface plasmon resonance. Opt Quantum Electron 48:1–9

Wu JJ, Gao JX (2016) Temperature-dependent optical properties of defect mode in dielectric photonic crystal heterostructure containing a superconducting layer. Mater Chem Phys (Elsevier) 171:91. https://doi.org/10.1016/j.matchemphys.2015.12.003

Xie Q, Chen Y, Li X, Yin Z, Wang L, Geng Y, Hong X (2017) Characteristics of D-shaped photonic crystal fiber surface plasmon resonance sensors with different side-polished lengths. Appl Opt 56(5):1550–1555. https://doi.org/10.1364/AO.56.001550

Yablonovy E (1993) Photonic band-gap structures. J Opt Soc Amer B 10:283–295

Yun BJ, Kwon JE, Lee K, Koh W-G (2019) Highly sensitive metal-enhanced fluorescence biosensor prepared on electrospun fibers decorated with silica-coated silver nanoparticles. Sens Actuators B: Chem 284:140–147. https://doi.org/10.1016/j.snb.2018.12.096

Chapter 13
Detection and Diagnosis of Cervical Cancer Using Machine Learning Models

P. Amsini, Amrita Rai, and V. Shanmugasundaram

Abstract Cervical cancer is a dangerous disease that occurs in advancing sovereignty where public awareness is low. Screening Papnicolaou, or Pap test, is the most common test to detect cancer, which develops in the uterus and occurs in most women. Image processing algorithms play a leading role in the segmentation of the cancerous region in cervical images. The cancer region is segmented in cervical cancer images by fuzzy support vector machine (FSVM) algorithm. The cervical cancer region was separated from background images. The k-means classification algorithm is an existing algorithm applied to cervical cancer images. The existing and proposed segmentation results are compared using quality measurement techniques such as accuracy and precision. From this evaluation, proposed algorithm is provided the highest accuracy (98%) when compared to the previous algorithm. In this chapter, we described about the machine learning models, determination of the cancer region and collected the data from the cervical images for diagnosis of cancer level using support vector machine (SVM) classifier and k-nearest neighbours (KNN) classifier and classify it is a benign or malignant stages. The accuracy, sensitivity, and specificity are which calculated for classification accuracy. The accuracy is 97 percentages is gained using SVM compared to the KNN.

Keywords Cervical cancer · Image segmentation · FSVM · Image quality · K-means · Machine learning · SVM

P. Amsini (✉)
Department of Computer Science, Padmavani Arts & Science College for Women, Salem, India
e-mail: amsini1621992@gmail.com

Amrita Rai
Department of Electronics & Communication Engineering, LLOYD Institute of Engineering and Technology, Greater Noida, India

V. Shanmugasundaram
Department of Electrical and Electronics Engineering, Sonna College of Technology, Salem, India
e-mail: shanmugasundaram@sonatech.ac.in

A. Rai et al. (eds.), *Convergence of Artificial Intelligence, Machine Learning, and the Internet of Things in Industry 4.0 Applications*, Transactions on Computer Systems and Networks, https://doi.org/10.1007/978-981-96-5269-3_13

13.1 Introduction

Cervical cancer and precancerous cells that may eventually become malignant can both be found using screening tests. At the begin stage, the cells in the cervix region might be developed as cervical cancer and then lowermost portion of the uterus that is joined at the vagina region which is called as cervix. The human papillomavirus (HPV) is culpable for cervical malignancies and approached in various strains.

A recurrent infection increase throughout intercourse is HPV and it causes of cervical cancer at cervical tissue and ultimately malignancy. If cervical cancer is discovered earlier and addressed quickly, it can be cured. In the decades to come, nations worldwide are striving to expedite the eradication of cervical cancer, including a set of three goals that must be accomplished by 2030. To detect the cervical cancer at the earlier stages, this leads to curable of cervical cancer. The deep learning methods, artificial intelligence, image analysis, and data analytics methods are helps detection of benign or malignant. A central initial step is recognizing indication and getting medical help to address any concerns. To consult the doctor when these kind of symptom will be noticed by the patient. The symptoms of cervical cancer are given below:

- Anomalous vaginal discharge.
- Chronic back, leg, or persistent of pelvic pain.
- A typical bleeding amid periods behind menopause.
- They get weight loss, exhaustion, and appetite loss.
- Discomfort throughout vaginal swell up legs.

In this chapter, we discuss the image acquisition, image pre-processing, and segmentation of cancer regions using fuzzy-based support vector machines and quality measurement for the segmentation algorithm. The computer-aided diagnosis (CAD) tools are urbanized to distinguish cancer in the cervix region, where portable technology is needed to identify aberrant cervix alterations that may lead to invasive carcinoma at a low cost. The developed tools are assisting with secondary examinations for expert assessments for early detection. The proposed work consists of picture collection to obtain cervical cancer data for examination of the cancer region. Then pre-processing is one of the most efficient ways to improve the cancer digital image. The goal of pre-processing is to enhance the picture data by suppressing unwanted distortions or enhancing certain visual elements that are relevant for later processing. The proposed methods of fuzzy-based support vector machine (FSVN) and k-means classification were used to segment the cancer region from the image background. This algorithm quality was tested in terms of precision and accuracy.

13.1.1 Use Cases of Cervical Cancer

A specific kind of cancer called cervical cancer arises in the cells of the cervix, the area of the uterus (womb) that links to the vagina. The cervical cancer is caused by primary sexually where transmitted infection as human papillomavirus (HPV) and HPV infections also developed a cancer cells and some are low-risk and do not, while others are high-risk and can develop abnormalities in the cervical tissue and ultimately malignancy. To gather information from patient reports of cervical cancer in order to prevent the disease's progression, either combine information from imaging or ensure that the data is balanced.

Numerous techniques, including big data, data analytics algorithms, and machine learning models, are available for diagnosing the illness. This disease may be occurred for both female and male. Early detection of the disease one of key point for improves the patient safety. Every nation has committed to ending cancer of the cervical cavity as a problem affecting the public. The World Health Organization worldwide approach has three goals to be met by 2030 in order to put every nation on track towards eradication, which is defined as bringing the annual incidence of new infections down to Four or below per million women. In 2024, HPV vacancies will be available worldwide for the prevention of cervical cancer diseases. When administered by intramuscular injection, antibodies are produced, which is thought to stop the emergence of new infections specific to a certain kind and the subsequent development of cervical intraepithelial neoplasia (Arbyn and Xu 2018).

Cervical cancer is developed on the cervix region of women body; it causes death so early detection is most important to prevent the cancer and improve patient outcomes. Cervical cancer develops in cells found in the cervix. The cervix is the bottom straight end of the womb anatomy shown in Fig. 13.1. The cervix functions as an intermediary connecting the womb and the birth canal. The cervical cancer frequently grows on slowly over time in the cervix. When cancer develops in the uterus, the cells experience a transition phase known as dysplasia, during which cells that are abnormal begin to emerge in the cervical epithelium. If the abnormal cells are not eliminated, they will eventually develop into carcinoma cells that continue to multiply and propagate deeper into the cervix region and adjacent areas. The HPV vaccination and Pap can both help prevent cervical cancer. Diagnostic of cancer patient is an examination for physical and the inspection of the uterus and region of the vagina to determine whether infection has occurred or not. Then a Pap test is conducted to collect the cells from the cervix region to determine whether cancer is present or not. There are two types of cervical cancer such as squamous cell carcinoma and adenocarcinoma carcinoma. The squamous cell is developed at ectocervix from cells. An adenocarcinoma is growth on the glandular cells on the endocervix.

Cervical cancer has several important use cases, ranging from prevention and screening to treatment and support. HPV vaccination is a crucial use case in preventing cervical cancer. Vaccination can protect against infection which causes the cervical cancer with HPV types and produced risk. It is recommended for preteen boys and girls, ideally before they become sexually active, to provide a protection

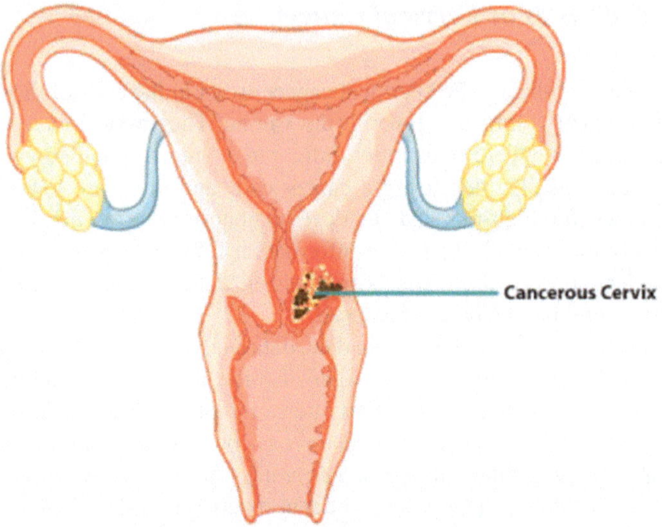

Fig. 13.1 Anatomy of cervical cancer

alongside for HPV-based diseases such as cervical cancer. Screening for cervical cancer is another important use case. The Pap test is one of the regular screening tests. The HPV deoxyribonucleic acid (DNA) tests are used to perceive the pre-cancerous changes in the cervix sooner than it was developed as cancer cells. Detection of cancer at earlier stages via screening allows for early treatment, intervention, and prevent from the cancer lesions into invasive cancer. Screening for cervical cancer is another important use cases.

Cervical cancer diagnosis involves the use of different examination and actions to substantiate occurrence of cancer, establish its level (stage), and guide treatment decisions. Diagnostic tests may include colposcopy, biopsy, imaging tests such as magnetic resonance imaging (MRI) or computer tomography (CT) scans, and sometimes, lymph node biopsy to assess whether the cancer has spread beyond the cervix. Treatment planning is a critical use case in cervical cancer management.

Once the cancer is diagnosed and staged, healthcare professionals develop a personalized treatment plan tailored to the individual patient's condition. Treatment options could comprise private clinic, emission therapy, chemotherapy, targeted rehabilitation, or an amalgamation of these modalities which depends on the period of the cancer, the female largely wellbeing, in addition to erstwhile factors. After treatment for cervical cancer, regular follow-up and surveillance are essential to monitor for recurrence or any potential complications. Follow-up visits typically involve physical examinations, imaging tests, and other assessments to ensure that the patient remains cancer-free and to address any ongoing needs or concerns. Cervical cancer survivors may require ongoing supportive care to manage treatment side effects, cope with emotional challenges, and address any long-term effects of cancer and its treatment. Survivorship care plans may include guidance on maintaining overall health, managing late effects of treatment, and accessing support services and resources. Early detection is most important in cancer diagnosis which is improved the patients safety and outcomes.

13.2 Machine Learning Algorithms Are Applied to Diagnosis of Cervical Cancer

One of the principal causes of cancer-related demise in women internationally is cervical cancer, which behaves epidemiologically like a low-infectious venereal illness (Zhang Shaokai et al. 2020; Zur Hausen 1989). Human papillomavirus (HPV) 16/18 along with smouldering are presently significant aspect in the multifaceted, progressive carcinogenic theory in the cervix uteri. As a result, it is advised to implement screening programmes, HPV vaccination campaigns, and society-based preventive and control measures (Zhang Shaokai et al. 2020; Zur Hausen 1989). The small segment of the uterus to relations to the vagina is called the cervix, and here is where cervical cancer starts. Image segmentation is used extract the region of cancer so medical image segmentation is used for together therapeutic and diagnosis reason. The region of interest, classification, threshold methods are applied in a medical imaging where identified with the use of image segmentation. This phase is essential for further analysis because it enables the emphasis on a particular region that is pertinent to the screening of cervical cancer. Cervical abnormalities or lesions are common presentations of cervical cancer. These lesions can be located and identified via image segmentation.

Healthcare practitioners can identify regions that could need more investigation with the use of automated segmentation techniques. In the event that an anomaly or tumour is present, segmentation aids in defining its limits. Precise segmentation facilitates comprehension of the tumour form and size, which is important for treatment planning. Segmentation aids in defining the limits of any tumours or abnormalities that may be present. For the purpose of treatment planning, accurate segmentation helps determine the size and form of the tumour. Multiclass support vector machine (multiclass-SVM) was applied for cervical images which detect and segment the abnormal region as object from the background region (NCCC 2010). Sometimes, the image features are extracted from the cancer images where SVM classification is applied for segment the cancer regions. Before used of SVM, select the kernel type leads to better classification. There are important standard kernel types such as polynomial kernel, radial basis kernel, and Gaussian and linear kernel (Scholkopf and Smola 2002).

This chapter is focused on cancer detection utilizing image processing algorithms to find the cancer nucleus region. Most machine learning approaches are used to identify cancer such as SVM, CNN architect (convolutional neural network), and it has been one of the autonomous processes that predict cancer at an early stage,

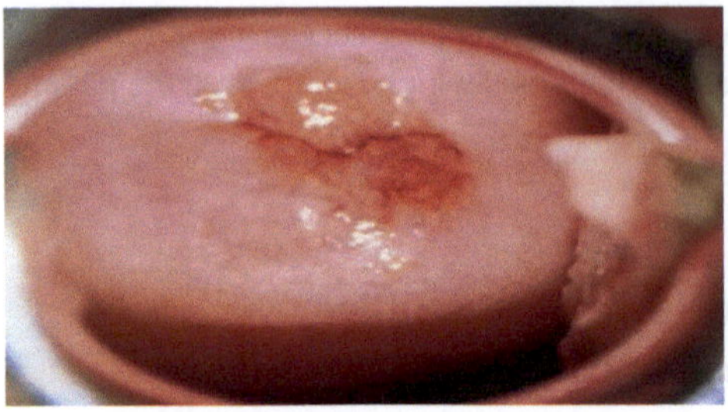

Cervical cancer image

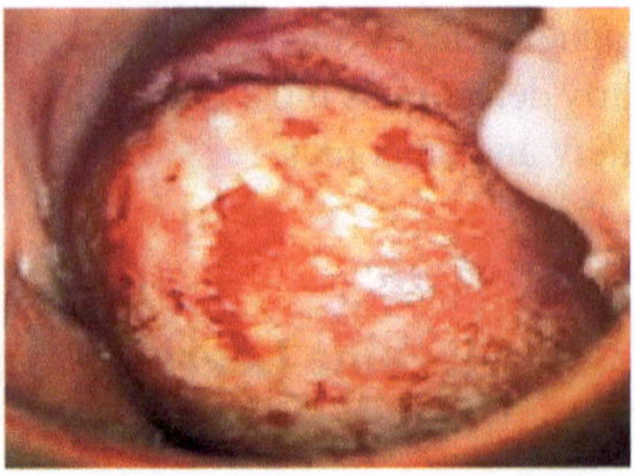

Cervical cancer image(at stage 2nd)

Fig. 13.2 Illustration of cervical cancer images

improving patient safety (Wei et al. 2020). The k-nearest neighbour (KNN) algorithm is provided classification precision of eighty-four percentage for segment the cancer cervical images (Sharma and Singh 2016). The standard k-means algorithm is applied for colposcopy images with accuracy 87.25% (Bing et al. 2018). This literature study reveals that many common algorithms are utilized to segment cancer regions in cervical images, and this research raises awareness about the need for experimental work in medical-based cancer research. Follow-up scans and other serial imaging techniques can be used to track the advancement or regression of malignancy. Observing changes in tumour features and size over time is crucial for assessing how well a treatment is working. Image segmentation helps to segment the cancer region from the cervical cancer images. The National Cancer Institute (NCI) generated the Guanacaste dataset in 1997, and this collection of cervical photos includes photographs of patients with cancer and those without any abnormal abnormalities. It contains ground-truth images that have been examined by the experts (Jaya et al. 2018). Figure 13.2 illustrates the cervical cancer image1 and image2.

13.3 Detection of Cancer Region Using Machine Learning Model

Firstly, computers process digital images and characterize them numerically. Digital image processing is used to process digital images on computers. The general purpose of research work consists of image acquisition, pre-processing, image segmentation, feature extraction, and then finally, applied classification method based on machine learning is shown in Fig. 13.3.

The digital image processing algorithms help improve the image, or if the image contains errors, they reduce noise and enhance the brightness or luminance. In this chapter, the red, green, and blue (RGB) image is rehabilitated to hue, saturation, and value. The support vector machine (SVM) is used for segmentation tasks, including cancer image analysis and then medical imaging. SVM is one of the most important

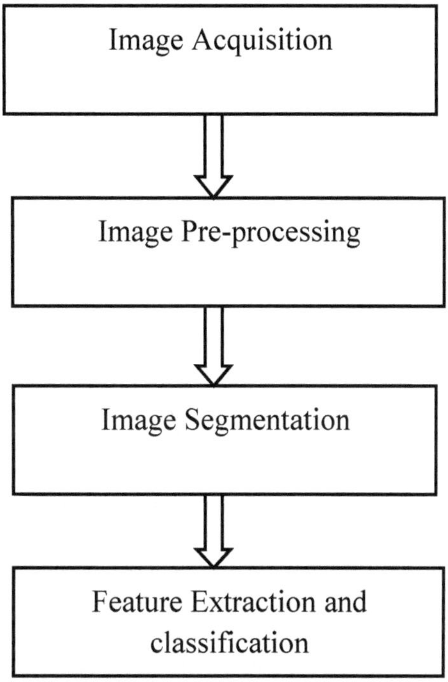

Fig. 13.3 Proposed method for image analysis and applied machine learning for classification

supervised machine learning methods that is applied for both section and categorization tasks. The specific region object is extracted from the background image and assigned to a specific category or based on classes.

The SVM is based on structural risk minimization principle model and present space complexity with n number of dimension and training data. It work based on hyperdimensional so used for cancer cervical images for image segmentation purposes. The support vector machine method is most recommend by the scientific authors and researchers for segmentation and classification purposes such as classification, segmentation and cancer stages predication, remote sensing images and medical applications and reorganization fields. The sub-pixel classification is cannot handled by the SVM so it is extent through fuzzy set theory.

The SVM is introduced by vapnix followed that Lin Wang is introduced by Lin and Wang. There various kinds of kernel are used in SVM classifier such as linear, polynomial, and radial basis function. When uncertainty occurs in data or image, it is handled by fuzzy-based support vector machine. A classifier is learned the training set as input and measured the membership function from the medical images which is consist uncertainty. Fuzzy pixel-based sub-classification helps to classifier the image. The pixels belong to the membership functions when used fuzzy-based SVM classifier and considered as training pixels. The image training pixels have N number of samples with correlated pixels in the input image. The regularization parameter is applied to the kernel functions.

The image processing technique uses cervical pictures to identify cancer in the cervix. In the current environment, cervical cancer can be detected at the terminal phase, resulting in untimely death. In numerous uses, people do not experience any discomfort or soreness in the cervix region, and no symptoms develop. If the cancer is diagnosed early, the patient's life can be preserved (Jaya et al. 2018; Kashyap et al. 2016). Researchers and scientists spent more time detecting cancer in its early stages and improved the automatic tools for segmentation and prediction.

A collection of specified pixels is all that makes up a picture. The similar pixels are clustered together with meaningful information from the original image. The pixel-by-pixel mask was created for each item in the image, and the object was divided from the image at a deeper and more detailed level. The Pap smear test examines the cells in the cervix area and diagnoses the nuclei sections. Image segmentation is important assignment to distinguish the cancer-affected region from the cancer cervical images. We discussed the applied segmentation algorithm and the comparison of quality based on measurement. The k-means segmentation is one of the standard methods: randomly assign the data points and then calculate the cluster centre point. Then calculate the Euclidean distance between the centre points. This process continuously produces data points that cannot be changed. It is a very complex and time-consuming process to handle cancer image data. To begin with, the original image is converted to an HSV image, and the resultant image is adjured for a segmented process. The SVM has the capability to segment the images, and the intelligence approaches are an SVM and neural network. SVM is more sensitive and one of the standard methods, besides. The SVM method has one complication, which is sensitivity to pixel initialization, which can be overcome by applying fuzzy-based SVM (Cai 2007). From the proposed algorithm, we measured the hue, saturation, and values, extracted them, and applied fuzzy and SVM for cancer in the cervical region. The fuzzy-based SVM yields better segmentation results compared to the existing k-means clustering algorithm. The algorithm steps are given below:

- To set the initial parameter and then the quantity of clusters.
- The numeral of colour pixels is clustered through the membership function.
- Classifying the pixels in the image by membership value.

To create the whole training set, combine the training samples from each cluster. As the test set, retain the remaining picture pixels. Then train the SVM classifier (with a train set with FCM class labels by SVM). Experimental results for cancer region detection using k-means classification and FSVM. The proposed and existing methods applied for cervical cancer images to extract the cancer region are shown in Fig. 13.4. Comparing the existing and proposed methods, the cancer region is better segmented from fuzzy-based SVM.

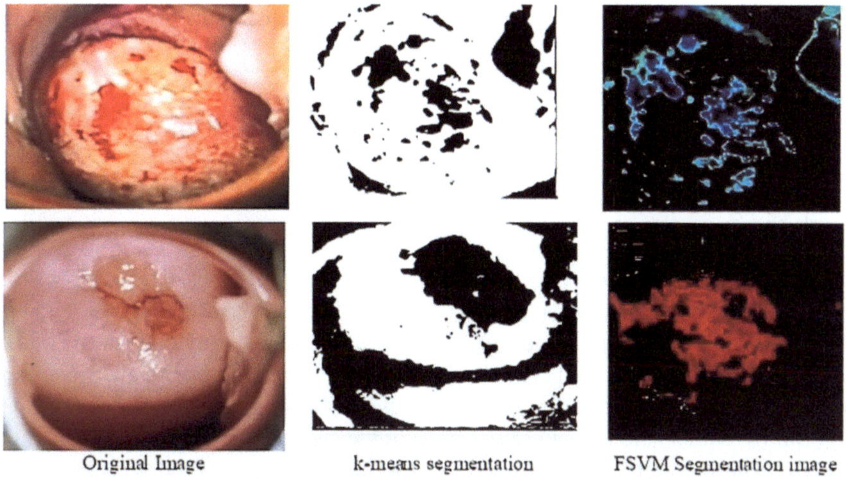

Fig. 13.4 Proposed algorithm applied for cancer region detection

13.3.1 Quality Measurements Between Existing and Proposed Algorithms

Quality measurement is a crucial way for determining if the diagnosis procedure was completed correctly or not. The recommended cervical cancer segmentation technique is evaluated in terms of precision and accuracy, as shown in Tables 13.1 and 13.2.

When compared to the current k-means segmentation approach, the FSVM-based cancer area segmentation method performs well based on quality measurements. When applied k-means for cervical cancer image 1, the precision value is 87%, followed by accuracy of 96.2%. As a result, the suggested FSVM segmented measurement value of accuracy is 98.6%, while the precision value is 89%. When k-means is used for cervical cancer image 2, accuracy is 94.4% and precision is 89%. As a

Table 13.1 Quality measurement value for cervical cancer_Image 1

Algorithms	Precision (%)	Accuracy (%)
K-means	87	96.2
FSVM	89	98.6

Table 13.2 Quality measurement value for cervical cancer_Image 2

Algorithms	Precision (%)	Accuracy (%)
K-means	89	94.4
FSVM	97	98.0

result, the suggested FSVM segmented measurement value of accuracy is 98%, and precision is 97%.

13.4 Disease Diagnosis Using Machine Learning Algorithms

The cervical cancer data is derived from the segmented pictures. There are several machine learning algorithms available for diagnosing cancer, whether malignant or benign. Machine learning models are employed in medical applications to continually identify or forecast cancer. The nuclei cell detection is one of the important one to perceive the cancer that manner image processing-based machine learning algorithms, which help to analyse the stage of the cancer after applied segmentation process. The researchers should be faced lot of challenges when detect the cancer nuclei and stages found out by applied machine learning algorithms. There are lot of benchmark dataset is available at the network era. In this chapter, we selected thousands of images and then applied feature selection algorithm applied to analyse the stage of the cancer. The grey-level co-occurrence matrix (GLCM) features and intensity histogram are extracted from the segmented images. The selected features from segmented images are listed in given Tables 13.3 and 13.4.

The features values are selected from the cancer region segmented images and then collecting the features from the segmented images using GLCM and histogram intensity. We are randomly selected the features from the segmented images which are invoked for SVM classification. The SVM classification is used for binary classification. The SVM classifier yields almost ninety-seven percentage of accuracy compared to the k-nearest neighbours (KNN) classifier. The KNN classifier is supervised learning algorithm. It is one of the easiest methods of classify the data which

Table 13.3 Features are extracted from segmented Image 1 _cervical cancer

S.No	GLCM features name	Feature value cervical cancer Image 1
1	Contrast	1.80
2	Energy	0.102
3	Correlation	0.51
4	Entropy	2.6
5	Homogeneity	2.4
6	Shade	1.23
7	Prominence	2.32
8	Mean	0.81
9	Variance	0.32
10	Skewness	0.12
11	Kurtosis	0.213

Table 13.4 Features are extracted from segmented Image 1_cervical cancer

S.No	GLCM features name	Feature value cervical cancer image 2
1	Contrast	1.78
2	Energy	0.132
3	Correlation	0.4
4	Entropy	1.23
5	Homogeneity	1.4
6	Shade	1.03
7	Prominence	0.32
8	Mean	0.72
9	Variance	0.45
10	Skewness	0.12
11	Kurtosis	0.213

Table 13.5 Results of classification

Classifier	SVM	KNN	Definition
Sensitivity (TP/(TP + FN))	96.4	89.521	The cancer is classified correctly
Specificity (TN/(FP + TN))	92.1	86.31	Correctly diagnosed the cancer
Accuracy (TP + TN)/(P + N))	97.1	87.42	Rate for both cancer or non-cancer stages

have similar points that can be originate by near another one. Euclidean distance is calculated in the KNN. This machine learning algorithm is overfed for large dataset. Both algorithm classification works are compared through accuracy, sensitivity, and specificity. Table 13.5 shows the classification result.

From the results, the SVM classifier is performed well compared to the KNN classifier. Diagnosis of cancer stages based on features extraction from the segmented cancer images. The machine learning algorithms are improving the classification accuracy.

13.5 Conclusion

Cervical cancer is the second majority malignancy among women worldwide. This cancer originates in the cervix where forms closest to the uterus. Since cervical cancer patients typically exhibit few symptoms, early diagnosis is crucial to detecting the disease. In these cases, automated detection employing image processing algorithms is used to identify the malignancy. In this chapter, we select the cervical cancer images and detect the cancer region by using image processing algorithms such as existing method of k-means and then proposed FSVM. The FSVM method is performed well compared to the existing one. Based on quality assessments, the FSVM-based cancer area segmentation method outperforms the existing k-means segmentation strategy. Using k-means to cervical cancer images, the accuracy is 96.2% and the precision is 87%. Consequently, the accuracy value of the proposed FSVM segmented measurement is 98.6%, while the precision value is 89%. For picture 2 of cervical cancer, k-means yields accuracy of 94.4% and precision of 89%. This leads to a proposed FSVM segmented measurement value of 98% for accuracy and 97% for precision. Early detection of cancer was improved the patients outcomes. We selected the GLCM and histogram features from the segmented images for cancer-level diagnosis using support vector machine (SVM) and k-nearest neighbours (KNN) classifiers, as well as determining whether the stage is benign or malignant. To calculate the accuracy, sensitivity, and specificity are used for classification accuracy. We gained 97% accuracy while using SVM compared to the KNN machine learning algorithm.

References

Arbyn M, Xu L (2018) Efficacy and safety of prophylactic HPV vaccines. a Cochrane review of randomized trials. Expert Rev Vaccines 17(12):1085–1091. https://doi.org/10.1080/14760584.2018.1548282

Bai B, Liu P-Z, Du Y-Z, Luo Y-M (2018) Automatic segmentation of cervical region in colposcopic images using K-means. Australas Phys EngSci Med 41(4):1077–1085

Cai W, Chen S, Zhang D (2007) Fast and robust fuzzy c-means clustering algorithms incorporating local information for image segmentation. Pattern Recogn 40(3):825–838

Jaya BK, Kumar SS (2018) Image registration based cervical cancer detection and segmentation using ANFIS classifier. Asian Pacific J Cancer Prevent: APJCP 19(11):3203–3209

Kashyap D, Abhishek S, Jatin S et al (2016) Cervical cancer detection and classification using independent level sets and multi SVMs. In: Vienna: 39th international conference on telecommunications and signal processing (TSP), pp 523–528

Mattiuzzi C, Lippi G (2020) Cancer statistics: a comparison between World Health Organization (WHO) and Global Burden of Disease (GBD). Eur J Public Health 30:1026–1027

NCCC (2010) Cervical tumor. http://www.nccc-online.org/index.php/cervicaltumor

Scholkopf B, Smola AJ (2002) Learning with kernels: support vector machine, regularization, optimization, and beyond. MIT Press, Cambridge, Mass

Sharma M, Singh SK, Agrawal P, Madaan V (2016) Classification of clinical dataset of cervical cancer using KNN. Indian J SciTechnol 9(28). ISSN:0974-5645

Soumya MK, Sneha K, Arunvinodh C (2016) Cervical cancer detection and classification using texture analysis. Biomed Pharmacol J 9:123–134

Wei LZ, Mustafa WA, Jamlos MA, Idrus SZ, Sahabudin MH (2020) Cervical cancer classification using image processing approach: a review. IOP Conf Series: Mater SciEng 917:012068

Zhang S, Xu H, Zhang L, Qiao Y (2020) Cervical cancer: epidemiology, risk factors and screening. Chin J Cancer Res 32(6):720–728

Zur Hausen H (1989) Papillomaviruses in anogenital cancer as a model to understand the role of viruses in human cancers. Cancer Res 49:4677–4681

Chapter 14
AI/ML-Based Approaches in the Healthcare Industry

Abhishek Das, Saumendra Kumar Mohapatra, and Mihir Narayan Mohanty

Abstract Research on signal and image processing is a multi-disciplinary application that includes academia, engineering, industry, and medicine. It is progressing day by day. Out of all these applications, the input/desired data is either signal, image, or both. However, the work discussed in this chapter is medical image analysis for cancer detection. Recent technical advancements support healthcare through modern equipment for diagnosis. In most disease diagnosis cases, imaging has a significant role in representing the patients' physiological parameters. Different techniques need to be developed for accurate detection and diagnosis. Simultaneously, it has an essential role for patients who have cancer. AI technology handles most problems effectively and grows exponentially in this digital world. The critical point of this technology is machine learning techniques, represented as intelligent techniques. The growth in this area paves the way for many applications. In this chapter, the models are developed for the classification using modified variants of machine learning techniques. These are deep learning algorithm-based convolutional neural networks and ensemble neural networks.

Keywords Artificial intelligence · Machine learning · Medical signal processing · Image processing

---

A. Das
School of Engineering & Technology, Centurion University of Technology and Management, Paralakhemundi, Odisha, India

S. K. Mohapatra
Faculty of Engineering and Technology, Sri Sri University, Cuttack, Odisha, India

M. N. Mohanty (✉)
Department of Electronics and Communication Engineering, ITER, Siksha 'O' Anusandhan (Deemed to be University), Bhubaneswar, Odisha, India
e-mail: mihir.n.mohanty@gmail.com

A. Rai et al. (eds.), *Convergence of Artificial Intelligence, Machine Learning, and the Internet of Things in Industry 4.0 Applications*, Transactions on Computer Systems and Networks, https://doi.org/10.1007/978-981-96-5269-3_14

14.1 Introduction

By facilitating more precise diagnoses, more tailored treatment programs, and better patient outcomes, AI and ML have utterly transformed the healthcare industry. It aids in medical imaging, predictive analytics, and patient engagement. However, challenges like concerns of data privacy, compliance with regulations, and ethical issues need to be handled for responsible implementation in healthcare. Modern medical equipment in healthcare provides accurate results quickly for analysis, diagnosis, detection, and better treatment. Only physiological signals may not be sufficient to analyze, whereas medical images can be more helpful. Also, the need for only pictorial representation is practical for certain diseases. This tends to image analysis through computational techniques. Many systems are developed from X-rays to acquire physiological parameters in the form of images. According to WHO, 10 million deaths occurred in 2020, which is nearly one in six deaths (http://gco.iarc.fr/; https://gco.iarc.fr/today/online-analysis-table). Any abnormal tissue growth is termed a tumor and can be hidden from the immune system. These may be of either benign or malignant types. Benign tumors may grow into large sizes without spreading to nearby tissues and are non-cancerous, whereas malignant tumors are spreading types that lead to cancer. Imaging techniques play a significant role in cancer analysis and diagnosis. Cancer is a genetic abnormality caused by mutations in the genes that control how our cells behave, specifically how they grow and divide. It accumulates several chromosomal alterations, such as chromosome duplications and deletions. This disease damages DNA due to several causes including radiation, such as harmful UV rays from the sun (initiating skin cancer) (https://www.cancer.gov/about-cancer/understanding/what-is-cancer; Gefeller and Diehl 2022; Gruijl 1999). Research in medicine, with the support of technology, is growing exponentially. Observing the present scenario of society, like heart disease, cancer diagnosis needs to be studied well.

In recent decades, the use of machine learning techniques has improved the efficiency of medical imaging effectively. Considerable developments in deep learning-based technologies in medical image examination are found. This motivated me to carry forward the research in biomedical image classification through variants of machine learning techniques for automated and intelligent diagnosis systems.

- Detection of tumors and cancers from gene expression has little application. Performance improvement is another issue for disease detection from gene expressions. This area of research needs to be explored more.
- Though the statistical machine learning models have been proposed earlier, the scope is to design a suitable robust classifier. Similarly, the raw data for processing needs to be well organized for faster processing with better accuracy.

This chapter considers three types of cancers for detailed analysis and discussion.

(i) Brain malignant tumor detection for brain MRI.
(ii) Breast cancer detection from histopathology images.
(iii) Skin cancer detection from dermatoscopic images.

These three types of cancers are analyzed from their gene expressions. For this purpose, the one-dimensional (1D) data is considered to be a two-dimensional (2D) image, as the models are based on a convolutional neural network (CNN). The following methods for cancer detection are adopted in this chapter, and the explanation and discussion are detailed.

(i) Gene analysis for three different types of cancer.
(ii) Use deep learning techniques to design a suitable cancer detection and classification model. CNN is utilized for three different cancers.
(iii) Design of ensemble classifier through adaptable algorithms.
    In this approach, the models are chosen with homogeneous and heterogeneous types.
(iv) Fuzzy ensemble classifier developments with the deep learning approach.
    Neural network systems are ensembled with fuzzy rule-based models. The final stage classification model is utilized with either neural network or fuzzy models.

Initially, the most informatic gene data for cancer detection is analyzed with image representation. Further, cancer with and without tumors is considered for analysis in the proposed techniques. Therefore, three different cancers are considered: brain cancer through tumor detection and breast cancer through cyst detection. The non-cystic type of cancer is considered skin cancer. Each cancer was experimented with through each proposed model, and substantial improvement was found in detection and diagnosis.

Deep learning-based models provide efficient results for image data found in earlier works. Gene data is of 1D type and is converted to two-dimensional data for image representation using the Convex Hull algorithm and three data reduction techniques named multi-dimensional scaling (MDS) along with t-distributed Stochastic neighbor embedding (t-SNE) and locally linear embedding (LL-Embedding). The synthetic images are further analyzed through single and ensemble CNN models for cancer recognition. Ensemble learning is used in designing models for brain tumor detection rather than single models. Modification in ensemble models is done by considering both neural network and fuzzy-based models meta-classifiers. Multilayer perceptron (MLP) and fuzzy min–max (FMM) models are stacked to the ensemble models. The development of heterogeneous and homogeneous ensemble learning algorithms for breast cancer diagnosis using histopathology pictures is suggested. In this analysis, MLP and fuzzy ARTMAP are used as meta-classifiers. The recurrent ensemble learning model classified the breast histopathology images accurately. Class imbalance in the dermatoscopic image dataset is overcome using data augmentation and SMOTE. A fuzzy-based ensemble model is proposed, considering ARTMAP and FMM as base classifiers, whereas a neuro-fuzzy classifier (NEFCLASS) is used to classify skin cancer detection. Performance improvement using ensemble learning models in combination with fuzzy models paves a new way of analyzing medical image processing for cancer detection (Kumar et al. 2010).

Conversion of gene data to images and processing by CNN models is described in Sect. 14.2. Section 14.3 thoroughly analyzes the models designed for brain tumor recognition. A detailed analysis of the performance obtained using homogeneous and heterogeneous ensemble models in breast cancer detection is provided in Sect. 14.4. Skin cancer detection using is discussed in Sect. 14.5. Section 14.6 concludes the work and suggests the scope for future research.

14.2 Gene Analysis for Cancer Classification Using Image Processing Techniques

Disease detection from images in an efficient way for analysis and diagnosis. Medical imaging, especially for cancer analysis, has a high value. Also, the intelligent model chosen for this work performs better than the signal input. The 1D data is the root of the disease for genetic prediction. Therefore, the corresponding gene is considered for this work.

Further, it is converted to represent through images for faster and better analysis. Due to this purpose, the objective is to convert 1D–2D data. Three data reduction techniques, including the Convex Hull algorithm, are used for such purposes.

Deep learning models require massive data for their training. The size of the 1D genes dataset is not enough to train a deep learning model. The dataset increase task is done using the following procedures. Each sample of data is reduced using techniques such as t-SNE, MDS, and LLE. The Convex Hull algorithm creates boundaries, as illustrated in Fig. 14.1 using t-SNE.

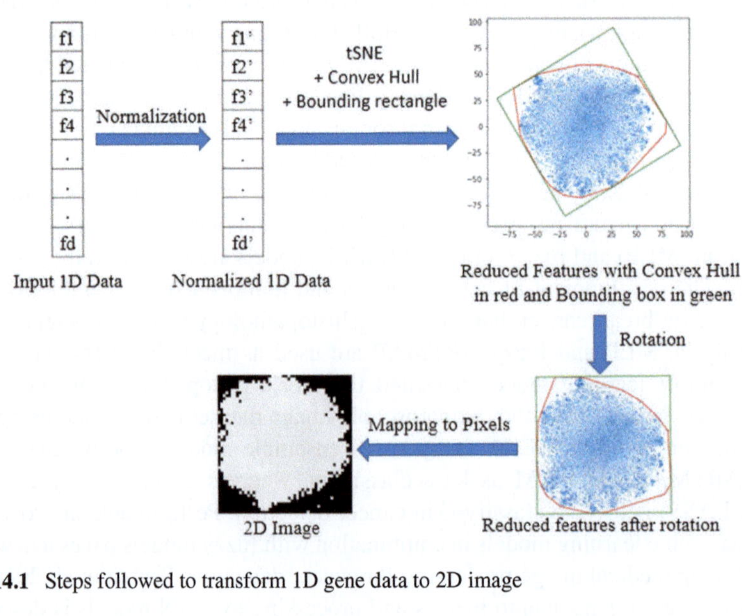

Fig. 14.1 Steps followed to transform 1D gene data to 2D image

Other data reduction techniques mentioned earlier are applied similarly to generate images from gene data. Mean normalization (Eq. 14.2.1) is applied to normalize 1D data.

$$f' = \frac{f - \text{average}(f)}{\max(f) - \min(f)}. \tag{14.2.1}$$

14.2.1 Brain Gene Data to Image Conversion and Classification Using Flexible Deep Learning Techniques

For brain tumor classification, the proposed stacked ensemble model is fed with images produced from 1D data of gene expressions. Three CNNs are trained in parallel by the synthetic image datasets. Multilayer perceptron (MLP) is trained using the results from an ensemble of three CNN models, and it performs the final classification in our work. Figure 14.2 depicts the schematic portrayal of the suggested model.

The architecture of each CNN model is given in Table 14.1 for a two-class classification.

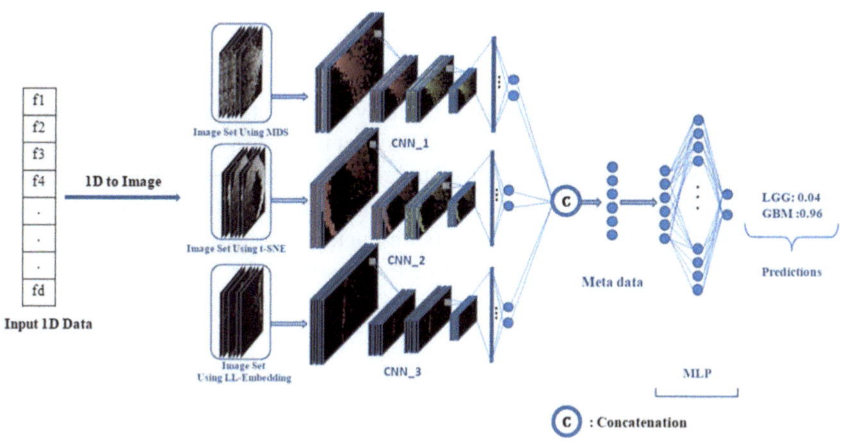

Fig. 14.2 Architecture of the proposed stacked ensemble model

Table 14.1 Parameters of proposed CNN architecture

Layer (type)	Output shape
Convolutional (2D)	(50 × 50, 32)
Max-pooling (2D)	(25 × 25, 32)
Convolutional (2D)	(25 × 25, 64)
Max-pooling (2D)	(12 × 12, 64)
Flatten	9216
Dense	2

14.2.1.1 Dataset Description

Deep learning-based models require training on considerable datasets to achieve high detection, prediction, or classification accuracy. A gene expression dataset accessible to the public is considered to classify low-grade gliomas (LGG) and glioblastoma multiforme (GBM). The dataset includes TCGA-GBM and TCGA-LGG gene data in the National Cancer Institute's Genomic Data Commons (GDC) site (https://portal.gdc.cancer.gov/repository). There are 166 RNA sequences in the TCGA-GBM dataset and 512 in the TCGA-LGG dataset. Each gene expression contains 60483 profiles.

14.2.2 Breast Cancer Detection from Gene Data with an Ensemble Learning Method

The suggested method improves performance by employing both the dataset and the decomposed forms of the same. Empirical wavelet transform (EWT) and variational mode decomposition (VMD) are utilized to break down the sample images and train the models at the lower level. There are two stages of this classification process. Three CNNs serve as the fundamental classifiers for the initial stage of classification. MLP is employed in the second stage as the meta-classifier. The proposed steps with a sample image prediction are shown in Fig. 14.3.

14.2.2.1 Breast Cancer Gene Dataset Description

This work is developed to process gene data related to breast cancer (Xie et al. 2016). There are a total of 590 samples. Five hundred twenty-nine images are cancerous, and the remaining 61 images belong to the normal category. Each sample of the gene dataset includes 17814 genes.

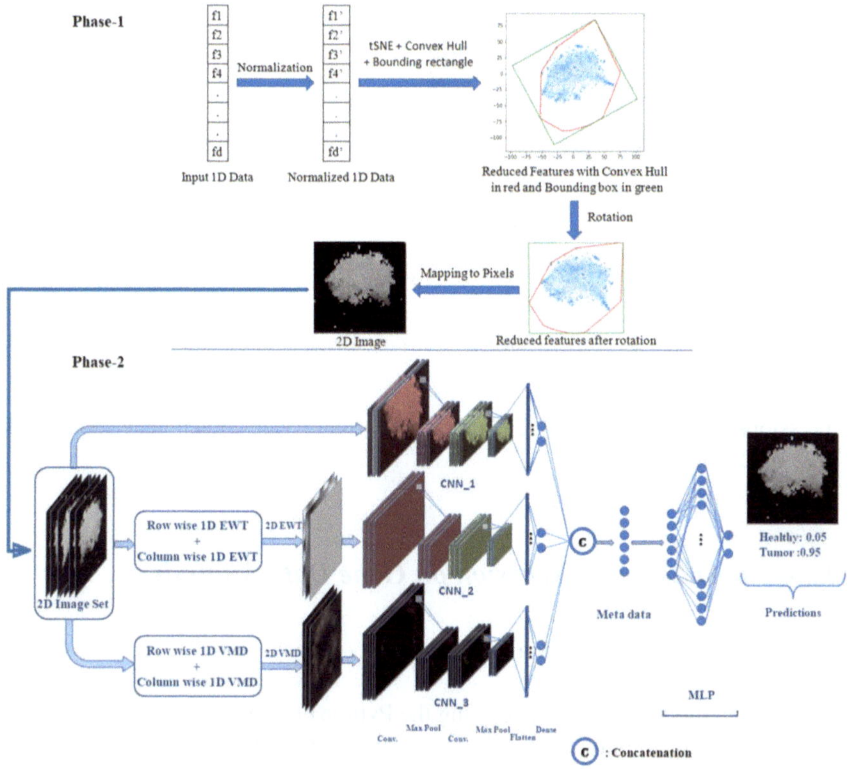

Fig. 14.3 Workflow diagram of the proposed 1D–2D conversion and classification model

14.2.3 Converting One-Dimensional Data Into a Two-Dimensional Image Through a Stochastic Mapping Technique for the Secure Classification of Skin Lesions

Merkel cell carcinoma, squamous cell carcinoma, and basal cell carcinoma are the three main types of skin malignancies, including melanoma and non-melanoma varieties. The architecture of the suggested technique is shown in Fig. 14.4 together with a sample of data and prediction from the model.

14.2.3.1 Dataset Description

The NCBI's Gene Expression Omnibus (GEO) database (Jee et al. 2015) is used to detect non-melanoma skin cancers from gene expressions. The dataset has 16 gene

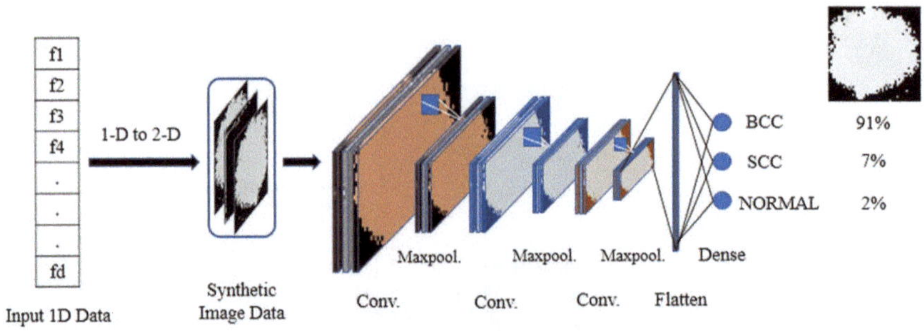

Fig. 14.4 Workflow of the proposed method, including a sample data prediction

expressions for BCCs, 5 for SCCs, and 5 for normal tissues. Each gene sequence is 47323 base pairs in length.

14.2.4 Results and Discussion on Gene Data Analysis

14.2.4.1 System Specification

The proposed models are designed using the Python environment on Google Colaboratory, providing an online graphics processing unit (GPU) for faster execution.

14.2.4.2 Results Obtained from Gene Analysis for Brain Tumor Detection

Three distinct data reduction methods are employed to produce three varieties of synthetic image datasets. The results are, therefore, different. These reduced data points are marked with blue color. After applying the data reduction methods, Convex Hull is used to wrap the data points within a polygon and marked as the red color boundary. The image shape is given to the data points using a rectangular bounding box marked as green. The corresponding sample results for brain gene expressions are shown in Fig. 14.5.

The coordinates created using the abovementioned steps are assigned pixel values, and the results obtained in this step are the required images. The samples of synthetic images are shown in Fig. 14.6 for brain gene data, respectively.

Three CNN models are trained with the synthetic datasets. The MLP does the final classification taught with the outcomes of these three CNNs. The accuracy obtained in the detection of brain tumors is 97.38%.

The proposed models are evaluated using certain constraints such as accuracy, recall, precision, F1-score, sensitivity, and specificity using Eqs. (14.2.2)–(14.2.7).

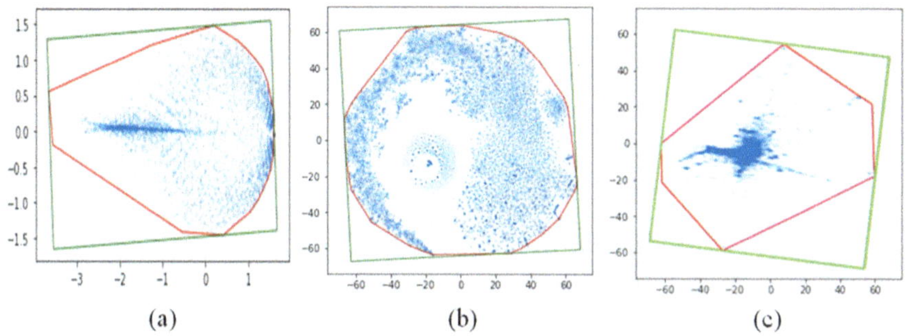

Fig. 14.5 Outputs of convex Hull algorithm on **a** MDS, **b** t-SNE, and **c** LL-embedding of gene expressions

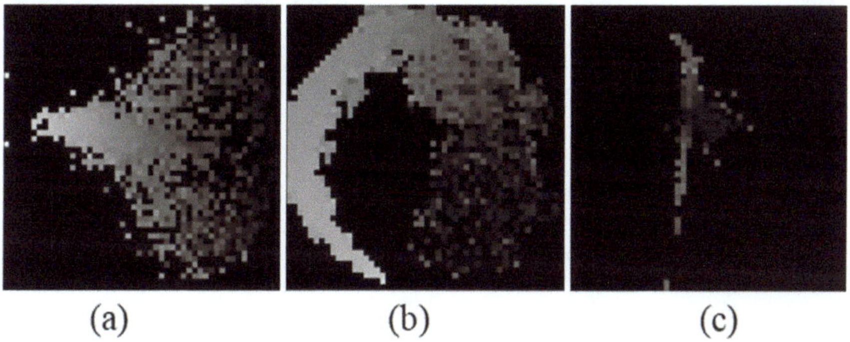

Fig. 14.6 Outputs of gene expression using **a** MDS, **b** t-SNE, and **c** LL-embedding

$$\text{Accuracy} = \frac{\text{TP} + \text{TN}}{\text{TP} + \text{TN} + \text{FP} + \text{FN}} \times 100 \tag{14.2.2}$$

$$\text{Precision} = \frac{\text{TP}}{\text{TP} + \text{FP}} \times 100 \tag{14.2.3}$$

$$\text{Recall} = \frac{\text{TP}}{\text{TP} + \text{FN}} \times 100 \tag{14.2.4}$$

$$F1\text{Score} = 2 \times \frac{\text{Recall} \times \text{Precision}}{\text{Recall} + \text{Precision}} \tag{14.2.5}$$

$$\text{Sensitivity} = \frac{\text{TP}}{\text{TP} + \text{FN}} \tag{14.2.6}$$

$$\text{Specificity} = \frac{\text{TN}}{\text{TN} + \text{FP}}. \tag{14.2.7}$$

These equations evaluate the proposed models described in all the sections.

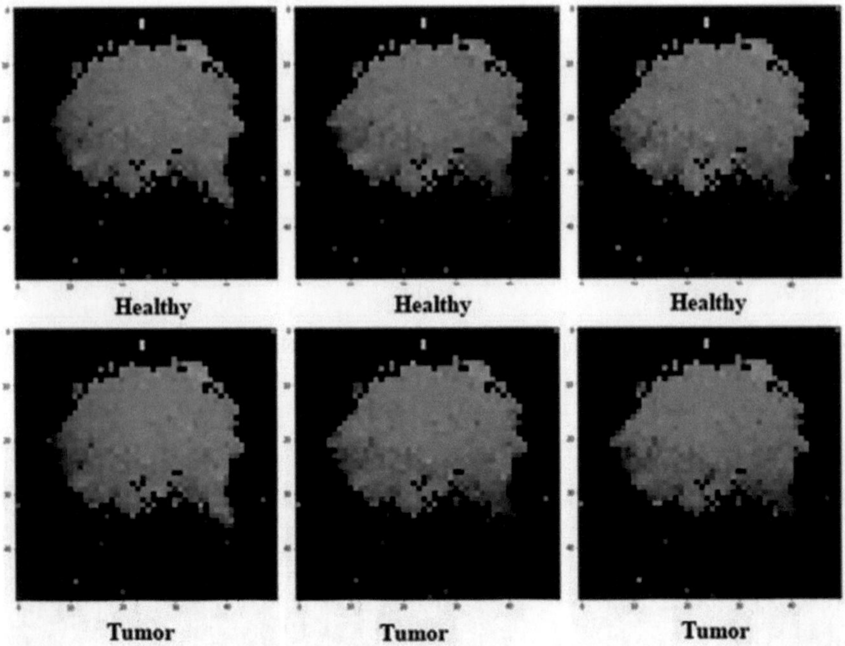

Fig. 14.7 Examples of the synthetic gene image dataset

14.2.4.3 Results Obtained from Gene Analysis for Breast Cancer Detection

To test how well the suggested stacked ensemble model works, we use artificial image datasets made from breast cancer-related one-dimensional gene expression data. Figure 14.7 displays examples of each image type in the synthetic gene image dataset. 80:20 split is used to create a train set and test set from the image dataset.

Different modes resulted using EWT are shown in Fig. 14.8.

Figure 14.9 displays the different modes VMD generates from the gene data.

The findings derived from the synthetic gene image dataset are outlined in Table 14.2.

14.2.4.4 Results Obtained from Gene Analysis with Skin Data

Gene data conversion into images is achieved by utilizing three distinct data reduction techniques in the DeepInsight framework. Figures 14.10, 14.11, and 14.12 depict the images created using t-SNE, LLE, and MDS, respectively.

The CNN model is trained using images obtained via three distinct data reduction strategies. The Model had 96.32% training accuracy and 92.13% testing accuracy.

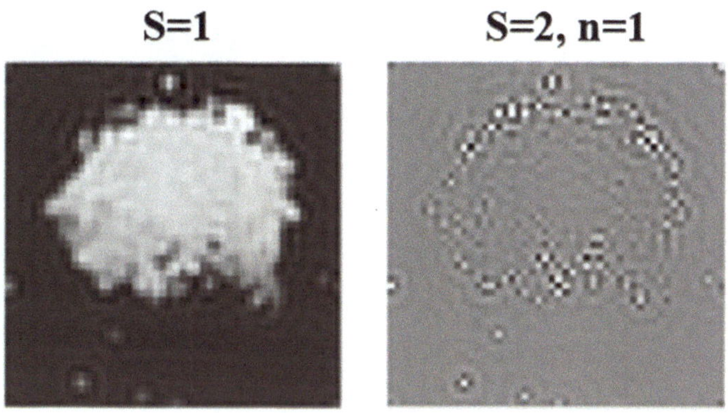

Fig. 14.8 Decomposed modes obtained through empirical wavelet transform (EWT) from synthetic gene images

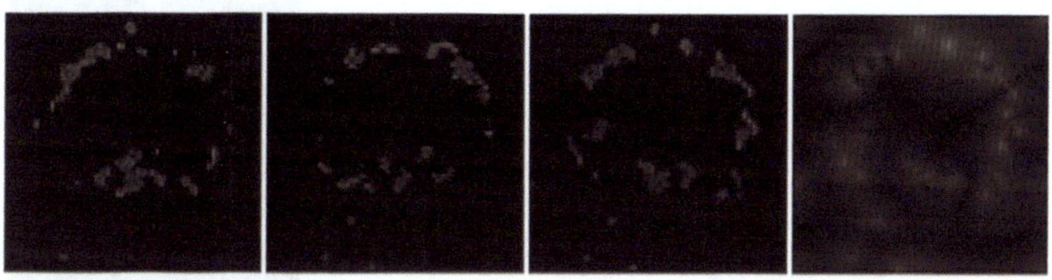

Fig. 14.9 Modes decomposed through VMD from synthetic gene images

Table 14.2 Evaluating the performance of the suggested model by utilizing two datasets

Dataset	Accuracy (%)	F1-score	Recall	Precision	Sensitivity	Specificity
Synthetic gene dataset (tumor vs. healthy)	95.52	0.9444	0.9455	0.9432	0.9444	0.9583

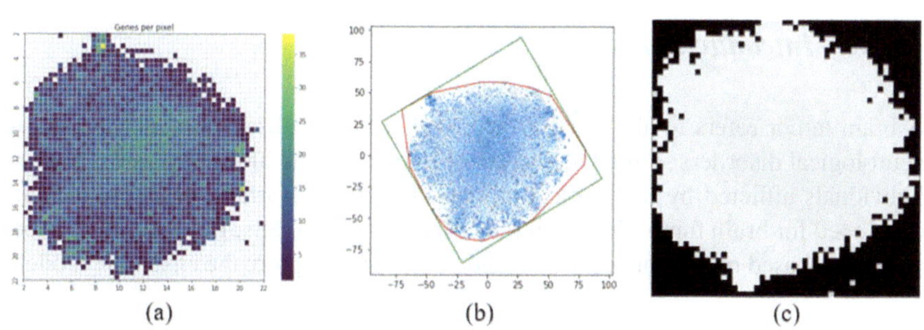

Fig. 14.10 Outputs of **a** t-SNE, **b** Convex Hull, and **c** image

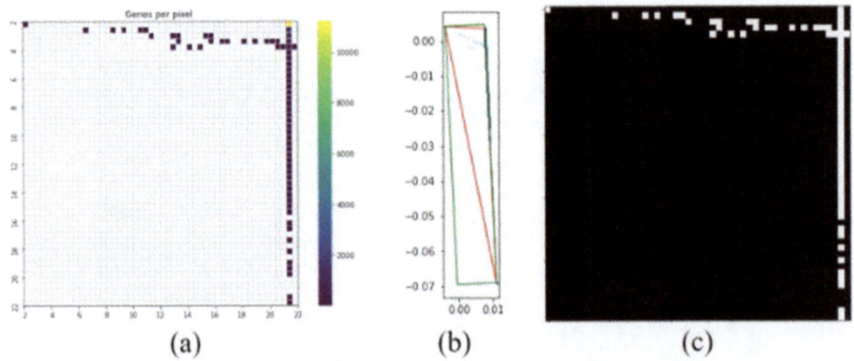

Fig. 14.11 Outputs of **a** LLE, **b** Convex Hull, and **c** image

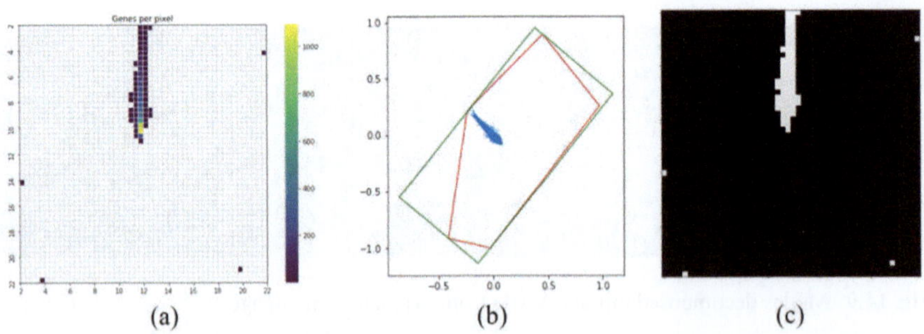

Fig. 14.12 Outputs of **a** MDS, **b** Convex Hull, and **c** image

14.3 Brain Tumor Detection Using Ensemble Deep Learning Technique

14.3.1 Introduction

A brain tumor refers to abnormal growth within brain cells, potentially leading to neurological disorders. Timely detection of tumors is crucial for saving the lives of individuals afflicted by this critical ailment. Ensemble learning-based models are developed for brain tumor detection rather than single models. Both neural network and fuzzy-based models are proposed and observed to explore the ensemble models.

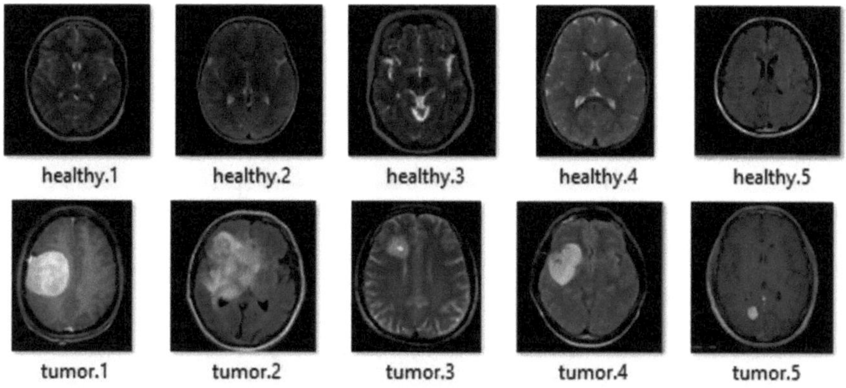

Fig. 14.13 Sample images of the brain MRI dataset

14.3.2 Dataset Description

A dataset is essential for training and testing the models to distinguish between brain tumors and healthy brain images. This dataset is sourced from Kaggle and is publicly accessible (Chakrabarty 2019). A glimpse of the training dataset is depicted in Fig. 14.13. The included magnetic resonance imaging (MRIs) includes a mix of T1-weighted, T2-weighted, and flair-type images. The training process takes up 80% of the dataset, while model testing takes up the other 20%. Also, there are 80:20 splits between the training set and the validation set.

14.3.3 The Utilization of an Optimized Extreme Gradient Boosting Ensemble Classifier for Brain Image Classification

The Extreme Gradient Boosting (XGBoost) model is configured with several decision trees. By tuning the number of trees and learning rate to various values, the model is effectively optimized for classifying brain MRI image. The block diagram illustrating the proposed model is presented in Fig. 14.14.

14.3.3.1 XGBoost Classifier

In XGBoost, trees are sequentially added to the parent model to refine predictions and rectify errors from previous trees. The performance of the XGBoost classifier is assessed based on cross-entropy loss, also referred to as log loss. A lower log loss value indicates a more accurate classification.

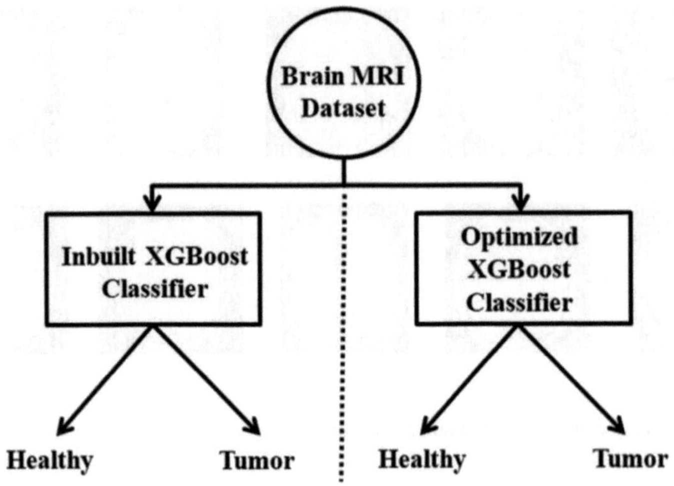

Fig. 14.14 Diagram depicting the proposed method's blocks

$$\text{log loss} = y\ln(p) + (1 - y)\ln(1 - p), \tag{14.3.1}$$

where $= \frac{1}{1+e^{-x}}$, y is the real value in $\{0, 1\}$, and p is the probability of matching evaluated by the Model.

14.3.3.2 Optimized XGBoost Classifier

The objective function ($L^{(m)}$) of XGBoost is tuned.

$$L^{(m)} = \sum_{i=1}^{k} l(y_i, \widehat{y_i}^{(m-1)} + f_m(x_i)) + \Omega(f_m), \tag{14.3.2}$$

where $\Omega(f) = \gamma T + \frac{1}{2}\lambda||\omega||^2$, l is the loss function, m represents the number of trees, y_i and $\widehat{y_i}^{(m-1)}$ are the target and predictions, respectively, Ω represents the Model's complexity, which is contingent on the sampling rate and the number of features, T is the number of leaves in a tree, λ is a constant, ω represents leaf weights

14.3.4 Classification of MRIs of the Brain Using an Ensemble Model

CNN, RNN, and LSTM models serve as base classifiers for feature extraction and initial classification stages. The predictions generated by these base classifiers are

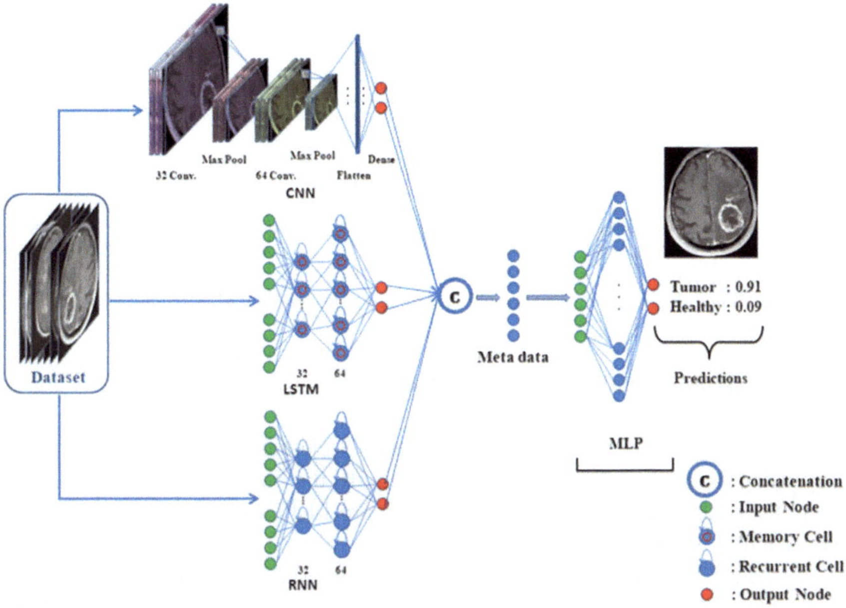

Fig. 14.15 System architecture of the proposed model

then forwarded to an MLP for the next training phase, and Fig. 14.15 depicts the layout of the suggested model, accompanied by a sample prediction.

This method employs a CNN model that has two layers, one with 32 filters and the other with 64 filters. The activation function known as the rectified linear unit (ReLU) operates on these levels. Each layer is followed by a max-pooling layer to shrink the filtered features to half with stride 2. The flattened layer combines all the features as a single linear vector. The last dense layer is designed with two nodes triggered by the softmax activation function for a two-class classification. RNN and LSTM models are designed with the same structure as CNN without any max-pooling layer. Model error functions are evaluated with the help of the binary cross-entropy loss function, and the model is optimized with the help of the Adam optimizer. The number of epochs is set to 20 as models provide the highest accuracy in this range.

14.3.5 Development of a Deep Ensemble Model with a Fuzzy Decision Technique for Brain MRI Image Classification

The diagram in Fig. 14.16 shows the block diagram of the proposed method.

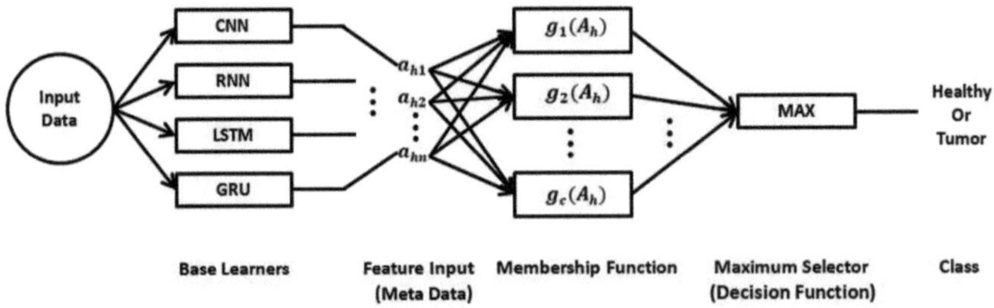

Fig. 14.16 Block diagram of the suggested ensemble model with FMM as meta-classifier

14.3.5.1 Training of Base and Meta-Classifiers

As shown in Fig. 14.18, the fuzzy model takes an 8-element vector as input and trains itself using the results of the basic models. Finally, the FMM classifier's last layer is where the actual classification takes place. Figure 14.17 displays the FMM model's layer configurations.

When the FMM model is being trained, the feature vector Ah is formed by merging the output vectors from the basis learners. This is done in order to train the model. The feature vector consists of label vectors y_i and value vectors Z_i, where Z_i ranges from 0 to 1 (i.e., $Z_i \in (0,1)$). Using these features, the FMM model training begins, and the relevant layer FB is updated to include new hyperbox nodes. Each hyperbox, denoted as $B_j = \{V_j, W_j, y_j\}$, consists of the minimum point (V_j), maximum point (W_j), and the corresponding class label (y_j). The hyperbox is established based on the membership function b_j, as delineated by Eq. (14.3.3).

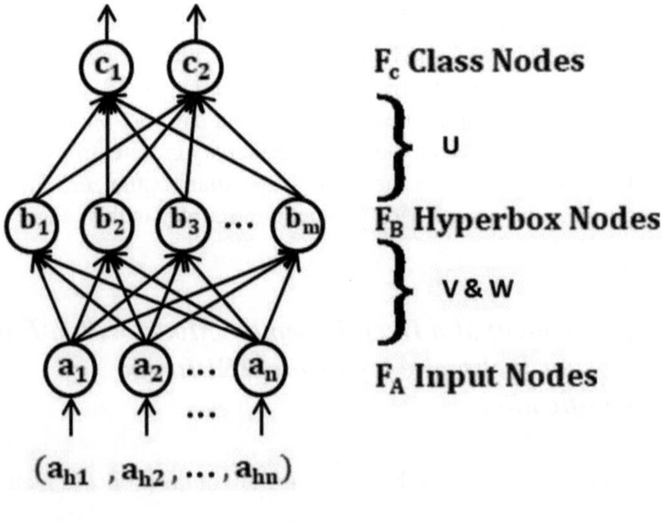

Fig. 14.17 Three-layer fuzzy neural network

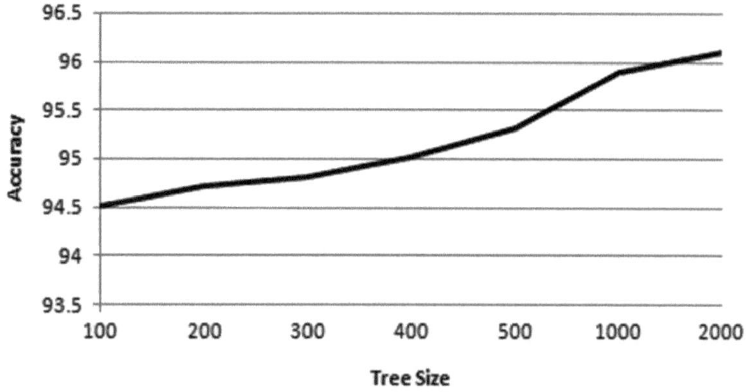

Fig. 14.18 Plot illustrating the accuracy with varying tree sizes

$$b_j = \frac{1}{2n}\sum_{i=1}^{n}[\max(0, 1 - \max(0, \gamma \min(1, a_{hi} - w_{ji})))$$
$$+ \max(0, 1 - \max(0, \gamma \min(1, v_{ji} - a_{hi})), \qquad (14.3.3)$$

where $V_j = (v_{j1}, v_{j2} \ldots, v_{jn})$ represents the minimum points and $W_j = (w_{j1}, w_{j2}, \ldots w_{jn})$ denotes the maximum values for the hyperbox B_j, A_h and B_j's distance determines the variance in the membership function, and the sensitivity value γ that governs this variation is denoted by. In order to train the FMM classifier, these two-dimensional hyperboxes are used.

Using the expansion criterion given in Eq. (14.3.4), the hyperbox is extended to define the boundary between two classes.

$$n\theta \geq \sum_{i=1}^{n}(\max(w_{ji}, x_{hi}) - \min(v_{ji}, x_{hi})). \qquad (14.3.4)$$

If the criterion above is satisfied, the expansion of the minimum and maximum points is executed according to Eqs. (14.3.5) and (14.3.6).

$$v_{ji}^{\text{updated}} = \min(v_{ji}^{\text{priveous}}, x_{hi}) \text{ for, } i = 1, 2, \ldots n \qquad (14.3.5)$$

$$w_{ji}^{\text{updated}} = \max(w_{ji}^{\text{old}}, x_{hi}). \qquad (14.3.6)$$

14.3.5.2 Method of Classification

Closeness to the membership function determines the class prediction, b_j. The components of the vector U, which represents the links between the last two stages of the FMM, are represented by

$$u_{jy} = \begin{cases} 1 \text{ if } b_j \in \text{ class } y \\ 0 \text{ Otherwise.} \end{cases} \tag{14.3.7}$$

In order to get the anticipated class of the matching data, the final layer of FMM, F_c, is used. The F_c layer's nodes include the values acquired by,

$$C_y = \max_{j=1}^{m} b_j u_{jy}. \tag{14.3.8}$$

Similar to layer FB, the classification technique is described in Algorithm 14.1, where m is the amount of hyperbox nodes.

Algorithm 14.1 Test data classification

1. **Input**: $T_i = \{x_i, y_i\}_{i=1}^{n}$, the test image
2. Generate A_h
3. Build Hyperbox
4. Compare with trained hyperbox
5. Calculate the class by applying Eqs. (14.3.7) and (14.3.8)
6. **Output**: Predicted class

The next part presents and elaborates upon the experimental results obtained using the suggested technique.

14.3.6 Results and Discussion

14.3.6.1 System Specifications

Latest version of Python on the Google Colaboratory is used to create the suggested models, which allows for convenient access to online GPU/TPU resources.

14.3.6.2 Outcomes of the Proposed Optimized XGBoost Classifier

Classification by XGBoost Classifier

The accuracy achieved concerning the number of trees is depicted in Fig. 14.18.

As the number of decision trees in the XGBoost model increases, classification performance improves. Achieved accuracies are 94.5%, 95.3%, and 96.1% for 100,

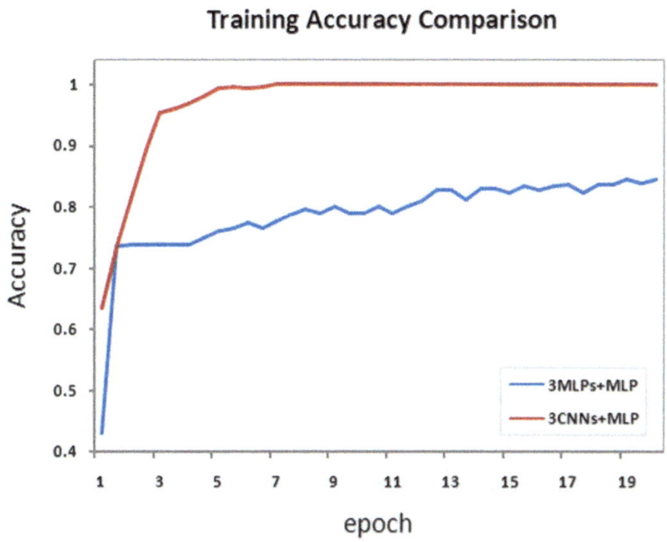

Fig. 14.19 Training accuracy comparison

500, and 2000 tree sizes, respectively. Optimal combinations of tree number and learning rate are considered for enhancing the XGBoost classifier's performance.

14.3.6.3 Results Obtained from Deep Ensemble Model with Meta-Classifier MLP

Preprocessing

The images within the dataset vary in size. To mitigate computational complexity and reduce processing time, all images are resized to dimensions of 50 × 50. Labels are then assigned to these resized images for the train and validation sets. The test dataset contains no labels to verify the proposed Model.

The training accuracies obtained from both models are shown in Fig. 14.19.

The training and validation accuracies provided by the base models CNN, RNN, LSTM, and stacked ensemble model are shown in Fig. 14.20.

Table 14.3 presents the accuracy, F1-score, precision, recall, sensitivity, and specificity.

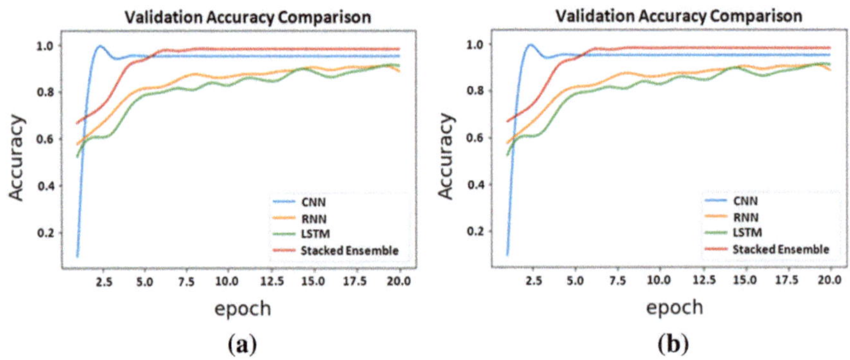

Fig. 14.20 Comparison plot for **a** training accuracy, **b** validation accuracy

Table 14.3 Evaluation parameters of the proposed model

	Precision	Recall	F1-score	Support	Sensitivity	Specificity	Accuracy
Healthy (0)	0.93	1.00	0.96	14	1.00	0.96	0.97
Tumor (1)	1.00	0.96	0.97	28			

14.3.6.4 Results Obtained from Deep Ensemble Model with Meta-Classifier FMM

Hyperboxes are created using various values of the hyperparameter θ, and a subset of these hyperboxes generated by the FMM during training over brain MRI scans is illustrated in Fig. 14.21.

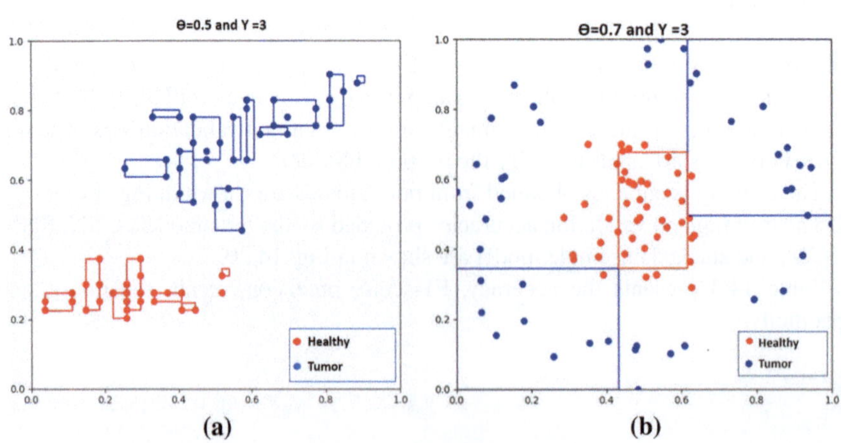

Fig. 14.21 Hyperbox formation for **a** $\theta = 0.5$ and **b** $\theta = 0.7$

Increase in θ increases the number of points in the hyperplane. The hyperplane points are notably low when θ is set to smaller values, such as 0.1 and 0.3. However, with a θ value of 0.5, the classification accuracy is observed to be higher. When θ is set to 0.7, even though the quantity of points in the X–Y plane continues to increase, there is also noticeable overlapping, adversely impacting performance. Conversely, at $\theta = 0.5$, clusters are distinct without overlap, leading to accurate class separation. The proposed model achieves a test accuracy of 97.62% in brain tumor classification using the Kaggle dataset.

14.3.6.5 Comparison of Performance with State-of-the-Art Models

Evaluating parameters of the proposed models are summarized with a comparison to earlier works in tumor detection from brain MRI in Table 14.4.

Table 14.4 shows that the proposed models provide better classification in terms of accuracy, recall, F1-score, sensitivity, and specificity for brain image classification compared to state-of-the-art models.

14.4 Summary

Improved model design with fewer preprocessing steps is the main moto of this research. This is achieved with the ensemble learning technique. By obtaining greater performance in comparison with individual base classifiers that use single architectures, the hybrid model that was presented in this work demonstrates that ensemble learning is an effective method of learning. The effect of some deep learning-based classifiers is analyzed while designing the ensemble model. Two different meta-classifiers, MLP and FMM, are stacked to aggregate the variance among the base classifiers. Combining deep learning with fusion techniques using a fuzzy min–max classifier represents a novel approach to biomedical image classification. Section 14.4 proposes and discusses the utilization of ensemble learning models for histopathology image classification in breast cancer detection.

14.5 Breast Cancer Detection Via Ensemble Learning Technique

14.5.1 Introduction

Exponential growth in tumors within the living body may lead to cancer. Breast cancer is one of the examples of this category. CT scans, mammograms, and X-rays provide the structure of the tumor inside the breast. However, histopathology images

Table 14.4 Comparison between the proposed method and state-of-the-art approaches

References	Model architecture	Precision	Recall (in %)	Accuracy (in %)	F1-score	Sensitivity	Specificity
Özyurt et al. (2019)	Randomly generated graph-based CNN	–	–	95.49	–	–	–
Ge et al. (2020)	3D CNN	0.818	–	–	0.952	0.947	–
Huang et al. (2020)	Transfer learning based on CNN	0.940	0.887	–	0.913	–	–
Ker et al. (2019a)	Ensemble of 2 CNNs	0.94	–	96	–	0.98	0.94
Ker et al. (2019b)	Fuzzy-sure entropy + CNN + SVM	–	–	95.62	–	0.9625	0.95
	Fuzzy-sure entropy + CNN + KNN	–	–	90.62	–	0.90	0.9125
Ertosun and Rubin (2015)	GAN for dataset enlargement + CNN for classification	–	–	88.2	–	0.7999	0.9217
Huda et al. (2016)	Decision tree + bagging	0.658	–	–	0.648	–	–
Proposed methods	Optimized XGBoost model	0.92	0.94	96.1	0.95	1.00	0.945
	{CNN, RNN, LSTM} + MLP	**0.965**	0.98	97	**0.965**	1.00	0.96
	{CNN, RNN, LSTM, GRU} + FMM	0.9286	**1.00**	**97.62**	0.96	**1.00**	**1.00**

provide a molecular view, and they help distinguish the types of cancer and the grade of severity by observing the cell conditions. Tissue architecture, cellular distribution, and structural inconsistencies in individual cells are all factors considered for the evaluation. Manual disease diagnosis is less accurate, depending on the pathologist's experience. Breast histopathology images validate the classification models developed for brain MRI images. These models could not perform well due to their structural complexity compared to brain MRI images described in Sect. 14.3. Accurate, reliable, and early cancer detection is proposed based on ensemble learning that overcomes the limitations of manual diagnosis, including X-ray imaging techniques.

14.5.2 Dataset Description

The breast histopathology image dataset used for different proposed models is taken from the Kaggle database (Janowczyk and Madabhushi 2016). The details of the data are described in this section. The collection includes 198,738 photos that are not classified as IDC and 787,860 images that are classified as IDC. Class values for Non-IDC and IDC are denoted by 0 and 1, respectively. Samples from each category are illustrated in Fig. 14.22. With 80% of the dataset allocated for model training, the remaining 20% is reserved for the purpose of testing. Furthermore, the training set is subdivided into training and validation subsets, maintaining an 80:20 ratio.

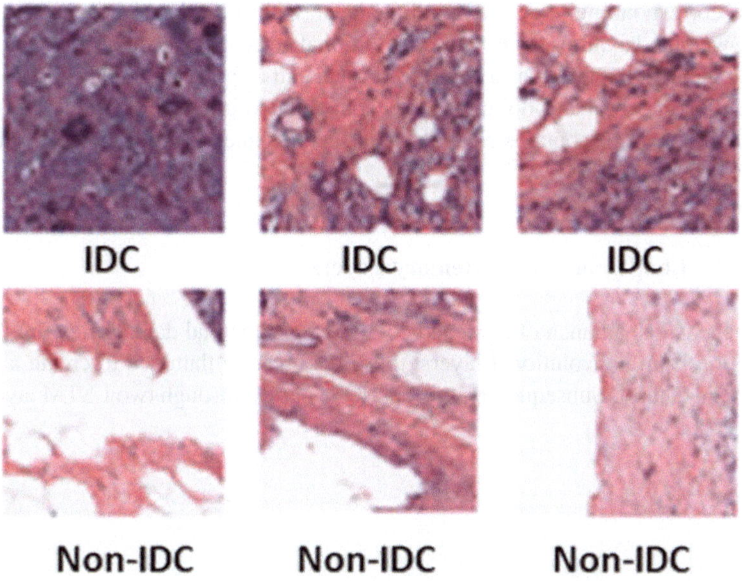

Fig. 14.22 Samples of the breast histopathology image dataset

14.5.3 *Detection of Breast Cancer Through a Concatenated Deep Learning Model*

Single models process the images using a single learning method. The generated features become limited, and decision-making is poor. Series and parallel are the two possible ways of combining different models. A hybrid architecture comprising convolutional and LSTM layers is proposed and arranged sequentially that extracts both spatial and temporal features, enhancing the classification accuracy. The structure of the proposed method is depicted in Fig. 14.23, alongside a sample prediction. In this instance, the sample input image is classified as non-IDC, as the node assigned to non-IDC exhibits the highest prediction match.

14.5.3.1 Convolutional Layers

The input images undergo processing through convolutional layers, where they are filtered. A total of 32, 64, and 32 filters are utilized in each of the three convolutional layers, correspondingly. A max-pooling layer with a stride of 2 is used to pool highly weighted features from each convolutional layer. This successfully reduces the filtered pictures to half their original size. Each and every feature that was produced by the most recent max-pooling layer is consolidated into a single linear vector by a flattening layer. It contains all the spatial features and is further processed by the LSTM layers.

14.5.3.2 Long Short-Term Memory Layers

An image is just a bunch of rows and columns of sequential data. Following feature extraction from convolutional layers, these features are flattened to create a single sequential dataset. Subsequently, this dataset is passed through two LSTM layers for

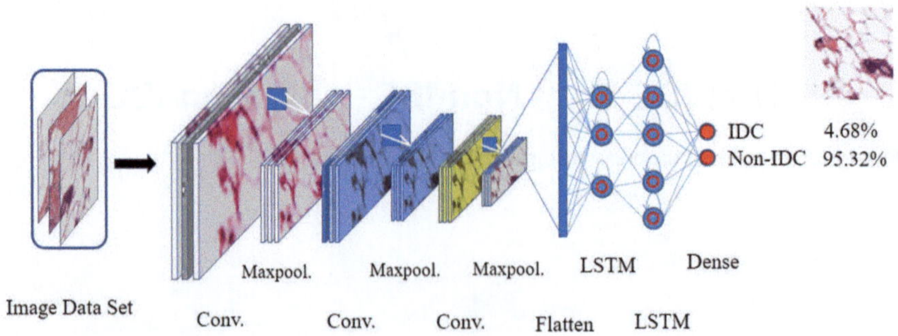

Fig. 14.23 Architecture of the proposed technique, along with a forecast of some sample data

temporal feature extraction. In the two consecutive LSTM layers, 32 and 64 LSTM cells are employed, respectively. Classification occurs in the output layer; a dense layer is formed by activating two nodes using the softmax activation algorithm. The model employs binary categorical cross-entropy as its loss function and utilizes the Adam optimizer to optimize the mathematical model.

14.5.4 Breast Cancer Detection Using LSTM-Based Ensemble Learning Techniques

While CNN models are prevalent in biomedical image processing and classification tasks, there is a need to explore alternative deep learning techniques for breast cancer detection. In this context, two different models are considered for classifying breast histopathology images. Initially, a single LSTM model is designed for training and classification. However, the classification accuracy achieved by the single model was deemed insufficient. Subsequently, parallel training of two LSTMs, along with a stacked MLP, is explored. The workflow diagram illustrating the proposed model is presented in Fig. 14.24.

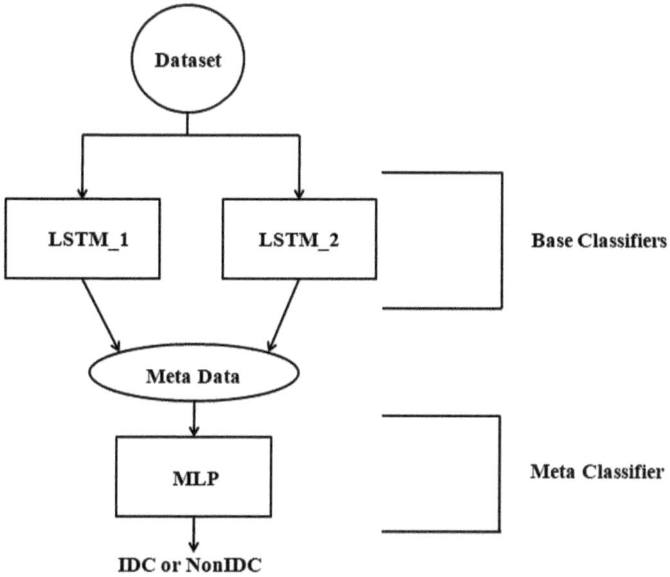

Fig. 14.24 Block diagram of the proposed method

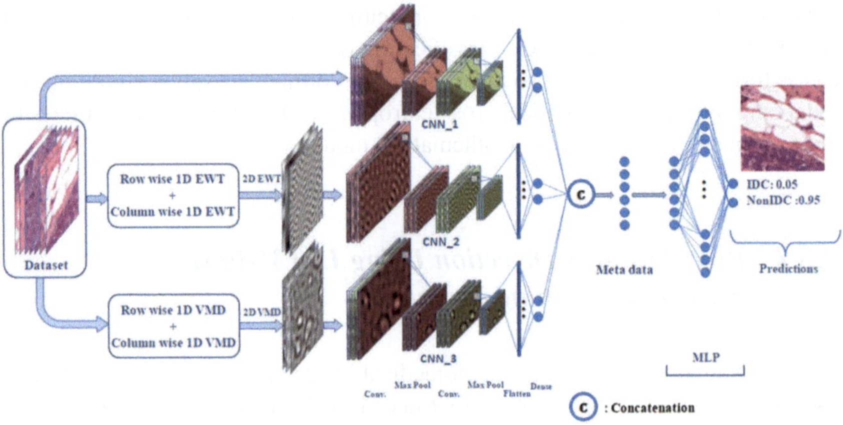

Fig. 14.25 Proposed classification model's system architecture

14.5.5 *Detection of Breast Cancer Via Ensemble Deep Learning Approach with Decomposed Features*

This subsection outlines the formulation of the problem. Instead of relying solely on the dataset, decomposed modes generated using empirical wavelet transform (EWT) and variational mode decomposition (VMD) are considered, as described in Sect. 2.2. Coefficients from EWT and VMD are images and supplied straight into the base classifiers. No step is used to extract statistical features from these coefficients. CNN models are base classifiers as they perform well with image input. To finalize the classification process, an MLP module is employed, which is subsequently positioned atop the parallel connection of three CNNs. The architecture of the proposed classification model is depicted in Fig. 14.25, accompanied by an example image prediction. This figure also provides a detailed description of the model.

14.5.6 *Ensemble Recurrent Model with Stacked Fuzzy ARTMAP*

The base models consist of RNN, LSTM, and GRU architectures. Variance among these base models is managed by incorporating Adaptive Resonance Theory Mapping (ARTMAP). If the winning criteria are met, new nodes are introduced to the F_2^a Layer, enhancing the robustness of the ARTMAP model for improved classification. The proposed method is depicted as blocks in Fig. 14.26.

Each block of the suggested model is elucidated in the subsequent subsections.

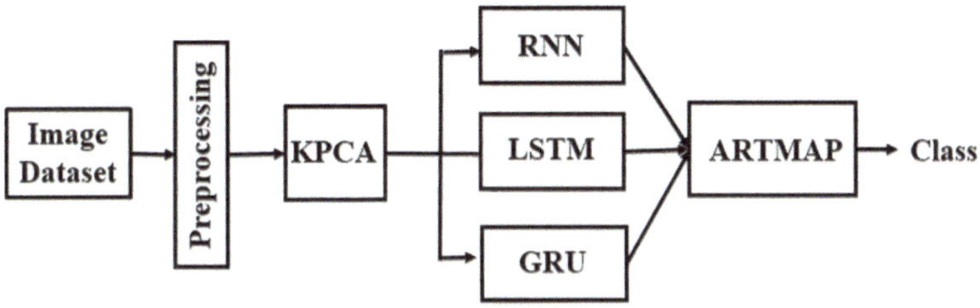

Fig. 14.26 Workflow diagram of the suggested model

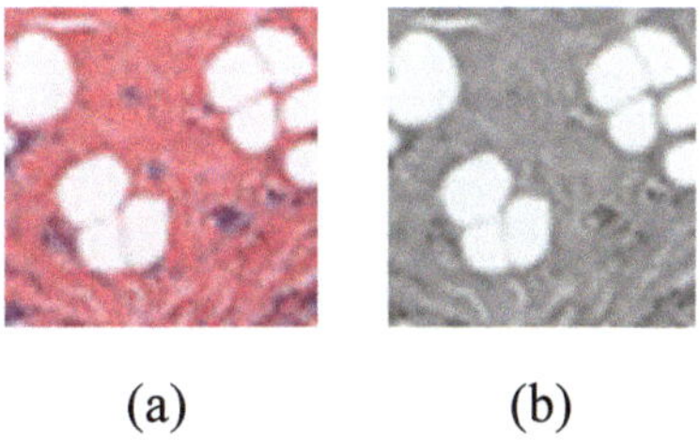

Fig. 14.27 **a** Dataset image of size 50 × 50 × 3, **b** images of size 50 × 50 after conversion

14.5.6.1 Preprocessing

To limit the amount of computational work that is required, color images of breast histopathology are converted into grayscale images. Figure 14.27 illustrates a sample image post-conversion from color to grayscale.

The grayscale images and their attributes (f) undergo normalization using mean normalization (14.4.1).

$$f' = \frac{f - \text{average}(f)}{\max(f) - \min(f)}. \quad (14.4.1)$$

14.5.6.2 Image Denoising and Feature Selection

Each sample in the data set contains 2500 pixels, resulting in substantial data volume and increased processing time. Additionally, noise in the input data contributes to data complexity. To mitigate these effects, kPCA selects the 500 highest eigenvalues. The cosine kernel is chosen due to its ability to measure similarity between documents regardless of size.

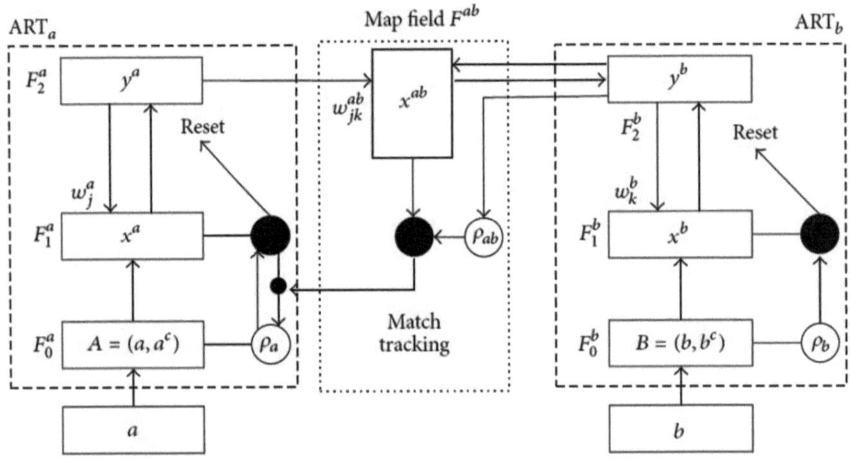

Fig. 14.28 Structure of the ARTMAP

14.5.6.3 Training

For the purpose of training all of the base classifiers, the denoised data is utilized. There are four layers in all the base models. The softmax activation function is used to activate the outer most layers, which comprises two nodes, as there are two classes for classification. As depicted in Fig. 14.26, the base classifiers indirectly generate features for ARTMAP. The structure of ARTMAP is shown in Fig. 14.28.

Fuzzy ARTMAP comprises two ART models, ART_a and ART_b, interconnected by a map field (Carpenter et al. 1992), allowing for the creation of maps connecting clusters produced in the ART_a input domain with the ART_b output domain. Figure 14.28 shows the two domains' normalization, input, and recognition layers, labeled as F_0, F_1, and F_2, respectively.

14.5.7 Results and Discussion

14.5.7.1 Result Analysis of the Proposed Concatenated Model

In terms of test accuracy, a single CNN model with three separate layers achieved 93.12%, whereas a single LSTM method with two layers achieved 89.5%. The method of classification is enhanced, and the concatenated model achieved 97% training and 95.32% testing accuracy levels, respectively.

Table 14.5 Performance parameters of the proposed model

	Precision	Recall	F1-score	Accuracy	Support
Non-IDC (0)	1.00	0.93	0.96	0.98	14
IDC (1)	0.97	1.00	0.98		28

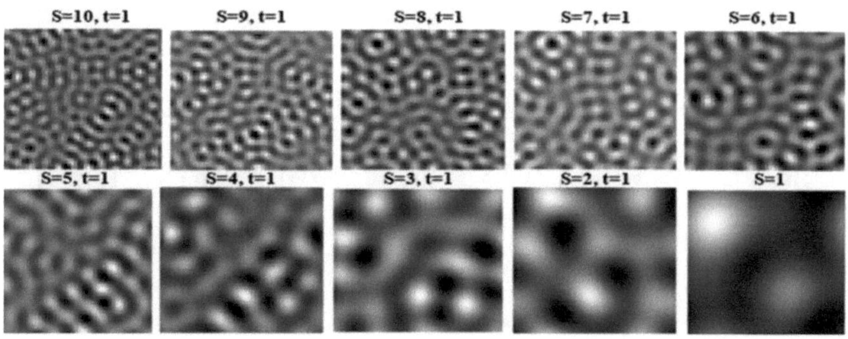

Fig. 14.29 Decomposed modes resulted from EWT from the breast histopathology image dataset

14.5.7.2 Classification Results Obtained from the LSTM-Based Ensemble Learning

The classification of breast histopathology images by the single LSTM model is insufficient. This task is improved by designing an ensemble model. The deep ensemble model is stacked with MLP for final classification. The classification was enhanced, and testing accuracy reached 98%. Accuracy, F1-score, recall, and precision are the measures that are utilized in the evaluation of the recommended technique. Table 14.5 provides the exact values that are associated with these parameters.

14.5.7.3 Performance Analysis of Ensemble Deep Learning Model Trained with Decomposed Features

Decomposition Results

To extract intrinsic features related to spectral modes, the dataset is decomposed by EWT and VMD.

Results of EWT

Figure 14.29 displays the various modes decomposed by EWT from the image datasets.

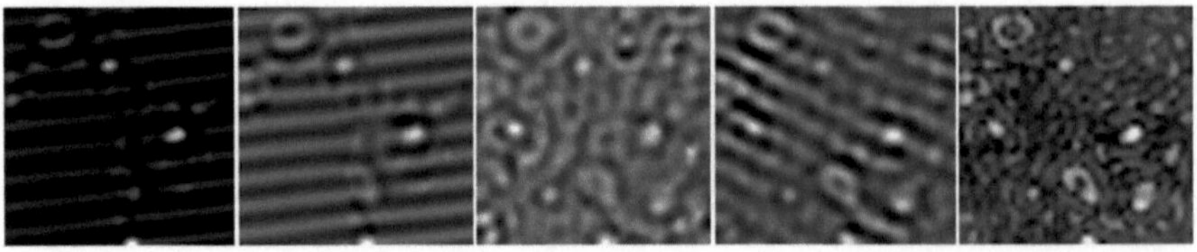

Fig. 14.30 Decomposed modes resulted from VMD from the breast histopathology image dataset

Table 14.6 Evaluating parameters obtained from the proposed model

Dataset	Accuracy (%)	F1-score	Recall	Precision	Sensitivity	Specificity
Breast cancer dataset (IDC vs. Non-IDC)	98.08	0.988	0.9920	0.9841	0.9920	0.9355

The outcomes are derived for a variety of scale (*s*) and rotation angle (*t*) values. By holding the rotation angle *t* constant (i.e., at a value of 1) and altering s from 1 to 10, the EWT findings are displayed in Fig. 14.29.

Results of VMD

Observe how VMD employs the breast histopathology dataset to generate multiple modes, as depicted in Fig. 14.30.

Parameters Evaluation

When compared to CNNs that were trained alone with the dataset, CNNs that were trained utilizing EWT and VMD coefficients exhibit considerably better performance. In accordance with the findings shown in Table 14.6, the combination of these CNNs and MLP led to the achievement of correct classification.

14.5.7.4 Classification Results of the Ensemble Recurrent Model Stacked with ARTMAP

Test and Training Features with Denoising

Following denoising of each dataset image, we utilize kPCA to condense the values from 2,560 pixels into 500 high-priority values. This transformation is crucial as deep learning models tend to underperform when presented with fewer features.

Classification Results

We train three recurrent models on the denoised data and keep the best-fitting ones for further stacking. The testing accuracy for the suggested model was 98.72%.

14.5.7.5 Performance Comparison with State-of-the-Art Models

When tested on a dataset of breast histopathology images, the suggested models outperform the current gold standard in terms of classification accuracy. In this respect, Table 14.7 presents an in-depth examination. The bolded results stand for the top scores in their respective groups.

14.5.8 Summary

Breast cancer diagnosis using histopathology images is the topic of discussion in this part. The suggested models that make use of hybrid deep learning and ensemble learning techniques are the primary emphasis of this discussion. Ensemble learning-based models have been shown to exhibit considerably better detection accuracy when compared to single models. The fusion of different deep learning models with fuzzy decision-making proved to be a better combination for classification. The proposed models for histopathology images did not fit well with dermatoscopic images for skin cancer detection due to no regular pattern and color variation. The classification methods proposed for skin cancer detection are described in Sect. 14.5.

14.6 Skin Cancer Detection Using Intelligent Computing Technique

14.6.1 Introduction

The atypical proliferation of skin cells often leads to skin lesion issues. Timely detection can prevent certain forms of cancer such as melanoma and basal cell carcinoma. Identifying and categorizing cancerous skin growths in the early stages pose significant difficulties for academics and medical practitioners. The varied shapes and diverse appearances of skin lesions contribute to the complexity and time-consuming nature of clinical manual dermatoscopic or biopsy screening. This complexity arises from the fact that patients exhibit different types of skin lesions. Modern technology, such as machine learning, is advancing in its ability to surpass the time-consuming conventional approaches.

Table 14.7 Comparative analysis of the performance of the suggested model with related works

Work	Method used			Accuracy (%)	F1-score
	Preprocessing	Feature extraction	Classification		
Chiu et al. (2020)	Data reduction using PCA	MLP	SVM	86.97	–
Rashmi et al. (2020)	Color normalization	Texture feature	SVM + MLP	–	0.83
Li et al. (2020)	–	RCA	AdaBoost	93.30	–
Kausar et al. (2020)	Segmentation using A-FCNN	MS-RCNN		–	0.902
Das et al. (2020)	–	MI-CNN		96.63	
Alom et al. (2020)	Segmentation using R2U-net	IRRCNN		–	0.878
Brancati et al. (2019)	–	Residual-CNN		89.57	–
Wang et al. (2019)	–	CNN	ELM	86.50	–
Li et al. (2019)	–	SVM	CNN	76	–
Qi et al. (2018)	–	Pre-trained AlexNet		92.84	–
Sebai et al. (2020)	–	Ensemble of 2 CNNs		–	0.698
Sharma and Kumar (2022)	Magnification	Xception	SVM R	96.25	0.96
Proposed works	–	Hybrid CNN + LSTM		95.32	0.95
	–	Ensemble of 2 LSTMs	MLP	98	0.975
	EWT and VMD	Ensemble of 3 CNNs	MLP	98.08	0.98
	Data denoising using kPCA	Ensemble of RNN, LSTM and GRU	Fuzzy ARTMAP	98.72	0.985

14.6.2 Dataset Description

Freely accessible HAM10000 dataset from Kaggle (Tschandl et al. 2018) is utilized in this part. The dataset consists of 10,015 dermoscopic images categorized into seven groups: melanocytic nevi (Nv), melanoma (Mel), benign keratosis-like lesions (Bkl), basal cell carcinoma (Bcc), actinic keratoses (Akiec), vascular lesions (Vasc), and dermatofibroma (Df). The number of images in each category is as follows: 6,075 for Nv, 1,113 for Mel, 1,099 for Bkl, 514 for Bcc, 327 for Akiec, 142 for Vasc, and 115 for Df. The dataset is partitioned into training and testing sets with an 80:20

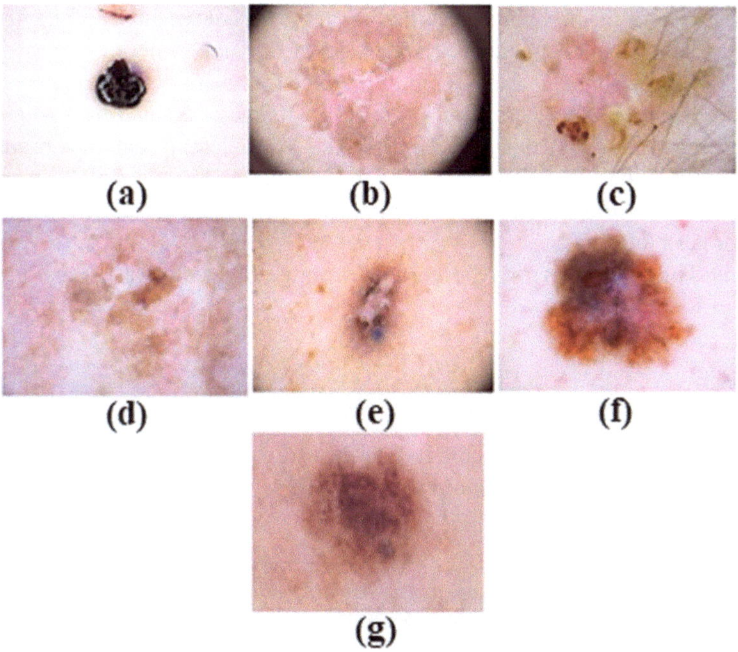

Fig. 14.31 Sample images from every seven classes: **a** Vascular lesions, **b** actinic keratoses, **c** basal cell carcinoma, **d** benign keratosis-like lesions, **e** dermatofibroma, **f** melanoma, **g** melanocytic nevi

ratio, meaning that the training set consists of 8012 photos and the testing set consists of 2003 images. Figure 14.31 displays sample photos of each kind of skin cancer.

14.6.3 A Concatenated Model with Deep Learning Technique for Skin Lesion Detection

In order for any object identification system to function properly, it is necessary to extract color as a critical component. The mean, standard deviation, skewness, and kurtosis are the four statistical parameters that are produced as a result of the analysis of each of these channels independently. The definition of a picture is comprised of twelve moments, consisting of four moments for each channel. The symbol is a representation of the probability distribution of the *i*th color channel of the *j*th pixel. In mathematics, a mathematical variable is denoted by the symbol 〖P〗 _ij. In order for any object identification system to function properly, it is necessary to extract color as a critical component. The mean, standard deviation, skewness, and kurtosis are the four statistical parameters that are produced as a result of the analysis of each of these channels independently. The definition of a picture is comprised of twelve moments, consisting of four moments for each channel. The symbol is a representation of the probability distribution of the *i*th color channel of the *j*th pixel. In mathematics, a mathematical variable is denoted by the symbol (P_{ij}).

Each of these channels is analyzed to produce four statistical parameters: mean, standard deviation, skewness, and kurtosis. An image is defined by 12 moments, namely, four moments for each channel. The symbol represents the probability distribution of the *i*-th color channel of the *j*-th pixel. The symbol P_{ij} represents a mathematical variable. The calculation of the four statistical characteristics based on color is as follows:

Mean: The mean of an image refers to the average value of its various color values.

$$\mu_i = \frac{1}{N}\sum_{j=1}^{N} P_{ij}. \tag{14.5.1}$$

Standard Deviation: The square root of the estimated variance is used to characterize the degree of color sharing in an image. This value is obtained by performing a calculation.

$$\sigma_i = \sqrt{\frac{1}{N}\sum_{j=1}^{N} \left(P_{ij} - \mu_i\right)^2}. \tag{14.5.2}$$

Skewness: Skewness is the measure of the extent to which the color distribution deviates from symmetry. It represents the characteristic of uneven color distribution, referred to as asymmetric color distribution, which is determined by analyzing the form of the distribution and computed using,

$$s_i = \sqrt[3]{\frac{1}{N}\sum_{j=1}^{N} (P_{ij} - \mu_i)^3}. \tag{14.5.3}$$

Kurtosis: Kurtosis is a statistical measure that quantifies the form of the color distribution, especially the degree of peakedness or smoothness in relation to a normal distribution. It is comparable to skewness and is calculated by

$$k_i = \sqrt[4]{\frac{1}{N}\sum_{j=1}^{N} (P_{ij} - \mu_i)^4}. \tag{14.5.4}$$

The feature set is meant to include statistical moments up to the 4th order for each color channel individually.

$$m_R = \left\{\mu_R, \sigma_R, s_{R,} k_G\right\} \tag{14.5.5}$$

$$m_G = \left\{\mu_G, \sigma_G, s_{G,} k_G\right\} \tag{14.5.6}$$

$$m_B = \{\mu_B, \sigma_B, s_{B,}k_B\}. \tag{14.5.7}$$

A feature set of moments is designed by the union of m_R, m_G, and m_B as

$$f_m = m_R \cup m_G \cup m_B = \{\mu_R, \sigma_R, s_{R,}k_G, \mu_G, \sigma_G, s_{G,}k_G, \mu_B, \sigma_B, s_{B,}k_B\}. \tag{14.5.8}$$

Utilizing the gray-level co-occurrence matrix (GLCM) to extract features.

These attributes stem from second-order statistical texture features, with the following characteristics computed from the gray-level co-occurrence matrix (GLCM).

Contrast: Contrast pertains to the variations in the gray-level co-occurrence matrix (GLCM) between different regions.

$$C = \sum_{i,j=0}^{N-1} P_{i,j}(i-j)^2. \tag{14.5.9}$$

Correlation: Correlation refers to the frequency of occurrence of pixel pairs that are being analyzed together.

$$C_r = \sum_{i,j=0}^{N-1} P_{i,j}\frac{(i-\mu)(j-\mu)}{\sigma^2}. \tag{14.5.10}$$

Energy: To quantify the total of squared components that are contained inside the GLCM, energy, which is also referred to as angular second moment or homogeneity, is utilized.

$$E = \sum_{i,j=0}^{N-1} \left(P_{i,j}\right)^2. \tag{14.5.11}$$

Homogeneity: A measure of homogeneity is the degree to which the values that are distributed along the diagonal of the gray-level co-occurrence matrix are equivalent to one another.

$$H = \sum_{i,j=0}^{N-1} \frac{P_{i,j}}{1+(i-j)^2}. \tag{14.5.12}$$

Dissimilarity: In the context of a particular region of interest, the term "dissimilarity" refers to the measurement of the distance between adjacent pairs of objects, more precisely pixels.

$$D = \sum\sum |i-j|P_{i,j}. \tag{14.5.13}$$

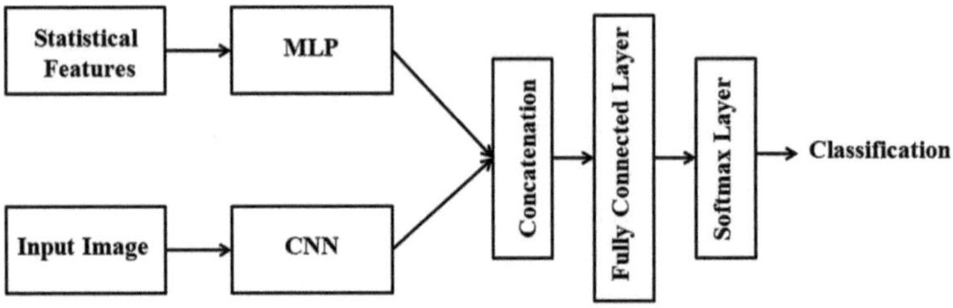

Fig. 14.32 Block diagram of the proposed model

A feature set comprising these textural properties is formulated as

$$f_{\text{GLCM}} = \{C, C_r, E, H, D\}. \tag{14.5.14}$$

For the purpose of classifying skin lesions, the concatenated deep learning model is developed by integrating the capabilities of handmade feature mining with CNN learning. Figure 14.32 presents a visual representation of our suggested technique.

The convolutional neural network (CNN) model and the multilayer perceptron (MLP) model are the components that make up the hybrid model that is being presented. Two types of inputs are taken into consideration by the system: the feature set, which consists of RGB color pictures, GLCM characteristics, and even color moments. A fully linked model with four layers is employed in order to categorize the qualities that have been stated, employing a nonlinear mapping function $f(\bullet)$ to translate x into y. The mapping is defined according to the reference (Chiu et al. 2020) and is presented in Eq. (14.5.15).

$$y = f(wx + b) \tag{14.5.15}$$

The weight matrix is denoted by the letter w, while the bias vector is denoted by the letter b.

14.6.4 Cancer Detection Using Hybrid Fuzzy Ensemble Model

Ensemble learning and the class balancing technique are two methods that have been recommended for use in the process of diagnosing skin cancer. There are three basic models and one meta-classifier that make up the ensemble version of the model. The integration of both homogeneous and heterogeneous ensemble learning approaches is accomplished by use of two ARTMAPs and one fuzzy min–max (FMM) model. The ultimate decision is determined by the layered NEFCLASS model. Figure 14.33 displays the process diagram for the suggested technique.

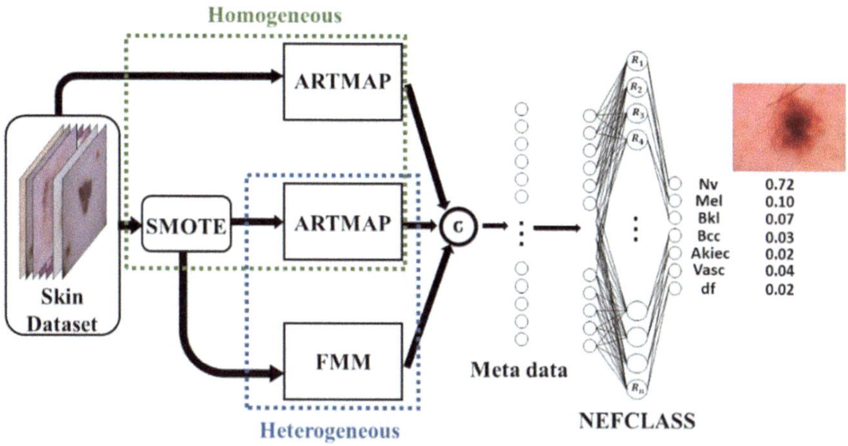

Fig. 14.33 Architecture diagram of the proposed model

14.6.5 Class Balancing Technique—SMOTE

The skin dermatoscopic image collection achieves class balance by utilizing the SMOTE technique. The test set remains unchanged in order to evaluate performance both before and after the implementation of SMOTE.

14.6.6 Ensemble Model Design

One FMM model and two ARTMAPs make up the basic classifier. Finding the correlation between precision and foresight becomes much easier when one employs fuzzy analytic methods, especially neuro-fuzzy ones. Consequently, we propose a hybrid ensemble learning model, NEFCLASS, that combines three standard classifiers with a fuzzy-based meta-classifier. Both kinds of data are used directly to train the foundation models. For the purpose of training the meta-classifier, the predictions are utilized. The basic classifiers' outputs (Y_1, Y_2, Y_3. .) form the basis of the final decision. Thus, the rules (R) created from the variables dictate the output of the stacked meta-classifier NEFCLASS. Y_1, Y_2 and Y_3.

$$y_{\text{stacking}}(x) = \text{NEFCLASS}((Y_1, \dot{Y}_2 Y_3, R), \tag{14.5.17}$$

where x represents the input data.

Figure 14.34 displays the NEFCLASS model along with its corresponding rules.

The rule layer often consists of the rules utilized to make decisions. The NEFCLASS has a total of three inputs. Therefore, the number of alternative rules is 8, which can be calculated as 2^3. The NEFCLASS algorithm utilizes the "Best rule

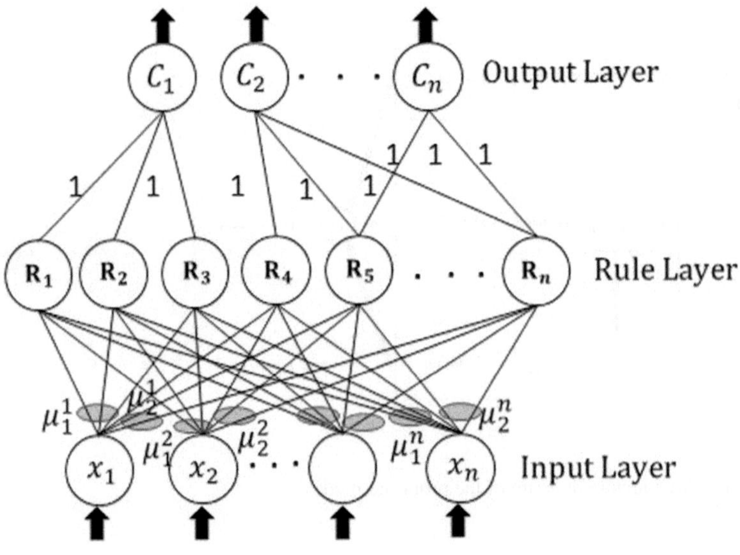

Fig. 14.34 Architectural diagram depicting the planned NEFCLASS paradigm

learning" technique to select just four rules with the highest cumulative activation for class *c*. The most effective rules utilized in this chapter are as follows.

If *y*1 is *c*, *y*2 is *c*, and *y*3 is X, then *x* belongs to class *c*
If *y*1 is *c*, *y*2 is X, and *y*3 is *c* then *x* belongs to class *c*
If *y*1 is X, *y*2 is *c*, and *y*3 is *c* then *x* belongs to class *c*
If *y*1 is *c*, *y*2 is *c*, and *y*3 is *c* then *x* belongs to class *c*

Where *x* represent the input image, and *y*1, *y*2, and *y*3 represent the outputs of the basic classifiers. *X* denotes any class other than *c*.

14.6.7 Results Analysis of the Proposed Methods in Skin Cancer Detection

14.6.7.1 Results of Ensemble CNN Model

The hybrid model outperforms the single CNN model in terms of classification accuracy. The second model exhibits greater precision, recall, and F1-score compared to the single CNN for each unique class. The test loss has decreased from 0.477 to 0.052, and the test accuracy has increased from 86.1 to 98.1%, as indicated in Table 14.8.

Table 14.8 Results in terms of performance metrics for the CNN model and concatenated deep learning model

	Performance metrics using a single CNN			Performance metrics using a concatenated deep learning model		
Classes	Precision	Recall	F1 score	Precision	Recall	F1 score
Vascular skin lesions	0.79	0.59	0.67	0.96	0.99	0.98
Actinic Keratoses	0.89	0.93	0.91	1.00	1.00	1.00
Basal cell carcinoma	0.92	0.77	0.84	0.99	1.00	1.00
Benign keratosis	0.84	0.52	0.65	0.96	0.99	0.97
Dermatofibroma	0.98	1.00	0.99	1.00	1.00	1.00
Melanocytic nevi	0.95	0.56	0.70	0.99	0.88	0.93
Melanoma	0.99	0.99	0.99	1.00	1.00	1.00
Test loss	0.477			0.052		
Test accuracy	0.861			0.981		

14.6.7.2 Results of the Hybrid Ensemble Model

The training set is created by selecting 80% of the whole dataset, with the class "Nv" having the maximum number of photos, specifically 5378. The issue of class imbalance was resolved by implementing the Synthetic Minority Over-sampling Technique (SMOTE), which led to an equal distribution of 5,378 photos in each class. Figure 14.35 illustrates the representation of the balanced class.

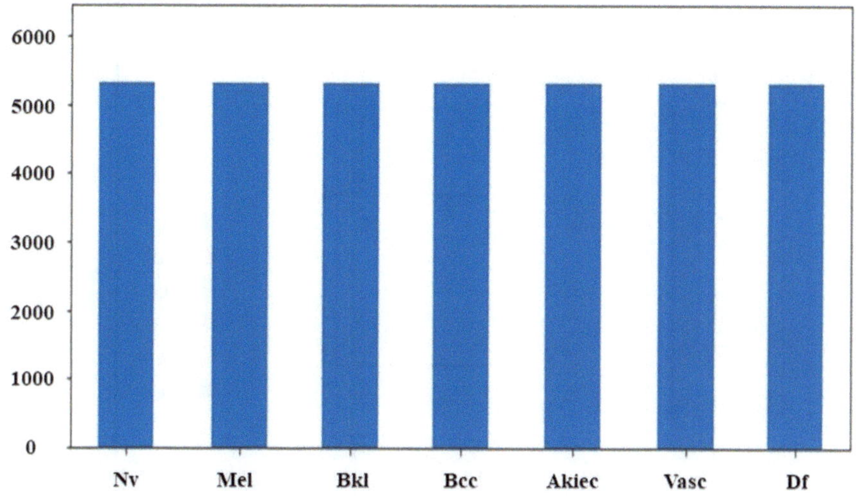

Fig. 14.35 Visualization of the balanced classes of the dataset

Table 14.9 displays the scores obtained from training, validation, and testing for each combination of data and model. Specifically, both the unbalanced and balanced datasets were used to verify the performance of ARTMAP and FMM.

The examples shown in Table 14.9 demonstrate how the performance of the models is greatly impacted by class imbalance. On the other hand, when compared to utilizing unbalanced data, the accuracy of training fuzzy classifiers is much improved when the dataset is balanced. The generation of hyperboxes for each class is seen in Fig. 14.36. This was accomplished by employing the FMM model that was trained with the balanced dataset.

Table 14.9 Performance metrics of each model

		Training accuracy (in %)	Validation accuracy (in %)	Testing accuracy (in %)
Direct data	ARTMAP	88.06	85.17	81.53
	FMM	86.11	83.92	80.63
Balanced data	ARTMAP	93.05	90.28	89.47
	FMM	92.25	89.23	89.12

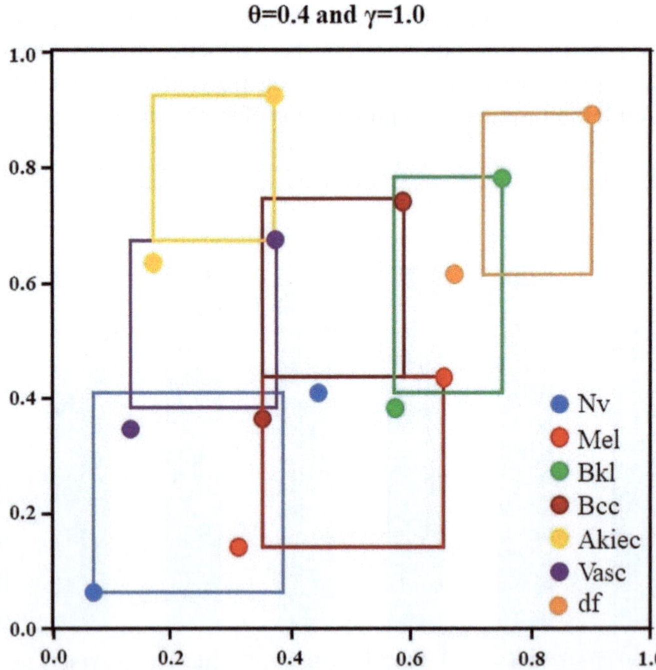

Fig. 14.36 Hyperboxes produced for each class are visualized using FMM

Despite implementing the balanced dataset, there is still a limited amount of overlap between the classes. The FMM model is tested with various values of θ, and the optimal performance is achieved when θ is set to 0.4 and γ is set to 1.

14.6.7.3 Performance Comparison with the State-of-the-Art Models

The suggested models achieve superior classification performance for the skin dermatoscopic image dataset compared to state-of-the-art approaches. Table 14.10 has a comprehensive examination on this matter. The bold results indicate the greatest values in their respective categories.

14.7 Summary and Future Scope

Technologies that are based on artificial intelligence and machine learning are a transformational force in the healthcare industry. These technologies provide great potential to improve diagnosis, treatment, and patient care. These technologies have the potential to revolutionize the delivery of healthcare, improve results, and ultimately save lives. They do this by utilizing sophisticated algorithms and massive volumes of medical data. However, in order to guarantee the appropriate and fair application of artificial intelligence and machine learning in the healthcare industry, it is essential to solve difficulties such as confidentiality of data, compliance with regulations, and ethical concerns. Artificial intelligence and machine learning have the potential to bring in a new era of personalized healthcare and precision medicine, provided that stakeholders work together and give due attention to the technology.

The primary objective of this chapter is to enhance the model design by reducing the number of preprocessing procedures. Accurate tumor and cancer detection is accomplished using the ensemble learning approach. The synthetic image dataset is generated using the one-dimensional gene data for the purpose of detecting cancer. The hybrid model provided demonstrates the concept of ensemble learning by delivering enhanced performance in comparison to individual base classifiers with a singular design. An analysis is conducted on the impact of certain classifiers based on deep learning when constructing the ensemble model. Two distinct meta-classifiers, namely the neural network-based MLP and the fuzzy-based FMM, ARTMAP, and NEFCLASS, are combined in a stacking manner to collectively capture the variation among the underlying classifiers. Fusion of deep learning with a fuzzy classifier introduces a novel approach to biological image classification.

The suggested methods for biomedical image classification are evaluated using three distinct datasets: brain MRI, breast histopathology, and dermatoscopic images. It has been noted that models tailored for certain datasets do not perform effectively on other datasets. The occurrence took place due to the presence of structural variances in the photographs of various categories. The complexity of breast histopathology images surpasses that of brain MRI imaging. Dermatoscopic images of the skin

Table 14.10 Comparative analysis of the suggested model with relevant studies in terms of performance

References	Preprocessing	Feature extraction	Classification	Testing accuracy (in %)
Alfed and Khelifi (2017)	–	Color features and location features	SVM-RBF	93
Ain et al. (2020)	–	–	DenseNet-169	92.25
Talavera-Martinez et al. (2020)	Noise removal using type II fuzzy	FSPO	Deep Q net	92.364
Tumpa and Kabir (2021)	Contrast enhancement	CNN features	ELM	93.4
Tan et al. (2018)	Segmentation	–	EncoderDecoder FCN	95.33
Fu et al. (2022)	–	Color features using LBP and Geometrical features	Ensemble of ML models	95.35
Monika et al. (2020)	Hair removal and segmentation	ABCD, GLCM, LBP	ANN	97.7
Saravanan et al. (2020)	Hair removal, noise reduction, segmentation by cropping, and binarization	Textural, shape, and color feature	KNN, SVM, and decision tree	96.68
Yu et al. (2021)	–	–	Optimized ensemble model using convolutional layers	77
Usmani et al. (2021)	–	–	BOA optimized (CNN + SVM)	90
Kwasigroch et al. (2020)	Segmentation using kernel fuzzy C-means	–	Red fox optimization-based MLP	90.5
Gu et al. (2019)	–	GAN	Progressive learning-based CNN	92.3
Kadampur and Al Riyaee (2020)	ROI image generation	–	Transfer learning DNN	97.2
Pham et al. (2020)	–	–	HMM	96
Biasi et al. (2021)	Localization using CNN	DenseNet201	ELM	88.39
Song et al. (2020)	–	Textural and color features	HG and HL	92.96

(continued)

Table 14.10 (continued)

References	Preprocessing	Feature extraction	Classification	Testing accuracy (in %)
Adegun and Viriri (2019)	–	–	Cloud-based CNN	96.27
Wei et al. (2020)	–	ABCD	Total dermatoscopic score	84
Wang et al. (2021)	Hair removal, smoothing, noise removal, edge detection	ABCD and GLCM	SVM	96.25
Adegun and Viriri (2020)	Augmentation	–	Transfer learning-based CNN	89.97
Kousis et al. (2022)	Augmentation, artifact removal, melanoma segmentation, image quality enhancement	Clinical feature	AlexNet	94
Giotis et al. (2015)	Image resize, zero normalization, and augmentation	FPN and RPN	Ensemble of three CNN	81.3
Kumar et al. (2022)	Augmentation and gamma correction	–	ResNet	82
Hsiao et al. (2021)	Noise reduction, resizing, normalization, and augmentation	–	CNN-based EncoderDecoder	95
Courtenay et al. (2022)	Semantic segmentation	–	Ensemble of 2 CNNs	96.2
Shahin et al. (2018)	–	–	Ensemble of ResNet and inception V3	89
Nahata and Singh (2020)	–	–	Ensemble of ResNet and inception	91
Bajwa et al. (2020)	–	–	Ensemble of ResNet, DenseNet, SE-ResNext, and NASNet	96.57

(continued)

Table 14.10 (continued)

References	Preprocessing	Feature extraction	Classification	Testing accuracy (in %)
Thurnhofer-Hemsi et al. (2021)	Time-shifted forms of input images	–	Ensemble of CNN	83.8
Ichim and Popescu (2020)	–	LBP, color HOGs, ABCD, and deep features	Perceptron-based convolutional classifier	97.5
Proposed works	–	Mean, standard deviation, skewness, kurtosis, and GLCM features	Ensemble of CNN and MLP as base classifier and MLP as meta-classifier	98.1
	Target-based augmentation for class balancing	–	DeepSVM (Ensemble of 2 SVMs as a base classifier and another SVM as meta-classifier)	98.27
	Class balance using SMOTE	–	Ensemble of two ARTMAPs, one FMM as base models and NEFCLASS as meta-classifier	**98.4**

exhibit less color fluctuation and possess a higher degree of anatomical complexity compared to brain MRI and breast histopathology images. Table 14.11 provides a concise overview of this observation.

Cancer subclass classification and cancer stage detection are the other directions for analyzing biomedical images, and they are kept for future analysis with improved machine learning models. The application of AI/ML makes most tasks automatic and accurate when appropriately designed and trained with a sufficient quantity of data.

Table 14.11 Performance analysis in terms of testing accuracy (in %) of the proposed models with each dataset

Proposed models	Brain MRI image	Breast histopathology images	Dermatoscopic images
CNN	**92.1**	91.61	86.1
{CNN, RNN, LSTM} + MLP	**97**	94.52	93.12
{CNN, RNN, LSTM, GRU} + FMM	**97.62**	95.81	94.21
CNN + LSTM	95.21	**95.32**	94.04
(EWT + VMD) {3 CNNs} + MLP	96.98	**98.08**	95
kPCA + {RNN, LSTM, GRU} + ARTMAP	97.03	**98.72**	95.30
{CNN + MLP} + MLP	96.13	97.91	**98.1**
Fuzzy ensemble	97.52	98.18	**98.4**

References

Adegun AA, Viriri S (2019) Deep learning-based system for automatic melanoma detection. IEEE Access 8:7160–7172

Adegun AA, Viriri S (2020) FCN-based DenseNet framework for automated detection and classification of skin lesions in dermoscopy images. IEEE Access 8:150377–150396

Ain QU, Al-Sahaf H, Xue B, Zhang M (2020) Generating knowledge-guided discriminative features using genetic programming for melanoma detection. IEEE Trans Emerg Top Comput Intell 5(4):554–569

Alfed N, Khelifi F (2017) Bagged textural and color features for melanoma skin cancer detection in dermoscopic and standard images. Expert Syst Appl 90:101–110

Alom MZ, Aspiras T, Taha TM, Bowen T, Asari VK (2020) MitosisNet: end-to-end mitotic cell detection by multi-task learning. IEEE Access 8:68695–68710

Bajwa MN, Muta K, Malik MI, Siddiqui SA, Braun SA, Homey B, Dengel A, Ahmed S (2020) Computer-aided diagnosis of skin diseases using deep neural networks. Appl Sci 10(7):2488

Brancati N, De Pietro G, Frucci M, Riccio D (2019) A deep learning approach for breast invasive ductal carcinoma detection and lymphoma multi-classification in histological images. IEEE Access 7:44709–44720

Carpenter GA, Grossberg S, Markuzon N, Reynolds JH, Rosen DB et al (1992) Fuzzy ARTMAP: a neural network architecture for incremental supervised learning of analog multi-dimensional maps. IEEE Trans Neural Netw 3:698–713

Chakrabarty N (2019) Brain MRI images for brain tumor detection. https://www.kaggle.com/navoneel/brain-mri-images-for-brain-tumor-detection

Chiu HJ, Li THS, Kuo PH (2020) Breast cancer-detection system using PCA, multilayer perceptron, transfer learning, and support vector machine. IEEE Access 8:204309–204324

Courtenay LA, González-Aguilera D, Lagüela S, Pozo SD, Ruiz C, Barbero-García I et al (2022) Deep convolutional neural support vector machines for the classification of basal cell carcinoma hyperspectral signatures. J Clin Med 11(9):2315

Das K, Conjeti S, Chatterjee J, Sheet D (2020) Detection of breast cancer from whole slide histopathological images using deep multiple instance CNN. IEEE Access 8:213502–213511

De Gruijl FR (1999) Skin cancer and solar UV radiation. Eur J Cancer 35:2003–2009

Di Biasi L, Citarella AA, Risi M, Tortora G (2021) A cloud approach for melanoma detection based on deep learning networks. IEEE J Biomed Health Inform 26(3):962–972

Ertosun MG, Rubin DL (2015) Automated grading of gliomas using deep learning in digital pathology images: a modular approach with ensemble of convolutional neural networks. In: AMIA annual symposium proceedings, vol 2015. American Medical Informatics Association, p 1899

Fu Z, An J, Yang Q, Yuan H, Sun Y, Ebrahimian H (2022) Skin cancer detection using Kernel fuzzy C-means and developed red fox optimization algorithm. Biomed Signal Process Control 71:103160

Ge C, Gu IYH, Jakola AS, Yang J (2020) Enlarged training dataset by pairwise GANs for molecular-based brain tumor classification. IEEE Access 8:22560–22570

Gefeller O, Diehl K (2022) Children and ultraviolet radiation, vol 9. MDPI, p 537

Giotis I, Molders N, Land S, Biehl M, Jonkman MF, Petkov N (2015) MED-NODE: a computer-assisted melanoma diagnosis system using non-dermoscopic images. Expert Syst Appl 42(19):6578–6585

Gu Y, Ge Z, Bonnington CP, Zhou J (2019) Progressive transfer learning and adversarial domain adaptation for cross-domain skin disease classification. IEEE J Biomed Health Inform 24(5):1379–1393

Hsiao YP, Chiu CW, Lu CW, Nguyen HT, Tseng YS, Hsieh SC, Wang HC (2021) Identification of skin lesions by using single-step multiframe detector. J Clin Med 10(1):144

http://gco.iarc.fr/. Accessed 15th Feb 2022

https://portal.gdc.cancer.gov/repository

Huang Z, Du X, Chen L, Li Y, Liu M, Chou Y, Jin L (2020) Convolutional neural network based on complex networks for brain tumor image classification with a modified activation function. IEEE Access 8:89281–89290

Huda S, Yearwood J, Jelinek HF, Hassan MM, Fortino G, Buckland M (2016) A hybrid feature selection with ensemble classification for imbalanced healthcare data: a case study for brain tumor diagnosis. IEEE Access 4:9145–9154

Ichim L, Popescu D (2020) Melanoma detection using an objective system based on multiple connected neural networks. IEEE Access 8:179189–179202

IAFR Cancer. Global Cancer Observatory. https://gco.iarc.fr/today/online-analysis-table. Accessed 11th June 2022

Janowczyk A, Madabhushi A (2016) Deep learning for digital pathology image analysis: a comprehensive tutorial with selected use cases. J Pathol Inform 7

Jee BA, Lim H, Kwon SM, Jo Y, Park MC, Lee IJ, Woo HG (2015) Molecular classification of basal cell carcinoma of skin by gene expression profiling. Mol Carcinog 54(12):1605–1612

Kadampur MA, Al Riyaee S (2020) Skin cancer detection: applying a deep learning based model driven architecture in the cloud for classifying dermal cell images. Inform Med Unlock 18:100282

Kausar T, Wang M, Ashraf MA, Kausar A (2020) SmallMitosis: small size mitotic cells detection in breast histopathology images. IEEE Access 9:905–922

Ker J, Singh SP, Bai Y, Rao J, Lim T, Wang L (2019a) Image thresholding improves 3-dimensional convolutional neural network diagnosis of different acute brain hemorrhages on computed tomography scans. Sensors 19(9):2167

Ker J, Bai Y, Lee HY, Rao J, Wang L (2019b) Automated brain histology classification using machine learning. J Clin Neurosci 66:239–245

Kousis I, Perikos I, Hatzilygeroudis I, Virvou M (2022) Deep learning methods for accurate skin cancer recognition and mobile application. Electronics 11(9):1294

Kumar KS, Suganthi N, Muppidi S, Kumar BS (2022) FSPBO-DQN: SeGAN based segmentation and fractional student psychology optimization enabled deep Q network for skin cancer detection in IoT applications. Artif Intell Med 129:102299

Kumar A, Mohanty MN, Routray A (2010) Design of support vector machines with time frequency kernels for classification of EEG signals. In: 2010 IEEE students technology symposium (TechSym). IEEE, pp 330–333

Kwasigroch A, Grochowski M, Mikołajczyk A (2020) Neural architecture search for skin lesion classification. IEEE Access 8:9061–9071

Li X, Radulovic M, Kanjer K, Plataniotis KN (2019) Discriminative pattern mining for breast cancer histopathology image classification via fully convolutional autoencoder. IEEE Access 7:36433–36445

Li G et al (2020) Effective breast cancer recognition based on fine-grained feature selection. IEEE Access 8:227538–227555

Monika MK, Vignesh NA, Kumari CU, Kumar MNVSS, Lydia EL (2020) Skin cancer detection and classification using machine learning. Mater Today: Proc 33:4266–4270

Nahata H, Singh SP (2020) Deep learning solutions for skin cancer detection and diagnosis. In: Jain V, Chatterjee J (eds) Machine learning with health care perspective (Learning and analytics in intelligent systems), vol 13. Springer, Cham, Switzerland

National Cancer Institute. https://www.cancer.gov/about-cancer/understanding/what-is-cancer

Özyurt F, Sert E, Avci E, Dogantekin E (2019) Brain tumor detection based on convolutional neural network with neutrosophic expert maximum fuzzy sure entropy. Measurement 147:106830

Pham TC, Doucet A, Luong CM, Tran CT, Hoang VD (2020) Improving skin-disease classification based on customized loss function combined with balanced mini-batch logic and real-time image augmentation. IEEE Access 8:150725–150737

Qi Q et al (2018) Label-efficient breast cancer histopathological image classification. IEEE J Biomed Health Inform 23(5):2108–2116

Rashmi R, Prasad K, Udupa CBK, Shwetha V (2020) A comparative evaluation of texture features for semantic segmentation of breast histopathological images. IEEE Access 8:64331–64346

Saravanan S, Heshma B, Shanofer AA, Vanithamani R (2020) Skin cancer detection using dermoscope images. Mater Today: Proc 33:4823–4827

Sebai M, Wang T, Al-Fadhli SA (2020) PartMitosis: a partially supervised deep learning framework for mitosis detection in breast cancer histopathology images. IEEE Access 8:45133–45147

Shahin AH, Kamal A, Elattar MA (2018) Deep ensemble learning for skin lesion classi_cation from dermoscopic images. In: Proceedings of the 9th Cairo International Biomedical Engineering Conference (CIBEC), pp 150–153

Sharma S, Kumar S (2022) The Xception model: a potential feature extractor in breast cancer histology images classification. ICT Express 8(1):101–108

Song L, Lin J, Wang ZJ, Wang H (2020) An end-to-end multi-task deep learning framework for skin lesion analysis. IEEE J Biomed Health Inform 24(10):2912–2921

Talavera-Martinez L, Bibiloni P, Gonzalez-Hidalgo M (2020) Hair segmentation and removal in dermoscopic images using deep learning. IEEE Access 9:2694–2704

Tan TY, Zhang L, Neoh SC, Lim CP (2018) Intelligent skin cancer detection using enhanced particle swarm optimization. Knowl-Based Syst 158:118–135

Thurnhofer-Hemsi K, López-Rubio E, Domínguez E, Elizondo DA (2021) Skin lesion classification by ensembles of deep convolutional networks and regularly spaced shifting. IEEE Access 9:112193–112205

Tschandl P, Rosendahl C, Kittler H (2018) The HAM10000 dataset, a large collection of multi-source dermatoscopic images of common pigmented skin lesions. Sci Data 5(1):1–9

Tumpa PP, Kabir MA (2021) An artificial neural network based detection and classification of melanoma skin cancer using hybrid texture features. Sens Int 2:100128

Usmani UA, Watada J, Jaafar J, Aziz IA, Roy A (2021) A reinforcement learning algorithm for automated detection of skin lesions. Appl Sci 11(20):9367

Wang Z et al (2019) Breast cancer detection using extreme learning machine based on feature fusion with CNN deep features. IEEE Access 7:105146–105158

Wang Y, Louie DC, Cai J, Tchvialeva L, Lui H, Wang ZJ, Lee TK (2021) Deep learning enhances polarization speckle for in vivo skin cancer detection. Opt Laser Technol 140:107006

Wei L, Ding K, Hu H (2020) Automatic skin cancer detection in dermoscopy images based on ensemble lightweight deep learning network. IEEE Access 8:99633–99647

Xie H, Li J, Zhang Q, Wang Y (2016) Comparison among dimensionality reduction techniques based on Random Projection for cancer classification. Comput Biol Chem 65:165–172

Yu Z, Nguyen J, Nguyen TD, Kelly J, Mclean C, Bonnington P et al (2021) Early melanoma diagnosis with sequential dermoscopic images. IEEE Trans Med Imaging 41(3):633–646

Chapter 15
Artificial Intelligence Techniques for Sentimental Analysis During Pandemic

Ayush Thakur and Reshu Agarwal

Abstract This chapter introduces the concept and applications of sentiment analysis, a natural language processing task that aims to identify and extract subjective information from text. It covers various topics such as data collection and preparation, machine learning techniques, case studies, model evaluation, and future directions. It also provides examples of how sentiment analysis can be used to gain insights into public sentiment and emotional trends related to COVID-19. The chapter is intended for readers who are interested in learning more about sentiment analysis and its potential impact on various domains.

Keywords COVID-19 · Sentiment analysis · Machine learning · Deep learning · Data processing

15.1 Introduction

15.1.1 What is Sentimental Analysis?

In natural language processing, sentiment analysis is a common activity. Sentiment analysis aims to categorise the text according to the attitude or mindset that is conveyed in it, which might be neutral, negative, or positive.

Sentiment analysis is the process of categorising a text passage as neutral, negative, or positive (Manguri et al. 2020). Sentiment analysis seeks to understand people's attitudes in a way that might facilitate corporate growth. It emphasises feelings (such as happiness, sadness, anger, etc.) in addition to polarity (positive, negative, and

A. Thakur · R. Agarwal (✉)
Amity Institute of Information Technology, Amity University Noida, Noida, India
e-mail: ragarwal1@amity.edu

A. Thakur
e-mail: ayush.th2002@gmail.com

A. Rai et al. (eds.), *Convergence of Artificial Intelligence, Machine Learning, and the Internet of Things in Industry 4.0 Applications*, Transactions on Computer Systems and Networks, https://doi.org/10.1007/978-981-96-5269-3_15

neutral). It makes use of a variety of NLP methods, including automatic, rule-based, and hybrid.

As an illustration, what if we wanted to determine if a product was meeting consumer needs or if the market needed it? Sentiment analysis is a useful tool for tracking reviews of that product. When we wish to automatically tag and classify a big amount of unstructured data, sentiment analysis is a very effective tool. Surveys measuring Net Promoter Score (NPS) are widely used to learn how customers feel about a product or service. Sentiment analysis's ability to handle massive amounts of NPS answers and provide reliable findings fast led to its increased popularity.

15.1.2 Why Perform Sentiment Analysis?

Sentiment analysis is the process of interpreting words in their context to reveal a brand's social sentiment (Manguri et al. 2020). It also assists businesses in assessing if the product they are producing will find a market.

According to the report, 80% of global data is unstructured (Sofaer et al. 2019). Whether the data is in the form of emails, messages, papers, articles, or something else entirely, it must be analysed and organised.

- Sentiment analysis is necessary since it saves data in an economical and effective manner.
- Sentiment analysis is a real-time problem solver that can assist you with all real-time problems.

15.1.3 Types of Sentiment Analysis

1. Fine-grained sentiment analysis: As we explore the field of sentiment analysis, we come across a complex methodology that depends on the polarity spectrum (Zirn et al. 2011). These are carefully crafted into five different emotional categories: "very positive," "positive," "neutral," "negative," and "very negative." Users assess feelings using a 5-point rating system, where 5 represents extreme optimism, 2 indicates dislike, and 3 indicates neutrality.
2. Emotion Recognition: Sentiment analysis encompasses a wide range of emotions that take on various forms, such as "happy," "sad," "angry," "upset," "jolly," "pleasant," and more. This aspect of sentiment analysis, where the complex subtleties of emotions are decoded, is frequently referred to as the lexicon method.
3. Aspect-Centric Sentiment Analysis: This type of sentiment analysis focuses on particular characteristics or attributes for a more in-depth assessment (Vijay et al. 2020). Consider examining a cell phone's functionality in detail; this method breaks down features like battery life, screen quality, and camera performance.

4. Multilingual Sentiment Analysis: When dealing with multilingual content in the ever-expanding global landscape, sentiment analysis faces a multiplicity of challenges (Dashtipour et al. 2016). This work involves interpreting sentiment across a range of languages, which calls for the division of the data into "positive," "negative," and "neutral" categories. This project is extremely difficult and mentally taxing.

15.1.4 Why is Sentimental Analysis Important During COVID-19?

The COVID-19 pandemic has ushered in a new era of uncertainty and disruption. In these challenging times, understanding public sentiment has become more critical than ever. Sentiment analysis, the process of determining public sentiment through data analysis, plays a pivotal role in navigating the complexities of the pandemic.

- Public Health Monitoring—Sentiment analysis facilitates the tracking and understanding of the public's emotional reaction to the pandemic. Researchers and public health authorities can learn more about the general sentiment surrounding COVID-19 by examining social media posts, news articles, and other online sources. Using this data to inform public health interventions and strategies is invaluable.
- Crisis Management—There is a great deal of fear and uncertainty due to the pandemic. Authorities can identify areas with high levels of fear, anxiety, or misinformation by using sentiment analysis. Equipped with this understanding, they can respond to these issues with prompt and precise information. This promotes a more composed reaction to the crisis and lessens panic. • Business Impact—The pandemic has had an adverse effect on businesses. Businesses can use sentiment analysis to measure customer sentiment, modify their approach, and make well-informed decisions in uncertain economic times. It is essential for businesses to comprehend customer sentiment regarding goods, services, and the general state of the economy in order to survive and adapt.
- Media Coverage—Responsible reporting is crucial during a crisis like COVID-19 since the media has a significant influence on public opinion. Media outlets may better understand audience sentiment with the use of sentiment analysis, which enables them to customise their coverage and present accurate and pertinent information. This guarantees the public receives reliable news by halting the spread of false information. • Mental Health Support—Globally, the epidemic has had a negative impact on people's mental health. Sentiment analysis tracks people's online expressions to identify those who are distressed. By identifying these indicators of concern, mental health professionals may provide prompt treatments and support, reducing the severity of the pandemic-exacerbated mental health crisis.
- Policy Decisions—To stop the spread of COVID-19, governments have had to enact a variety of measures. Sentiment analysis may be used to determine if the

public supports or opposes these initiatives. Authorities can make better judgements and modify policies to better represent the interests and concerns of the public by knowing popular sentiment.

One effective weapon in the fight against COVID-19 is sentiment analysis. We can better adapt to the constantly changing environment and adjust our activities to meet the demands, worries, and anxieties of the populace by keeping an eye on public mood. It is an essential tool in these difficult times for ethical journalism, public health, crisis management, corporate adaptability, mental health assistance, and well-informed policy decisions. Sentiment analysis is not only a data-driven procedure; it is an essential part of our group's reaction to the epidemic, guiding us through these uncharted seas with more purpose and clarity.

15.2 Overview of Machine Learning Techniques for Sentimental Analysis

Opinion mining, or sentiment analysis, is a natural language processing (NLP) (Jang et al. 2019; Kanakaraj and Guddeti 2015a) activity that entails identifying the sentiment or emotional tone conveyed in a document. This procedure has been automated in large part because to machine learning techniques, and sentiment analysis may be done in a number of ways. An outline of various popular machine learning methods for sentiment analysis is provided below.

15.2.1 Lexicon-Based Approaches

Lexicon-based techniques, which depend on predetermined lists of words and phrases linked to positive and negative sentiment, are a prominent approach in sentiment analysis (Onat Topal 2102). Because these techniques are easy to apply and work well in various situations, they are often utilised.

This method assigns sentiment ratings to words according to how positive or negative they are thought to be. Words like "amazing" and "love," for instance, may obtain high positive ratings, whereas "hate" and "terrible" might receive high negative values. These ratings are used to establish a text's overall emotion. Summing the scores might do this, or more intricate techniques that account for each word's context may be employed.

Lexicon-based sentiment analysis makes use of a wide variety of lexicons. The VADER and AFINN lexicons (Al-Shabi 2020) are two instances of this.

A set of pre-calculated English words with valence ratings ranging from minus five (negative) to plus five (positive) is called the AFINN lexicon (Medhat et al. 2014). Finn Arup Nielsen personally labelled the words between 2009 and 2011.

However, another well-liked option is the Valence Aware Dictionary and Sentiment Reasoner (VADER) vocabulary (Pennington et al. 2014). It is a sentiment analysis tool that is especially suited to the feelings expressed on social media, utilising a lexicon and rule-based approach. Text sentiment is scored by VADER using a mix of qualitative and quantitative techniques.

Lexicon-based methods have several drawbacks even if they can be quite successful. They may find it difficult to understand writings that contain rare words or phrases since they mostly rely on the calibre of the vocabulary employed. Furthermore, they could not adequately convey feeling in writings when the sentiment is conveyed through intricate language devices like irony or sarcasm.

15.2.2 Machine Learning Algorithms

Because machine learning algorithms can discover intricate patterns and correlations in data, especially supervised learning algorithms, have gained popularity in the field of sentiment analysis. Decision trees (Zharmagambetov and Pak 2015), support vector machines (SVM) (Miazga and Hachaj 2019), and Naive Bayes (Melo et al. 2021) are a few of the most often utilised algorithms.

The probabilistic classifier Naive Bayes uses strong independence assumptions between the features to apply the Bayes theorem (Taleb et al. 2018). It is quite easy to use and surprisingly successful for applications involving text categorisation, such as sentiment analysis. It is capable of handling many classes and performs well even with high-dimensional datasets.

Strong classifiers known as support vector machines operate by identifying a hyperplane in a high-dimensional space that divides several classes (Miazga and Hachaj 2019). SVMs may be very useful in sentiment analysis when the data is not linearly separable and when they are combined with an appropriate kernel function. They can also manage big feature spaces and are resistant to overfitting.

Another kind of classifier is a decision tree, which models judgements made based on characteristics in a manner like a tree (Zeng 2020). They can handle both category and numerical data and are simple to comprehend and analyse. They are, nevertheless, susceptible to overfitting, particularly in the case of noisy datasets.

Usually, labelled datasets with text combined with sentiment labels (such as positive, negative, or neutral) are used to train these algorithms. During the training phase, the algorithms identify patterns and correlations in the data, which they then utilise in the prediction phase to categorise fresh material according to sentiment.

It is crucial to remember that although these machine learning techniques can be quite successful, a significant amount of labelled training data is also needed. In addition, they could have trouble with cognitively demanding activities including irony, sarcasm, and context awareness. Machine learning is still a promising method for sentiment analysis in spite of these obstacles.

15.2.3 Recurrent Neural Networks

RNNs are a sort of artificial neural network meant to recognise patterns in data sequences such as text, genomes, handwriting, or spoken syllables. Because of this, they are quite helpful for tasks like sentiment analysis, in which a sentence's meaning may be significantly altered by the word order (Arras et al. 2017; Baktha and Tripathy 2017).

RNNs possess "memory" in the form of hidden states that store data about previous calculations. RNNs may utilise information in infinitely long sequences in principle, but in reality, the "vanishing gradient problem" restricts them to looking back just a few steps (Grefenstette 1999). Here, the gradients produced by the backpropagation method get smaller and smaller until they no longer significantly impact the network's ability to learn.

To solve this problem, variants such as gated recurrent units (GRU) (Roehrick 2020) and long short-term memory (LSTM) (Kanakaraj and Guddeti 2015b) units have been created. • A kind of RNN known as LSTM is immune to the vanishing gradient issue and is capable of learning and remembering across lengthy sequences. It employs a number of "gates" to do this. As the sequence is analysed, these many neural networks work together to determine which information should be retained and which should be deleted.

- In contrast, GRU is a condensed form of LSTM. It creates a single "update gate" by fusing the input and forget gates. In addition, it modifies a few more things and combines the concealed and cell states. As a result, there is a more straightforward model that, while not as well as LSTM, nonetheless performs on some tasks.

Because these versions are more resilient to the vanishing gradient problem and can capture longer-term relationships in sequence data than ordinary RNNs, they have shown to be more successful for many tasks.

These models can be very helpful in sentiment analysis for deciphering the context and semantic meaning of words. They are able to forecast sentiment more accurately because they comprehend that words used early in a phrase might influence the meaning of words used later.

15.2.4 Convolutional Neural Networks

The main use of convolutional neural networks (CNNs) is image analysis. They can automatically and adaptively learn the spatial hierarchies of features, which have led to their great success in tasks like picture identification and classification.

CNNs have, nonetheless, also been modified for text sentiment analysis, and the results have been encouraging. The fact that the convolutional kernels (filters) in CNNs may function as n-gram detectors is the foundation for the concept of employing CNNs for text (Nezhad and Deihimi 2022). A continuous series of n elements from a particular text or audio sample is called an n-gram. For instance, a 2-g, also known as a bigram, is a statement that consists of two words.

Convolutional layers are used by CNNs to extract high-level characteristics from embedded word vectors in text data. The network then uses these traits to categorise the text according to emotion.

The procedure is as follows: First, the text data is converted into a matrix, with each row denoting a word and each column a word embedding dimension. After that, this matrix is run through one or more convolutional layers, which extract features by swiping filters of various sizes over the whole matrix. After that, a fully linked layer processes these characteristics to provide the final sentiment categorisation.

CNNs are useful for sentiment analysis because they can detect significant n-grams that are predictive of sentiment and capture local dependencies in the text. Because of this, they work well on jobs like document categorisation where it is critical to comprehend the local semantic meaning.

It is important to keep in mind, though, that because CNNs lack sequential processing, they might not work as well as other models like RNNs or LSTMs on jobs that call for understanding context or longer-term dependencies.

15.2.5 Word Embeddings

Words with similar meanings can have comparable representations thanks to a kind of word representation called word embeddings (Bakarov 2018). They are a distributed representation of text that is perhaps one of the main innovations behind deep learning techniques' remarkable results on difficult natural language processing tasks.

A language's words are mapped to high-dimensional vectors using word embeddings. Numerous linguistic aspects of the words, including their syntax and semantics, are captured by these vectors. For instance, a well-trained word embedding model would show semantic similarity between synonyms if the cosine similarity between their vectors is high.

Word2Vec (Church 2017) and GloVe (Ouyang et al. 2015) are two well-liked methods for learning word embeddings.

- Google researchers created Word2Vec, a neural network-based solution that can train word embeddings using a continuous bag-of-words (CBOW) model or a skip-gram model. Whereas the CBOW model predicts a word given its context, the skip-gram model predicts the word given its context (Hodson 2022).
- Another technique for learning word vectors is called Global Vectors for Word Representation (GloVe), which was created by researchers at Stanford. Using global statistics of the corpus instead of local context windows is how it varies from Word2Vec. The fundamental tenet of GloVe is that the co-occurrence matrix may be used to infer semantic links between words (Ouyang et al. 2015).

These embeddings have the potential to greatly improve machine learning models' sentiment analysis task performance. They make it possible for models to recognise semantic similarities between words and to accurately depict the context in which they are used. As a result, sentiment forecasts become more precise.

Word embeddings, it should be noted, can capture many semantic intricacies of language, but they might not always be adequate to interpret sentiment. For instance, they could have trouble comprehending sarcasm or negations. As a result, they are frequently used with other sentiment analysis models or approaches.

15.2.6 Aspect-Based Sentiment Analysis

Beyond only analysing a text's overall sentiment, aspect-based sentiment analysis, or ABSA, offers more in-depth sentiment analysis. Rather than merely classifying a text as good, negative, or neutral, ABSA looks for particular elements or things referenced in the text and evaluates the feelings attached to them (Vijay et al. 2020).

This approach involves two main steps: aspect extraction and sentiment determination.

- Finding the precise aspects or things described in the text is known as aspect extraction. These might be characteristics of a company's services, a product, or any other particular entities. A review of a smartphone, for instance, would cover things like battery life, screen size, camera quality, etc.
- Determining the sentiment conveyed towards each recognised element is the process of sentiment determination. This might entail giving the feeling a sentiment score or categorising it as good, negative, or neutral.

ABSA is very useful for examining client comments and product reviews. Businesses may obtain more detailed insights into the thoughts and preferences of their customers by recognising feelings around particular features of a product or service. For instance, a business may utilise ABSA to find out that although consumers adore their product's look, they are dissatisfied with its robustness. Detailed feedback of this sort, which is not possible with generic sentiment analysis, can direct innovations and improvements.

But ABSA has its own set of difficulties. Because context and semantics must be understood, aspect extraction can be challenging. Furthermore, feelings can be nuanced and multifaceted; one sentence may convey distinct feelings on several topics. Notwithstanding these difficulties, sentiment analysis research on ABSA is moving into new and fascinating territory.

15.2.7 Deep Learning Models

Sentiment analysis has seen a radical change with the introduction of transformer models, specifically models such as Bidirectional Encoder Representations from

Transformers (BERT) (Alaparthi and Mishra 2007), Generative Pretrained Transformer (GPT) (Zhu and Luo 2022), and a Robustly Optimised BERT Pretraining Approach (RoBERTa) (Li and Qian 2016).

Google's BERT employs a bidirectional training methodology that enables it to comprehend a word's context by taking into account both the word's left and right surrounds. Without requiring significant changes to the task-specific architecture, the pre-trained BERT model may be improved with just one more output layer to provide state-of-the-art models for a variety of tasks, including question answering and language inference.

With 175 billion parameters (Tan et al. 2009), GPT is a massive transformer-based language model created by OpenAI. The most recent version of GPT, GPT-3, has demonstrated remarkable performance in tasks requiring the creation of language that is human-like. Poetry and short tales may be produced, essays can be written, lengthy papers can be summarised, languages can be translated, and even Python code can be produced.

Facebook AI's RoBERTa is a BERT version that employs an alternative training methodology. It trains with substantially bigger mini-batches and learning rates and does away with the next-sentence pretraining target. In addition, RoBERTa uses more data and trains for a longer period of time than BERT (Li and Qian 2016).

Because these transformer models have been pre-trained on big text data corpora, they are capable of learning a broad variety of linguistic patterns. After that, they might be improved upon for certain sentiment analysis jobs to produce cutting-edge outcomes. The method of fine-tuning entails utilising a smaller labelled dataset to train the model on a particular job, such as sentiment analysis.

These models' capacity to comprehend context via their attention processes is one of their main advantages. This enables them to pick up on subtleties in language that are important for precisely gauging emotion.

It is important to keep in mind, though, that even while these models can provide amazing outcomes, they also need a large amount of processing power for both training and inference. Additionally, because of the intricacy of the models, their projections might occasionally be difficult to understand.

15.3 Data Collection and Preparation

Sentiment analysis is a data-driven subject; therefore it is critical to understand that the type and quality of the data you use may have a significant and long-lasting influence on the outcomes you get. In this section, we will take an enlightening tour, exploring the different types of data sources that are frequently used for sentiment analysis, clarifying the various approaches to acquiring data, and highlighting the critical function that data preparation and meticulous cleaning procedures fulfil. These crucial elements are of utmost importance in this exciting and dynamic field since they jointly provide the cornerstone upon which sentiment analysis initiatives hinge.

15.3.1 Types of Data Sources for Sentiment Analysis

- Social media: Websites such as Facebook, Instagram, and Twitter are rich text data sources that convey a diverse spectrum of thoughts and feelings. • Product reviews: Because user-generated reviews and ratings are a feature of e-commerce websites like Yelp and Amazon, they are useful for sentiment analysis, especially when it comes to assessments of products and services.
- Surveys and feedback forms: Companies frequently use customer support interactions, feedback forms, and surveys to get input from their clients. This structured text data is then used for sentiment analysis. • News and stories: Text data from online news sources, blogs, and news stories is a useful tool for gauging public opinion on a range of subjects. • Forums and discussion boards: User-generated content may be mined for sentiment analysis from online forums such as stack exchange and Reddit. • Chat Logs: Real-time or historical sentiment analysis can be done using conversational data from chat programmes, emails, or customer support exchanges.
- Academic research: When benchmarking sentiment analysis models, datasets and corpora developed for academic sentiment analysis research are frequently utilised. Examples of these are the Twitter sentiment dataset and the IMDb movie reviews dataset.
- Bespoke data gathering: Depending on the demands of their particular study or company, organisations or researchers may use web scraping or other bespoke data gathering techniques to generate their datasets.

15.3.2 Data Collection Methods

- Web scraping: Data extraction from webpages is accomplished using a technique called web scraping. To obtain the necessary data, it entails sending HTTP requests to the URLs of particular websites, then processing the HTML result. This is a typical usage for Python libraries such as BeautifulSoup. When the data you want is unavailable via APIs or in structured forms such as CSV or JSON, web scraping may be especially helpful.
- APIs: Users may access their data programmatically through a variety of online platforms and services that offer Application Programming Interfaces (APIs). For instance, Twitter has APIs that let you get tweets from particular individuals, within particular regions, or with particular keywords. Understanding how to send HTTP requests and handle JSON replies is necessary in order to use APIs.
- Manual data entry: In certain situations, especially when working with offline sources like paper surveys or feedback forms, it might be necessary to manually enter data into a digital format so that it can be analysed. Although it can take a while, there are situations in which automated data collecting techniques might not be an option.

- Crowdsourcing: Crowdsourcing is the process of gathering information from a big number of individuals, usually online. Using services such as Amazon Mechanical Turk, you may use human annotators to gather and annotate data. This can be a rapid and economical approach to collect data, especially for activities like sentiment annotation that call for human assessment.
- Purchase of data: Certain organisations opt to buy datasets from data providers that have been labelled with sentiment. These professionally managed datasets may save a great deal of time when it comes to data collecting and preparation. They might not always have the precise data you want for your study, though, and they might also be costly.

15.3.3 Data Preparation and Cleaning

Data preparation and cleaning are essential to ensure the data used for sentiment analysis is of high quality and suitable for machine learning:

- Text preprocessing: This is the first step in any Natural Language Processing task. It involves preparing and cleaning the text data for analysis. This includes:
  - Tokenisation: This is the process of splitting the text into individual words or phrases, known as tokens (Fan et al. 2006). These tokens help in understanding the context or developing a model for prediction.
  - Lowercasing: Text data can often be inconsistent when it comes to case usage, so it is a common practice to convert all the text to lower case. This helps avoid having multiple copies of the same words.
  - Removing punctuation: Punctuation can provide grammatical context to a sentence which supports our understanding. But for our vectoriser which counts the number of words and not the context, it does not add value, so it is often removed.
- Stop word removal: Stop words are common words that are often removed in NLP because they do not carry much information or sentiment. Examples include "and," "the," and "is." Removing these words can help improve the performance of the model, as it reduces the amount of noise in the data.
- Normalisation: This is a systematic approach of decomposing and transforming complex words to their simplest form.
  - Stemming: This is a rudimentary rule-based process of removing suffixes ("ing," "ly," "es," "s," etc.) from a word (Balakrishnan and Lloyd-Yemoh 2014).
  - Lemmatisation: This takes into consideration morphological analysis of the words and transforms words to the root word (Balakrishnan and Lloyd-Yemoh 2014). For example, "running" would be transformed to "run."
- Handling missing data: Incomplete or missing data is a common issue in data analysis. There are several strategies to handle this:

  - Imputation: This involves filling missing values based on statistical measures such as mean, median, or mode.
  - Removing records: If a record has too many missing values, it might be best to entirely remove that record from the dataset.

- Dealing with noise: Noise is irrelevant or meaningless data that can distort your final results. It could be caused by data entry errors, poor transcription, etc. Strategies for dealing with noise include:

  - Filtering out noise: This involves identifying and removing noisy data. This could be irrelevant symbols or special characters in text data.
  - Data transformation: In some cases, noise can be reduced by transforming the data into another format.

- Balancing data: In machine learning, having an imbalanced dataset can lead to biased results, as the model will be more exposed to the majority class. Techniques like oversampling (increasing the number of minority class samples) or undersampling (decreasing the number of majority class samples) can be applied to balance the classes.
- Encoding labels: Machine learning models require numerical input. Therefore, sentiment labels like "positive," "negative," and "neutral" need to be converted into numerical values. This is typically done using techniques like one-hot encoding or label encoding.
- Data splitting: To evaluate a model's performance, the dataset is usually split into training, validation, and test sets. The training set is used to train the model, the validation set is used to tune parameters and prevent overfitting, and the test set is used to evaluate the final model's performance.
- Feature extraction: Text data needs to be transformed into a format that machine learning algorithms can understand. This involves extracting features from the text such as:

  - Term Frequency-Inverse Document Frequency (TF-IDF): This is a statistical measure used to evaluate how important a word is to a document in a collection or corpus (Christian et al. 2016).
  - Word embeddings: These are dense vector representations that capture semantic properties of words.

- Data quality assurance: Ensuring data quality is crucial for any machine learning task. This involves regularly monitoring and maintaining data quality to ensure that the dataset remains relevant and up-to-date for ongoing sentiment analysis tasks. It could also involve cleaning the data by removing duplicates, handling missing values, and correcting inconsistent entries.

15.4 Case Study

15.4.1 Case Study 1: Sentiment Analysis of COVID-19 Using Tweets

15.4.1.1 Introduction

Sentiment analysis, or opinion mining, is a branch of natural language processing that aims to identify and extract the subjective feelings and opinions expressed in text. This technique can be used to analyse the public sentiment and emotional reactions towards the COVID-19 pandemic by examining tweets posted by users from different countries and regions. By applying various methods and tools of sentiment analysis, such as lexicon-based, machine learning, and deep learning approaches, researchers can gain valuable insights into how people cope with the crisis, what are their main concerns and challenges, and how they perceive the actions and policies of governments and health authorities. In this case study (Chintalapudi et al. 2021; Liu et al. 2019; Verma and Thakur 2018), we will review some of the recent studies that have used sentiment analysis to explore the impact of COVID-19 on the social media discourse and the implications for public health communication and intervention.

15.4.1.2 Data Collection

The first step in this case study is data collection. For this study, we use the Twitter API to collect tweets related to COVID-19. To find relevant tweets, we may use terms like "COVID-19," "coronavirus," "pandemic," etc. It is crucial to remember that there are restrictions and use guidelines for Twitter's API that must be followed.

15.4.1.3 Preprocessing

Once the data is collected, it needs to be preprocessed for analysis. This involves steps like:

- Cleaning: Removing unnecessary information from the tweets like URLs, numbers, special characters, etc.
- Tokenisation: Breaking down the tweet into individual words or tokens.
- Stopword removal: Removing common words that do not contribute much information for analysis like "is," "the," "and," etc.
- Stemming/Lemmatisation: Reducing words to their base or root form (e.g. "running" to "run").

15.4.1.4 Sentiment Analysis

After preprocessing, the tweets are ready for sentiment analysis. There are several approaches for this:

- Lexicon-based methods: These methods calculate sentiment based on the presence of words from predefined lists of positive and negative words. • Machine learning methods: These methods involve training a classifier on a labelled dataset.

15.4.1.5 Results and Discussion

The results of the sentiment analysis can be visualised using graphs and charts to better understand public sentiment towards COVID-19. This can provide valuable insights for health organisations, governments, and policymakers.

15.4.1.6 Conclusion

Real-time insights into public opinion and emotional responses to events like as the COVID-19 outbreak may be obtained by sentiment analysis of tweets. Sentiment analysis should, however, be used in conjunction with other data sources for decision-making due to its limitations.

An overview of sentiment analysis as it relates to COVID-19 tweets is given in this case study. The real implementation would include more intricate procedures and could call for extensive understanding of machine learning and natural language processing.

15.4.2 Case Study 2: Sentimental Analysis of COVID-19 News Articles

15.4.2.1 Objective

To perform sentiment analysis on a collection of COVID-19 news articles to gain insights into public sentiment and emotional trends surrounding the pandemic.

15.4.2.2 Background

The COVID-19 pandemic, which emerged in late 2019, has been a global crisis with far-reaching impacts on society, economies, and public health. News media has played a crucial role in disseminating information and shaping public perception of the pandemic. This case study (Dashtipour et al. 2016; Yin et al. 2022) aims to analyse the sentiment of news articles related to COVID-19 to understand how the media has influenced public sentiment over time.

15.4.2.3 Data Collection

- Data Sources: A comprehensive dataset of COVID-19 news articles was collected from reputable news sources, including newspapers, online news websites, and magazines. The dataset spans from January 2020 to September 2023, encompassing different phases of the pandemic.
- Data preprocessing: The collected text data underwent preprocessing, which included text cleaning (removing HTML tags, special characters, and punctuation), tokenisation, stop word removal, and stemming to prepare it for sentiment analysis.

15.4.2.4 Methodology

Sentiment analysis was performed using natural language processing and machine learning techniques. The following steps were carried out:

- Sentiment lexicon: A sentiment lexicon containing positive and negative words was used to classify the sentiment of the articles. The lexicon was created from a large corpus of text data and assigned polarity scores to each word.
- Sentiment classification: Each news article was divided into sentences, and the sentiment of each sentence was determined based on the words it contained. The sentiment scores of all sentences in an article were then aggregated to calculate the overall sentiment of the article. • Time series analysis: Sentiment scores were analysed over time to identify trends and changes in public sentiment regarding COVID-19. The data was visualised using time series plots to illustrate sentiment shifts. • Topic modelling: To identify prevalent topics within the articles, topic modelling techniques like Latent Dirichlet Allocation (LDA) were applied. This helped to understand which aspects of the pandemic were most frequently discussed in positive or negative contexts.

15.4.2.5 Results

- Temporal sentiment analysis: The analysis revealed shifts in public sentiment over time. Early in the pandemic, news articles were predominantly negative, focusing on the spread of the virus and its consequences. As vaccination efforts progressed, sentiment became more positive due to the hope of controlling the virus.
- Topic analysis: Topic modelling identified themes such as public health measures, vaccine distribution, economic impacts, and misinformation. Positive sentiment was often associated with vaccination progress and scientific advancements, while negative sentiment frequently related to case surges and misinformation.

- Sentiment by region: Further analysis showed regional variations in sentiment, with some areas experiencing more negative sentiment due to local outbreaks or policy decisions.

15.4.2.6 Implications

- Media influence: The media plays a substantial role in shaping public sentiment. Understanding how news coverage affects public perception of the pandemic is crucial for public health officials and communicators. • Policy adjustments: Public sentiment can influence government policies and public compliance with health measures. Insights from this analysis can help policymakers tailor their messaging and strategies accordingly. • Monitoring misinformation: Identifying negative sentiment associated with misinformation can guide efforts to counter false information and promote accurate information.

15.4.2.7 Conclusion

Sentiment analysis of COVID-19 news articles provides valuable insights into how the pandemic has been portrayed in the media and how this portrayal has influenced public sentiment. By tracking sentiment over time and by region, organisations and policymakers can respond effectively to address public concerns and improve crisis communication strategies. This case study demonstrates the importance of analysing public sentiment in a dynamic and evolving crisis situation.

15.4.3 Case Study 3: Sentimental Analysis of COVID-19 Vaccine Reviews

15.4.3.1 Objective

To perform sentiment analysis on user-generated COVID-19 vaccine reviews from various sources to understand public sentiment and identify key factors influencing vaccine perception.

15.4.3.2 Background

The rapid development and distribution of COVID-19 vaccines have been a pivotal aspect of the global response to the pandemic. Understanding public sentiment regarding these vaccines is essential for public health campaigns, vaccine adoption, and addressing concerns and misconceptions (Mullen and Collier 2004; Yacouby and Axman 2020).

15.4.3.3 Data Collection

Data sources: A dataset of user-generated COVID-19 vaccine reviews was collected from various online sources, including social media platforms, review websites, and forums. The dataset covers reviews spanning from the initial vaccine rollout in late 2020 to the present day.

Data preprocessing: The collected data underwent preprocessing, including text cleaning, tokenisation, stop word removal, and stemming, to prepare it for sentiment analysis.

15.4.3.4 Methodology

Sentiment analysis was conducted using natural language processing and machine learning techniques. The following steps were taken:

- Sentiment classification: Each vaccine review was analysed for sentiment, classifying it as positive, negative, or neutral. Sentiment classification was based on the sentiment lexicon, which was developed specifically for COVID-19 vaccine reviews.
- Aspect-based sentiment analysis: To delve deeper, an aspect-based sentiment analysis was performed. This aimed to identify specific aspects or attributes mentioned in reviews (e.g. vaccine efficacy, side effects, and availability) and determine the sentiment associated with each aspect. • Topic modelling: To uncover common topics within the reviews, topic modelling techniques, such as Latent Dirichlet Allocation (LDA), were applied. This helped identify recurring themes or concerns in the reviews.

15.4.3.5 Results

Overall sentiment: The analysis revealed that the majority of vaccine reviews were positive, with users expressing relief, hope, and gratitude. However, a significant number of reviews were neutral, reflecting a lack of strong sentiment, while a smaller portion was negative, often citing concerns about side effects or vaccine hesitancy.

Aspect-based sentiment: Different aspects of the vaccine received varying sentiment. Efficacy and accessibility typically garnered positive sentiment, while side effects and vaccine distribution sometimes resulted in negative sentiment.

Common topics: Topic modelling unveiled several recurring themes, including vaccine safety, vaccination experiences, and information sources. Users' concerns often centred around vaccine safety and misinformation.

15.4.3.6 Implications

- Public health communication: Public health authorities can leverage the positive sentiment expressed in reviews to promote vaccine adoption. Addressing specific concerns and negative sentiment related to side effects or misinformation is crucial for effective communication.
- Vaccine development: The positive sentiment around vaccine efficacy reflects the success of vaccine development efforts, but this also highlights the importance of transparency and data-driven communication. • Addressing concerns: Understanding the negative sentiment and concerns expressed in reviews allows authorities to address these issues and work to build trust.

15.4.3.7 Conclusion

Sentiment analysis of COVID-19 vaccine reviews provides insights into how the public perceives and discusses vaccines. Positive sentiment can be leveraged for vaccination campaigns, while addressing negative sentiment is essential for building trust and improving vaccine adoption. The aspect-based analysis allows for a nuanced understanding of user concerns and preferences, aiding in tailoring public health campaigns and communication strategies. This case study underscores the significance of analysing public sentiment in the context of a global vaccination campaign during a public health crisis.

15.5 Evaluation of Machine Learning Models

Once machine learning models have been trained for sentiment analysis, it is crucial to evaluate their performance to assess how well they can predict sentiment in new, unseen data. In this section, well delve into the performance metrics used and the process of model selection and validation.

15.5.1 Performance Metrics

- Accuracy: One of the easiest measures to understand when analysing machine learning is accuracy. The percentage of correctly categorised cases relative to all instances is computed. It is crucial to remember, though, that precision can be deceptive when working with unbalanced datasets. A model that consistently predicts a class that comprises 95% of your data, for instance, will have an accuracy of 95%, but it may not be a strong model overall.
- Precision: The ratio of real positive predictions, or positively predicted occurrences that were made properly, to all positive predictions, including those that

were made wrong, is known as precision. When the cost of false positives is significant, this statistic becomes valuable. For instance, you would want to reduce the quantity of legitimate emails that are mistakenly categorised as spam in spam detection.

- Recall (Sensitivity): The percentage of true positive cases that are accurately detected is calculated using recall (Xing and Tsang 2022). When the cost of false negatives is significant, this statistic becomes extremely important. When diagnosing an illness, for example, you would like to reduce the proportion of sick patients who are mistakenly classified as healthy.
- F1 Score: A balanced measure between accuracy and recall, the F1 score is calculated as the harmonic mean of these two measures. It is especially helpful when handling unbalanced classes or when you worry about recall and precision equally (Xing and Tsang 2022).
- Confusion matrix: This table is frequently used to explain how well a classification model is performing (Xinran et al. 2020). True positives (TP), true negatives (TN), false positives (FP), and false negatives (FN) are all broken down in depth. The majority of classifier assessment metrics are based on these four results.
- ROC curve (receiver operating characteristic): This graphical representation shows how a binary classifier system's diagnostic capability changes with the discrimination threshold (Dey et al. 2016). Plotting the true positive rate (TPR), also known as recall, versus the false positive rate (FPR) at different threshold levels yields the curve. The classifier's total performance is summed up by a single scalar statistic called the area under the ROC curve (AUC-ROC). An AUC-ROC of 1 indicates a perfect classifier, whereas an AUC-ROC of 0.5 indicates a random classifier. • Area under the precision-recall curve (AUC-PR): In classification and assessment, AUC-PR (Santur 2019) is a crucial performance measure. AUC-PR emphasises the complex trade-off between precision and recall, which is a distinct aspect of model evaluation than AUC-ROC. This measure stands out as being especially helpful in the case of unbalanced datasets, which are defined by a low number of positive examples relative to negative ones. AUC-PR, which accounts for the delicate balance between the accuracy of positive predictions and the capacity to capture as many true positive instances as possible, offers a more nuanced and insightful measure of a model's efficacy in such scenarios. This can be crucial in domains where the costs associated with false positives or false negatives are not equal.
- Mean absolute error (MAE) and mean squared error (MSE): These are typical metrics for calculating the difference between expected and actual values in regression tasks (Hochreiter 1998). Without taking into account the direction of the mistakes, MAE calculates the average magnitude of the errors in a series of forecasts. It is the mean, with each individual difference carrying equal weight, of the absolute differences between the observed and predicted values over the test sample. While MSE and MAE are comparable, MSE penalises greater mistakes by squaring the difference before averaging them.

15.5.2 Model Selection and Validation

- Cross-validation: In the field of machine learning, cross-validation methods—like the popular k-fold cross-validation—are essential instruments. These techniques partition the dataset into several subsets, making it easier to assess a model's performance in a more thorough and reliable manner. Cross-validation increases our confidence in the efficacy and robustness of a model by iteratively rotating subsets for testing and training. This helps determine a model's capacity to generalise across various data divisions.
- Hyperparameter tuning: Adjusting hyperparameters (Bardenet et al. 2013) is a crucial step in the creation of a model. The goal of this project is to optimise a model's performance through methodical hyperparameter adjustments. To find the best configuration, methods like random search, which randomly samples hyperparameter combinations, and grid search, which examines a predetermined set of hyperparameters, are commonly used. To get the most possible performance out of the model, this careful tweaking procedure is necessary.
- Validation set: Using a validation set is essential to both training and evaluating models. In order to continuously track a model's performance as it learns, a piece of the data that is separate from the training and testing sets must be kept aside. This procedure is essential for identifying and addressing overfitting, a typical machine learning hazard. We may determine a model's applicability and efficacy during training by evaluating its performance on unknown data, which eventually improves model selection.
- Ensemble methods: Ensemble techniques offer a potent way to increase model resilience and performance, such as random forests and gradient boosting. By combining the predictions of several different models, these methods take use of their various advantages. Ensemble approaches can improve overall performance and generalisation by combining these predictions. They can smooth out individual model shortcomings and enhance their combined strengths, resulting in a more robust and accurate prediction model. They are especially useful when dealing with complicated issues or noisy datasets.
- Bias and fairness assessment: It is critical that sentiment analysis models undergo thorough evaluations for bias and fairness, especially when these models are used in sensitive contexts. Verifying that the model's predictions are impartial and fair to various demographic groupings is the goal. This involves a thorough analysis of the model's outputs in order to spot and address any discrepancies and make sure the model does not reinforce or magnify preexisting biases.
- Test on real-world data: Testing the most promising model with real-world data is essential once it has been chosen. The model's capacity to generalise to fresh, untested data in the application area is evaluated at this round of review. This stage is essential for determining a model's practical feasibility since real-world data may provide certain issues and complexities that are not fully captured during training.

- Continuous monitoring: For sentiment analysis models to continue working in changing situations, they frequently need to be continuously observed and retrained. It is critical to update the model to reflect changes in sentiment expressions and linguistic patterns over time. Constant observation guarantees that the model stays current and in step with the dynamic world of language and emotion.
- Interpretable models: Context should play a role in the model architecture selection process. Certain scenarios may call for using interpretable models—like decision trees or linear regression—instead of more complicated models—like deep neural networks. In situations where model interpretability is a top concern, interpretable models clearly have an advantage in terms of model comprehension and transparency, which can be crucial for regulatory compliance and decision-making procedures.

The process of selecting and validating models is iterative, and the selection of metrics is contingent upon the particular problem at hand as well as the significance of performance metrics like as accuracy and recall. Over time, sentiment analysis systems' accuracy and applicability depend on ongoing model performance monitoring and assessment.

15.6 Conclusion and Future Directions

15.6.1 Summary of Key Points

In this chapter, we covered:

- Data collection and preparation: The importance of sourcing relevant data, the methods for data collection, and the essential data preparation steps that ensure the quality and suitability of the data for sentiment analysis.
- Machine learning techniques: A comprehensive overview of the machine learning techniques and deep learning methods used for sentiment analysis, including lexicon-based approaches, supervised learning, recurrent and convolutions neural networks, and the influential role of transformer models.
- Evaluation of machine learning models: The performance metrics for assessing model accuracy and the model selection and validation processes crucial for determining the effectiveness of sentiment analysis models.

15.6.2 Conclusion

Sentiment analysis is an essential part of natural language processing and has proven to be useful for understanding and using sentiment in textual data in a variety of applications. Mood analysis has shown to be versatile across a range of disciplines, from enabling businesses with data-driven insights to tracking public mood during significant events like the COVID-19 epidemic. Sentiment analysis has evolved over time, moving from simple lexicon-based techniques to state-of-the-art deep learning models. This technology is set to develop in its ability to impact decision-making and provide a more thorough understanding of public emotion as it continues to broaden its scope. As sentiment analysis technology continues to progress and the variety of data sources grows, it is critical to be aware of new developments and prepared to adjust as natural language understanding and sentiment analysis become more and more of a developing field.

15.6.3 Future Directions

The subject of sentiment analysis is expected to develop and expand in tandem with the ongoing advancements in technology and data science. The following are some trends and future directions to be aware of:

- Multimodal sentiment analysis: In order to provide a more thorough understanding of sentiment in multimedia information, sentiment analysis will progressively combine several data modalities, including text, pictures, audio, and video.
- Emotion recognition: Sentiment analysis models will advance to identify and differentiate a wider range of emotions than only positive, negative, and neutral sentiment. This will improve the uses of these models in a variety of sectors, such as mental health and entertainment. • Real-time analysis: Sentiment analysis models will get increasingly responsive in response to the increasing need for real-time insights, allowing for quick sentiment tracking and decision-making.
- Fair and ethical sentiment analysis: Resolving biases in sentiment analysis models and guaranteeing openness, justice, and ethical concerns in their creation and application will receive more attention.
- Bespoke models and domain-specific sentiment analysis: In order to serve certain sectors or use cases, organisations are going to construct more and more bespoke models and datasets. This will lead to sentiment analysis that is more precise and specialised to a particular domain. • Interpretability: Explainable AI (XAI) methods will be used increasingly frequently in sentiment analysis models to improve the interpretability and reliability of their conclusions and decision-making procedures. • Sentiment analysis will be included into chatbots, virtual assistants, and conversational AI systems to improve their comprehension of user sentiment and emotions during interactions.
- Sentiment analysis for new languages: To enable a broader worldwide applicability, research and development initiatives will broaden sentiment analysis to encompass new languages and dialects.

References

Al-Shabi M (2020) Evaluating the performance of the most important lexicons used to sentiment analysis and opinions mining. IJCSNS 20(1):1

Alaparthi S, Mishra M (2020) Bidirectional encoder representations from transformers (BERT): a sentiment analysis odyssey. arXiv:2007.01127

Arras L, Montavon G, Müller KR, Samek W (2017) Explaining recurrent neural network predictions in sentiment analysis. arXiv:1706.07206

Bakarov A (2018) A survey of word embeddings evaluation methods. arXiv:1801.09536

Baktha K, Tripathy BK (2017) Investigation of recurrent neural networks in the field of sentiment analysis. In: 2017 international conference on communication and signal processing (ICCSP). IEEE, pp 2047–2050

Balakrishnan V, Lloyd-Yemoh E (2014) Stemming and lemmatization: a comparison of retrieval performances

Bardenet R, Brendel M, Kégl B, Sebag M (2013) Collaborative hyperparameter tuning. In: International conference on machine learning. PMLR, pp 199–207

Chintalapudi N, Battineni G, Amenta F (2021) Sentimental analysis of covid-19 tweets using deep learning models. Infect Dis Rep 13(2):329–339

Christian H, Agus MP, Suhartono D (2016) Single document automatic text summarization using term frequency-inverse document frequency (tf-idf). ComTech: Comput Math Eng Appl 7(4):285–294

Church KW (2017) Word2vec. Nat Lang Eng (1):155–162

Dashtipour K, Poria S, Hussain A, Cambria E, Hawalah AY, Gelbukh A, Zhou Q (2016) Multilingual sentiment analysis: state of the art and independent comparison of techniques. Cogn Comput 8:757–771

de Melo T, Figueiredo CM et al (2021) Comparing news articles and tweets about covid-19 in brazil: sentiment analysis and topic modeling approach. JMIR Public Health Surveill 7(2):e24585

Dey L, Chakraborty S, Biswas A, Bose B, Tiwari S (2016) Sentiment analysis of review datasets using Naive Bayes and k-nn classifier. arXiv:1610.09982

Fan J, Upadhye S, Worster A (2006) Understanding receiver operating characteristic (ROC) curves. Can J Emerg Med 8(1):19–20

Grefenstette G (1999) Tokenization. In: Syntactic wordclass tagging. Springer, pp 117–133

Hochreiter S (1998) The vanishing gradient problem during learning recurrent neural nets and problem solutions. Int J Uncertain Fuzziness Knowl-Based Syst 6(02):107–116

Hodson TO (2022) Root-mean-square error (RMSE) or mean absolute error (MAE): when to use them or not. Geosci Model Dev 15(14):5481–5487

Jang B, Kim I, Kim JW (2019) Word2vec convolutional neural networks for classification of news articles and tweets. PloS one 14(8):e0220976

Kanakaraj M, Guddeti RMR (2015a) NLP based sentiment analysis on twitter data using ensemble classifiers. In: 2015 3Rd international conference on signal processing, communication and networking (ICSCN). IEEE, pp 1–5

Kanakaraj M, Guddeti RMR (2015b) Performance analysis of ensemble methods on twitter sentiment analysis using NLP techniques. In: Proceedings of the 2015 IEEE 9th international conference on semantic computing (IEEE ICSC 2015). IEEE, pp 169–170

Li D, Qian J (2016) Text sentiment analysis based on long short-term memory. In: 2016 First IEEE international conference on computer communication and the internet (ICCCI). IEEE, pp 471–475

Liu Y, Ott M, Goyal N, Du J, Joshi M, Chen D, Levy O, Lewis M, Zettlemoyer L, Stoyanov V (2019) Roberta: a robustly optimized BERT pretraining approach. arXiv:1907.11692

Manguri KH, Ramadhan RN, Amin PRM (2020) Twitter sentiment analysis on worldwide covid-19 outbreaks. Kurdistan J Appl Res 54–65

Medhat W, Hassan A, Korashy H (2014) Sentiment analysis algorithms and applications: a survey. Ain Shams Eng J 5(4):1093–1113

Miazga J, Hachaj T (2019) Evaluation of most popular sentiment lexicons coverage on various datasets. In: Proceedings of the 2019 2nd international conference on sensors, signal and image processing, pp 86–90

Mullen T, Collier N (2004) Sentiment analysis using support vector machines with diverse information sources. In: Proceedings of the 2004 conference on empirical methods in natural language processing, pp 412–418

Nezhad ZB, Deihimi MA (2022) Twitter sentiment analysis from Iran about COVID 19 vaccine. Diabetes Metab Syndr: Clin Res Rev 16(1):102367

Ouyang X, Zhou P, Li CH, Liu L (2015) Sentiment analysis using convolutional neural network. In: 2015 IEEE international conference on computer and information technology; ubiquitous computing and communications; dependable, autonomic and secure computing; pervasive intelligence and computing. IEEE, pp 2359–2364

Pennington J, Socher R, Manning CD (2014) Glove: global vectors for word representation. In: Proceedings of the 2014 conference on empirical methods in natural language processing (EMNLP), pp 1532–1543

Roehrick K (2020) Valence aware dictionary and sentiment reasoner (vader)

Santur Y (2019) Sentiment analysis based on gated recurrent unit. In: 2019 international artificial intelligence and data processing symposium (IDAP). IEEE, pp 1–5

Sofaer HR, Hoeting JA, Jarnevich CS (2019) The area under the precision-recall curve as a performance metric for rare binary events. Methods Ecol Evol 10(4):565–577

Taleb I, Serhani MA, Dssouli R (2018) Big data quality assessment model for unstructured data. In: 2018 international conference on innovations in information technology (IIT). IEEE, pp 69–74

Tan S, Cheng X, Wang Y, Xu H (2009) Adapting Naive Bayes to domain adaptation for sentiment analysis. In: Advances in information retrieval: 31th European conference on IR research, ECIR 2009, Toulouse, France, April 6–9, 2009. Proceedings 31. Springer, pp 337–349

Topal MO, Bas A, van Heerden I (2021) Exploring transformers in natural language generation: GPT, BERT, and XLNET. arXiv:2102.08036

Verma B, Thakur RS (2018) Sentiment analysis using lexicon and machine learning-based approaches: a survey. In: Proceedings of international conference on recent advancement on computer and communication: ICRAC 2017. Springer, pp 441–447

Vijay T, Chawla A, Dhanka B, Karmakar P (2020) Sentiment analysis on covid-19 twitter data. In: 2020 5th IEEE international conference on recent advances and innovations in engineering (ICRAIE). IEEE, pp 1–7

Xing B, Tsang IW (2022) Out of context: a new clue for context modeling of aspect-based sentiment analysis. J Artif Intell Res 74:627–659

Yacouby R, Axman D (2020) Probabilistic extension of precision, recall, and f1 score for more thorough evaluation of classification models. In: Proceedings of the first workshop on evaluation and comparison of NLP systems, pp 79–91

Yin H, Song X, Yang S, Li J (2022) Sentiment analysis and topic modeling for covid-19 vaccine discussions. World Wide Web 25(3):1067–1083

Yu X, Zhong C, Li D, Xu W (2020) Sentiment analysis for news and social media in covid-19. In: Proceedings of the 6th ACM SIGSPATIAL international workshop on emergency management using GIS, pp 1–4

Zeng G (2020) On the confusion matrix in credit scoring and its analytical properties. Commun Stat-Theory Methods 49(9):2080–2093

Zharmagambetov AS, Pak AA (2015) Sentiment analysis of a document using deep learning approach and decision trees. In: 2015 Twelve international conference on electronics computer and computation (ICECCO). IEEE, pp 1–4

Zhu Q, Luo J (2022) Generative pre-trained transformer for design concept generation: an exploration. Proc Des Soc 2:1825–1834

Zirn C, Niepert M, Stuckenschmidt H, Strube M (2011) Fine-grained sentiment analysis with structural features. In: Proceedings of 5th international joint conference on natural language processing, pp 336–344

Chapter 16
Precision Oncology: Innovations in Computational Intelligent for Cancer Detection

Saumendra Kumar Mohapatra, Abhishek Das, and Mihir Narayan Mohanty

Abstract Precision oncology, a burgeoning field situated at the confluence of medicine, genomics, and computational intelligence, is poised to redefine cancer detection and treatment paradigms. This book chapter explores the transformative role of computational intelligence in advancing precision oncology, focusing on machine learning, deep learning, and other computational techniques applied to cancer detection. We begin by outlining the limitations of traditional cancer detection methods and how computational approaches offer new possibilities for early diagnosis and personalized treatment strategies. Genomics and bioinformatics are highlighted for their crucial role in analyzing vast biological datasets to identify cancer biomarkers and predict disease progression. Advancements in imaging technologies and the integration of artificial intelligence (AI) to interpret radiological images are discussed for their impact on tumor detection and characterization. Additionally, we explore digital pathology and its contribution to cancer diagnosis through automated analysis of tissue samples. Ethical considerations and challenges associated with implementing computational intelligence in clinical settings are addressed, emphasizing data privacy, algorithm transparency, and equitable access to technology-driven healthcare solutions. In summary, this book chapter provides a concise overview of how computational intelligence is revolutionizing precision oncology. By leveraging cutting-edge technologies, we aim to enhance cancer detection accuracy and develop personalized treatment plans that improve patient outcomes. The potential of computational methods to decipher complex biological data and optimize clinical decision-making underscores a promising future for precision oncology.

S. K. Mohapatra
Faculty of Engineering & Technology, Sri Sri University, Cuttcak, India
e-mail: saumendra.m@srisriuniversity.edu.in

A. Das
Department of Computer Science and Engineering, Centurian University of Technology and Management Paralakhemundi, Jatani, India

M. N. Mohanty (✉)
Department of ECE, ITER, Siksha 'O' Anusandhan (Deemed to be University), Bhubaneswar, Odisha, India
e-mail: mihir.n.mohanty@gmail.com

A. Rai et al. (eds.), *Convergence of Artificial Intelligence, Machine Learning, and the Internet of Things in Industry 4.0 Applications*, Transactions on Computer Systems and Networks, https://doi.org/10.1007/978-981-96-5269-3_16

Keywords Cancer · Machine learning · Artificial intelligence · Oncology · Deep learning

16.1 Introduction

Cancer, a disease that has haunted humanity for millennia, is at a crossroads. Traditional, one-size-fits-all treatment approaches are giving way to a new era of personalized medicine: Precision oncology. This revolutionary field tailors cancer care to the unique genetic and molecular makeup of each patient's tumor. At the heart of this transformation lies a powerful arsenal of tools—innovations in computational intelligence (Dlamini et al. 2020; Bhinder et al. 2021; Currie and Delles 2018; Sharma et al. 2020).

For decades, the fight against cancer has relied heavily on broad-spectrum therapies like chemotherapy and radiation. While these treatments have saved countless lives, they often come with a heavy burden of side effects and limited effectiveness due to tumor heterogeneity—the vast diversity within a single cancer type. Precision oncology seeks to address this challenge by creating a detailed profile of a patient's cancer, encompassing its genetic mutations, protein expression patterns, and the tumor microenvironment. This intricate picture allows oncologists to identify the specific drivers of the cancer and design targeted therapies that exploit their vulnerabilities. At the core of precision oncology lies the integration of computational intelligence and advanced analytics into the oncology field. This fusion of technology and medicine holds tremendous promise in unraveling the complexities of cancer biology, identifying novel biomarkers, predicting treatment responses, and ultimately improving patient outcomes. In recent years, the marriage of computational intelligence with oncology has led to remarkable breakthroughs, empowering clinicians with tools to make data-driven decisions and ushering in a new era of personalized cancer care. The essence of computational intelligence in precision oncology lies in its ability to process vast amounts of data—from genomics and proteomics to clinical records and imaging—in a manner that transcends human capabilities. Machine learning algorithms, artificial intelligence (AI), and data mining techniques have become indispensable allies in the quest to decode the intricacies of cancer, providing insights that are instrumental in early detection, accurate diagnosis, and the development of targeted therapies (Giansanti 2024; Beckmann and Lew 2016; Lin et al. 2023).

This book chapter delves into the forefront of innovations within computational intelligence specifically tailored for cancer detection. It explores a myriad of methodologies, from sophisticated machine learning models to deep neural networks, that are revolutionizing our understanding of cancer biology and clinical decision-making. By harnessing the power of big data and computational algorithms, researchers and clinicians are gaining unprecedented insights into tumor behavior, paving the way for more precise and effective treatment regimens.

The exponential growth of data in healthcare, fueled by advancements in genomics, medical imaging, and electronic health records, presents both a challenge and an opportunity. This vast amount of complex data is difficult to analyze with traditional methods. Here's where computational intelligence steps in, offering a suite of powerful techniques to unlock the secrets hidden within this data deluge. Machine learning algorithms, trained on massive datasets of cancer cases, can learn to identify subtle patterns and relationships that would escape the human eye. Deep learning, a subfield of machine learning inspired by the structure and function of the brain, allows computers to analyze complex images like mammograms or CT scans with unprecedented accuracy, aiding in early detection and tumor characterization (Khalighi et al. 2024; Mohapatra et al. 2023; Xu et al. 2019; Carrasco-Ramiro et al. 2017; Seyhan and Carini 2019).

This chapter delves into the exciting world of computational intelligence in cancer detection. We will explore how machine learning algorithms are being used to:

- **Analyze genetic data**: By identifying mutations and gene expression patterns associated with specific cancer types, machine learning can help classify tumors and predict their behavior.
- **Interpret medical images**: Deep learning algorithms can analyze mammograms, X-rays, and CT scans to detect cancerous lesions with high accuracy, leading to earlier diagnoses and improved treatment outcomes.
- **Integrate multi-modal data**: Precision oncology benefits from a holistic view of the patient. Machine learning can integrate data from various sources, including genomics, imaging, and clinical history, to create a comprehensive picture of the cancer and predict its response to therapy.

The impact of computational intelligence extends beyond detection. By analyzing vast clinical trial data, machine learning can help predict which patients are most likely to benefit from specific therapies, paving the way for personalized treatment plans and improved patient outcomes.

However, the road to a future powered by computational intelligence is not without its hurdles. Ethical considerations surrounding data privacy, security, and potential bias in algorithms need to be addressed. Additionally, ensuring equitable access to these advanced technologies for all patients remains a challenge.

Despite these hurdles, the potential of computational intelligence in cancer detection is undeniable. This chapter will explore the cutting-edge advancements, the ongoing challenges, and the future directions of this transformative field. As we harness the ever-growing power of computational intelligence, we stand on the threshold of a new era in cancer detection, one where personalized medicine becomes a reality for all, offering a brighter future for patients battling this formidable disease.

16.2 Basics of Precision Oncology

Precision oncology represents a transformative approach to cancer care that leverages advances in molecular biology, genomics, and targeted therapies to customize treatment strategies based on individual patient characteristics. This section provides an in-depth exploration of the fundamental principles, technologies, and challenges associated with precision oncology (Mohanty et al. 2019).

Precision oncology, also referred to as personalized or individualized medicine, is rooted in the concept of tailoring cancer treatments to the unique genetic, molecular, and clinical characteristics of each patient and their tumor. Unlike traditional one-size-fits-all approaches, precision oncology aims to optimize therapeutic outcomes by targeting specific molecular alterations driving cancer progression while minimizing adverse effects on normal tissues. The key principles guiding precision oncology include:

- **Molecular Profiling**: Comprehensive analysis of genetic mutations, gene expression patterns, and protein alterations within tumor tissue to identify specific molecular targets for therapy.
- **Targeted Therapies**: Development and use of drugs designed to selectively inhibit cancer-promoting molecular pathways identified through molecular profiling.
- **Patient Stratification**: Classification of patients into subgroups based on molecular profiles to optimize treatment selection and predict therapeutic response.
- **Treatment Personalization**: Customization of treatment regimens based on individual patient characteristics, including genetic makeup, tumor biology, and clinical history.

Precision oncology encompasses a shift toward a more nuanced understanding of cancer biology, recognizing that tumors can exhibit diverse molecular profiles even within the same cancer type. By elucidating these molecular aberrations, precision oncology enables the identification of actionable targets for therapeutic intervention, leading to improved patient outcomes and reduced treatment-related toxicity.

Central to precision oncology is the utilization of molecular profiling techniques to characterize tumors at a molecular level, enabling the identification of clinically relevant biomarkers and therapeutic targets. Molecular profiling involves the analysis of tumor tissue or circulating biomarkers (e.g., circulating tumor DNA) using advanced technologies such as next-generation sequencing (NGS), immunohistochemistry (IHC), and liquid biopsy. Key technologies used in molecular profiling include:

- **Next-Generation Sequencing (NGS)**: High-throughput sequencing methods for comprehensive analysis of cancer genomes, transcriptomes, and epigenomes, facilitating the detection of genetic mutations, gene fusions, and copy number alterations.
- **Immunohistochemistry (IHC)**: Protein-based assays performed on tumor tissue to detect specific biomarkers or assess protein expression patterns associated with cancer progression.

- **Liquid Biopsy**: Non-invasive testing of circulating tumor DNA, circulating tumor cells, or exosomes in blood or other body fluids, providing real-time insights into tumor dynamics and treatment response.

Based on molecular profiling results, targeted therapies can be tailored to inhibit specific molecular targets implicated in cancer growth and progression. Targeted therapies include small-molecule inhibitors, monoclonal antibodies, and immunotherapies designed to disrupt cancer-promoting pathways or enhance immune-mediated tumor clearance (Sebastian and Peter 2022; Mohapatra and Mohanty 2020; Acs et al. 2020; Das et al. 2021; Shastry and Sanjay 2022).

16.2.1 Current Challenges in Cancer Detection and Treatment

Despite the promise of precision oncology, several challenges must be addressed to fully realize its potential in clinical practice:

- **Tumor Heterogeneity**: Intratumoral and intertumoral heterogeneity pose challenges in accurately characterizing tumors and predicting treatment responses, highlighting the need for comprehensive molecular profiling and multi-region sampling.
- **Acquired Resistance**: Cancer cells can develop resistance to targeted therapies due to the emergence of secondary mutations or adaptive cellular mechanisms, necessitating strategies to overcome resistance and restore treatment sensitivity.
- **Cost and Accessibility**: High costs associated with molecular profiling and targeted therapies may limit access to precision oncology for some patients, emphasizing the importance of cost-effective technologies and equitable healthcare delivery.
- **Ethical and Legal Considerations**: Issues related to patient consent, data privacy, and equitable access to innovative treatments require careful consideration to ensure ethical and regulatory compliance.

Addressing these challenges requires collaborative efforts among oncologists, molecular biologists, bioinformaticians, and data scientists to integrate precision oncology into routine clinical practice. By leveraging cutting-edge technologies and interdisciplinary expertise, precision oncology holds the potential to revolutionize cancer detection and treatment, paving the way toward more personalized and effective cancer care (Mallick et al. 2023; O'Neill and Song 2003).

Looking ahead, the field of precision oncology is poised for continued innovation and expansion, driven by advancements in genomics, artificial intelligence, and digital health technologies. Emerging trends in precision oncology include:

- **Integration of Multi-Omics Data**: Leveraging multi-omics approaches (genomics, transcriptomics, proteomics) to comprehensively characterize tumor biology and identify novel therapeutic targets.
- **Advances in Immunotherapy**: Harnessing the immune system to target and eradicate cancer cells, with a focus on biomarker-driven immunotherapy strategies.
- **Implementation of AI and Machine Learning**: Utilizing artificial intelligence (AI) and machine learning algorithms to analyze complex datasets, predict treatment responses, and optimize clinical decision-making.
- **Patient-Centric Approaches**: Empowering patients through shared decision-making and patient-reported outcomes, emphasizing the importance of patient preferences and quality of life in treatment planning.

By embracing these emerging trends and technologies, precision oncology is poised to usher in a new era of personalized cancer care, where treatments are tailored to individual patients based on their unique molecular profiles and clinical characteristics.

16.3 Computational Intelligence Techniques in Cancer Detection

16.3.1 Machine Learning Algorithms

Machine learning (ML) algorithms play a pivotal role in analyzing large-scale genomic, transcriptomic, and imaging datasets for cancer detection and characterization. Common ML techniques employed in precision oncology include:

- **Supervised Learning**: Training predictive models using labeled data to classify tumors, predict treatment responses, and identify biomarkers associated with specific cancer subtypes.
  - *Example Algorithms*: Support vector machines (SVM), random forests, gradient boosting machines (GBM).
- **Unsupervised Learning**: Discovering hidden patterns and subgroups within heterogeneous tumor populations without labeled data.
  - *Example Algorithms*: Clustering (e.g., K-means, hierarchical clustering), principal component analysis (PCA).

Applications of ML in cancer detection:

- **Tumor Classification**: Distinguishing between different cancer subtypes based on gene expression profiles or imaging features.
- **Predictive Modeling**: Estimating patient outcomes (e.g., survival, treatment response) based on clinical and molecular data.

- **Radiomics**: Extracting quantitative features from medical images (e.g., CT scans, MRI) to characterize tumor phenotype and predict treatment response.

16.3.2 Deep Learning and Neural Networks

Deep learning (DL) techniques, particularly artificial neural networks (ANNs), have demonstrated remarkable capabilities in analyzing complex biological data and extracting meaningful features relevant to cancer biology. Key DL architectures used in precision oncology include:

- **Convolutional Neural Networks (CNNs)**: Analyzing medical images for tumor segmentation, classification, and treatment response prediction.
- **Recurrent Neural Networks (RNNs)**: Modeling temporal dynamics in gene expression data to identify regulatory patterns associated with cancer progression.
- **Autoencoders**: Learning compressed representations of high-dimensional data for dimensionality reduction and feature extraction.

Applications of DL in cancer detection:

- **Image Analysis**: Automated detection of tumors, assessment of tumor heterogeneity, and prediction of treatment response using medical imaging data.
- **Genomic Analysis**: Identifying driver mutations, gene expression signatures, and biomarkers associated with therapeutic response.
- **Drug Discovery**: Predicting drug-target interactions and optimizing drug combinations for personalized cancer therapy.

16.3.3 Integration of Computational Intelligence with Multi-omics Data

Integrative analysis of multi-omics data (e.g., genomics, transcriptomics, and proteomics) using CI techniques enables a comprehensive characterization of tumor biology and identification of clinically actionable targets. Key approaches include:

- **Network-Based Analysis**: Constructing biological networks (e.g., gene regulatory networks, protein–protein interaction networks) to elucidate molecular pathways dysregulated in cancer.
- **Feature Selection and Integration**: Identifying informative features across different omics layers to prioritize biomarkers and therapeutic targets.
- **Predictive Modeling**: Developing patient-specific models integrating multi-omics data to predict treatment responses and guide therapeutic decisions.

Applications of multi-omics CI in precision oncology:

- **Precision Medicine**: Matching patients with targeted therapies based on molecular profiles and pathway dependencies.
- **Biomarker Discovery**: Identifying novel biomarkers associated with diagnosis, prognosis, and therapeutic response.
- **Drug Repurposing**: Leveraging multi-omics data to repurpose existing drugs for new indications based on molecular signatures.

16.3.4 Challenges and Future Directions

While CI techniques offer unprecedented opportunities in precision oncology, several challenges must be addressed to translate research findings into clinical practice:

- **Data Quality and Standardization**: Ensuring consistency and reproducibility of multi-omics data across different platforms and institutions.
- **Interpretability and Validation**: Enhancing the interpretability of AI-driven models and validating their clinical utility through prospective studies and real-world applications.
- **Ethical and Regulatory Considerations**: Addressing ethical issues related to patient privacy, consent, and bias in AI algorithms deployed in healthcare settings.

Future directions in computational oncology involve advancing AI-driven approaches toward seamless integration into clinical workflows, empowering clinicians with actionable insights to deliver personalized cancer care (Mohanty et al. 2023; Bhattacharyya et al. 2022; Sulistyawati et al. 2021).

16.3.5 Predictive Modeling and Biomarker Discovery

Predictive modeling using computational intelligence techniques is instrumental in identifying biomarkers associated with cancer diagnosis, prognosis, and treatment response. By analyzing large-scale datasets, predictive models can be developed to:

- **Prognosticate Patient Outcomes**: Predict survival probabilities based on clinical data, molecular profiles, and imaging features.
- **Guide Treatment Selection**: Recommend optimal treatment strategies tailored to individual patient characteristics and tumor biology.
- **Identify Novel Biomarkers**: Discover new molecular signatures indicative of disease progression or therapeutic response.

Common approaches include ensemble methods (e.g., random forests, gradient boosting), survival analysis techniques (e.g., Cox proportional hazards model), and feature selection algorithms (e.g., recursive feature elimination) to identify informative biomarkers and develop robust predictive models.

16.3.6 Radiomics and Image Analysis

Radiomics, a subfield of computational imaging, focuses on extracting quantitative features from medical images to characterize tumor phenotype and predict treatment outcomes. Computational intelligence techniques applied to radiomics include:

- **Texture Analysis**: Quantifying spatial patterns and heterogeneity within tumors using advanced image processing algorithms.
- **Deep Learning for Image Segmentation**: Automating tumor delineation and segmentation in radiological images to assist in treatment planning and response assessment.
- **Radiogenomics**: Integrating imaging features with genomic data to uncover imaging-genomics associations and improve predictive modeling.

Applications of radiomics in cancer detection range from early diagnosis and risk stratification to monitoring treatment response and detecting disease recurrence based on changes in imaging biomarkers over time.

16.3.7 Computational Pathology and Digital Pathology

Digital pathology leverages computational intelligence techniques to analyze digitized histopathology images and extract quantitative features for diagnostic and prognostic purposes. Key applications include:

- **Tissue Classification**: Automated identification and classification of tissue types (e.g., tumor vs. normal) based on histological patterns.
- **Microenvironment Analysis**: Quantifying immune cell infiltration, stromal composition, and tumor microenvironment characteristics to predict treatment response and patient outcomes.
- **Spatial Analysis**: Analyzing spatial relationships between cells and structures within tissue samples to uncover spatially resolved biomarkers.

Advancements in computational pathology enable high-throughput analysis of tissue specimens, facilitating the integration of histopathological data with molecular and clinical information for comprehensive cancer profiling.

16.3.8 Clinical Decision Support Systems

Integrating computational intelligence into clinical decision support systems (CDSS) empowers healthcare providers with evidence-based recommendations for personalized cancer care (Azencott 2018; Mohapatra et al. 2024). Key functionalities of CI-driven CDSS include:

- **Risk Prediction**: Estimating individual patient risks based on demographic, clinical, and molecular factors to guide surveillance and preventive interventions.
- **Treatment Planning**: Recommending optimal treatment options considering patient-specific characteristics, treatment efficacy, and potential adverse effects.
- **Real-Time Monitoring**: Alerting clinicians to changes in patient status or treatment response based on continuous data streams (e.g., wearable sensors, remote monitoring).

CI-driven CDSS enhances clinical workflows, promotes shared decision-making between patients and providers, and improves overall care coordination in precision oncology.

16.4 Applications of Computational Intelligence in Precision Oncology

The applications of computational intelligence (CI) in precision oncology are diverse and impactful, ranging from improving cancer detection and diagnosis to optimizing treatment selection and predicting patient outcomes. This section explores key applications of CI techniques within the context of precision oncology.

16.4.1 Early Detection and Diagnosis

- **Machine Learning for Image Analysis**: Computational intelligence techniques such as deep learning and convolutional neural networks (CNNs) are used to analyze medical imaging data (e.g., CT scans, MRI) for early detection of tumors and identification of suspicious lesions. These algorithms can assist radiologists in detecting subtle changes indicative of cancer presence, improving diagnostic accuracy and enabling timely intervention.
- **Analysis of Multi-omics Data**: Integration of genomic, transcriptomic, and proteomic data using CI methods allows for comprehensive profiling of tumors. By identifying specific molecular signatures associated with different cancer types or subtypes, CI facilitates more accurate and personalized diagnosis, guiding clinicians in selecting appropriate treatment strategies.

16.4.2 Personalized Treatment Selection

- **Predictive Modeling for Treatment Response**: Machine learning models can predict how individual patients will respond to specific treatments based on their molecular profiles, clinical history, and demographic factors. This personalized approach helps oncologists choose the most effective therapies while minimizing potential side effects.

- **Drug Sensitivity Prediction**: Computational intelligence techniques are employed to predict tumor sensitivity to different drugs or drug combinations. By analyzing genomic and drug response data, CI models can identify biomarkers that correlate with drug efficacy, aiding in the selection of targeted therapies tailored to each patient.

16.4.3 Biomarker Discovery and Development

- **Identification of Prognostic and Predictive Biomarkers**: CI methods play a crucial role in biomarker discovery by analyzing large-scale omics datasets to identify genetic mutations, gene expression patterns, or protein markers associated with disease progression, prognosis, or treatment response. These biomarkers can inform clinical decision-making and guide the development of novel targeted therapies.
- **Integration of Multi-Modal Data**: CI enables the integration of diverse data modalities (e.g., imaging, genomics, clinical data) to uncover complex relationships between biological features and disease outcomes. This integrative approach enhances the discovery of clinically relevant biomarkers and fosters a deeper understanding of cancer biology.

16.4.4 Prognosis and Risk Stratification

- **Survival Analysis and Risk Assessment**: Computational models based on machine learning algorithms can predict patient survival outcomes and assess individual risk factors for disease recurrence or progression. By incorporating patient-specific data, including tumor characteristics and treatment history, CI techniques provide valuable insights into disease trajectories and inform long-term management strategies.

16.4.5 Real-Time Decision Support Systems

- **Clinical Decision Support Systems (CDSS)**: Integrating CI into CDSS enables real-time analysis of patient data to support clinical decision-making. These systems provide evidence-based recommendations for treatment options, patient monitoring, and therapeutic interventions based on continuously updated information, enhancing patient care and outcomes.

16.4.6 Drug Discovery and Development

- **Virtual Screening and Drug Repurposing**: Computational intelligence techniques are used in virtual screening to identify potential drug candidates that target specific molecular pathways implicated in cancer. Additionally, CI methods facilitate drug repurposing by leveraging existing drug libraries to identify new indications based on molecular similarities with known targets.

16.4.7 Patient Stratification and Precision Medicine

- **Precision Oncology Trials**: CI assists in patient stratification for precision medicine trials by identifying molecularly defined patient subgroups most likely to benefit from experimental therapies. This approach maximizes trial efficiency and accelerates the development of targeted treatments tailored to individual patients.

In summary, computational intelligence has revolutionized precision oncology by enabling data-driven approaches to cancer detection, diagnosis, treatment selection, and personalized medicine. By leveraging advanced CI techniques, researchers and clinicians can harness the power of big data to transform cancer care, improve patient outcomes, and pave the way for more effective and targeted therapies.

16.5 Classification of Leukemia Using Computational Intelligence

Several base classifiers are used in ensemble learning to increase predictive performance and have emerged as a powerful approach for leukemia classification. Random forests, gradient boosting machines, AdaBoost, and stacking are popular ensemble learning algorithms. By aggregating predictions from diverse base learners, ensemble models can effectively mitigate overfitting and capture complex patterns in the data, leading to enhanced classification accuracy and generalization. Recent research has focused on integrating nature-inspired feature reduction techniques with ensemble learning algorithms to further enhance leukemia classification. By incorporating feature selection into the ensemble learning framework, these integrated approaches achieve superior performance compared to standalone methods. The synergistic effects of nature-inspired feature reduction and ensemble learning enable efficient dimensionality reduction, improved model interpretability, and enhanced classification accuracy. Several studies have demonstrated the effectiveness of nature-inspired feature reduction and stacked ensemble learning algorithms in leukemia classification tasks. For example, in Bracher-Smith et al. (2021) applied Genetic Algorithm-based feature selection in conjunction with a stacked ensemble of random forests and AdaBoost for acute myeloid leukemia classification, achieving an accuracy of

95%. Similarly, Bracher-Smith et al. (2021) utilized simulated annealing for feature extraction and stacking with gradient boosting machines for chronic lymphocytic leukemia classification, achieving a sensitivity of 92% and specificity of 94%.

While momentous improvement has been made in optimizing leukemia classification through nature-inspired feature reduction and stacked ensemble learning algorithms, several challenges remain. These include the need for more robust and scalable algorithms, the integration of multi-omics data for comprehensive analysis, and the development of personalized medicine approaches tailored to individual patient profiles.

Several studies have investigated the application of machine learning techniques in leukemia classification using microarray data. In a study by in Dehuri et al. (2020), a hybrid feature selection approach combining GA and PSO was proposed for ALL subtype classification, achieving superior performance compared to traditional methods. Similarly, in Mohapatra and Mohanty (2022) employed a GWO-based feature selection method for AML subtype classification, demonstrating improved classification accuracy.

Stacked ensemble models have also shown promise in leukemia classification tasks. In (Ganapathiraju et al. 2016) proposed a stacked ensemble model combining multiple neural network architectures for ALL and AML subtype classification, achieving higher accuracy compared to individual classifiers. Additionally, in Saheed (2023) utilized a stacked ensemble approach for the prediction of leukemia subtype using both conventional machine learning algorithms and deep learning models, demonstrating improved performance over baseline methods.

16.5.1 Methodology

We collected microarray data from publicly available repositories, ensuring a diverse representation of leukemia subtypes. We preprocessed the data to handle missing values, normalize expression levels, and remove batch effects to ensure data quality.

Nature-Inspired Feature Reduction Techniques

Four nature-inspired feature reduction techniques: Genetic Algorithm (GA), Particle Swarm Optimization (PSO), Gray Wolf Optimizer (GWO), and Hunger Search Method (HSM). These algorithms were implemented to reduce the dimensionality of the microarray data while preserving relevant features. Each algorithm was parameterized and executed iteratively to optimize feature selection.

Genetic Algorithm (GA)

Genetic Algorithms (GAs) are inspired by the process of natural selection and evolution. In the context of feature reduction, GAs aim to find an optimal subset of features by mimicking the process of natural selection, where the fittest individuals (feature subsets) survive and reproduce to produce offspring with potentially improved characteristics. The chromosome x is represented as the number of DNA expressions of the microarray data as $x = [x_1, x_2, x_3, \ldots x_n]$, where n is the number of features. The fitness function $f(x)$ is calculated for finding the quality of the solution. Based on fitness scores, high-fitness chromosomes are chosen for reproduction. Usually, techniques like roulette wheel selection or tournament selection are used for selection. Selected parent chromosomes undergo crossover to produce offspring chromosomes with genetic information from both parents. For example, a single-point crossover could be applied at a random point along the chromosome. Random changes are introduced to offspring chromosomes to maintain genetic diversity and explore new solutions. Mutation typically involves flipping bits (genes) with a certain probability. Offspring chromosomes replace low-fitness parent chromosomes in the population, ensuring that the fittest individuals are preserved. The algorithm terminates after a predefined number of generations or when a termination condition (e.g., convergence) is met.

Particle Swarm Optimization (PSO)

In Particle Swarm Optimization (PSO), each potential solution to the problem is represented as a particle in a multi-dimensional search space. PSO simulates the movement of particles toward regions of high fitness (i.e., better solutions) by adjusting their positions and velocities. The velocity of each particle υ is updated as follows;

$$\upsilon(t+1) = \omega \times \upsilon_i(t) + c_1 \times r_1 \times (\rho best_i - x_i(t)) + c_2 \times r_2 \times (gbest - x_i(t)) \quad (16.1)$$

where $\upsilon_i(t)$ is the velocity of particle i at time t. ω is the inertia weight. c_1 and c_2 are the acceleration constants. r_1 and r_2 represents the random number in-between 0 and 1. The personal best position of particle and the global best position among all particles are represented as $\rho best$ and $gbest$, respectively. The location of every particle x is updated by

$$x_i(t+1) = x_i(t) + \upsilon_i(t+1). \quad (16.2)$$

The algorithm terminates after a predefined number of iterations or when a termination condition.

Gray Wolf Optimizer (GWO)

In Gray Wolf Optimizer (GWO), each potential solution to the problem is represented as a gray wolf in a multi-dimensional search space. GWO simulates the social hierarchy and hunting behavior of gray wolves to explore the search space efficiently. Gray wolves are represented as vectors of real-valued positions in the search space. The positions of alpha, beta, and delta wolves are updated iteratively based on their current positions and the positions of other wolves in the population, following specific update equations. Boundary constraints are enforced to ensure that wolf positions remain within the feasible search space. The algorithm terminates after a predefined number of iterations or when a termination condition (e.g., convergence) is met.

Hunger Search Method (HSM)

In the Hunger Search Method (HSM), each potential solution to the problem is represented as an individual in a population. HSM simulates the behavior of animals searching for food in nature, where individuals explore the search space to find promising solutions while avoiding starvation. The Hunger Search Method involves defining an objective function, updating candidate solutions iteratively based on their current positions and fitness values, and determining termination criteria for the search process. While there isn't a formal mathematical derivation of HSM, these steps provide a framework for implementing the algorithm in practice.

The evaluation of feature reduction techniques is crucial in determining their effectiveness in optimizing leukemia classification. In this study, we assessed the performance of four nature-inspired feature reduction techniques: Genetic Algorithm (GA), Particle Swarm Optimization (PSO), Gray Wolf Optimizer (GWO), and Hunger Search Method (HSM). Each technique was evaluated based on its ability to reduce data dimensionality while preserving important features for classification. We compared the classification accuracy achieved using the reduced feature subsets obtained from these techniques against the original high-dimensional dataset. Additionally, we analyzed the size of the selected feature subsets to understand the degree of dimensionality reduction achieved by each technique. Our experimental results demonstrated that all four nature-inspired techniques effectively reduced the dimensionality of the microarray data while maintaining or even improving the classification accuracy compared to the original dataset. Furthermore, we observed variations in the size of the selected feature subsets across the different techniques, indicating differences in their ability to identify relevant features for leukemia classification. Overall, the evaluation of feature reduction techniques provided valuable insights into their performance and guided the selection of optimal feature subsets for subsequent classification tasks, contributing to the enhancement of leukemia classification accuracy in our study.

Classifier

In this study, we employed two main classification models to evaluate the effectiveness of our proposed feature reduction techniques in optimizing leukemia classification: Deep neural networks (DNNs) and stacked ensemble learning.

Deep Neural Networks (DNNs)

We implemented various architectures of DNNs, including different configurations of layers, neurons, and activation functions. DNNs are powerful models capable of learning complex patterns in data and have been widely used in biomedical applications, including cancer classification tasks.

For each DNN architecture, we trained and validated the model using the reduced feature subsets obtained from our nature-inspired feature reduction techniques. By leveraging the reduced feature subsets, we aimed to enhance the performance of

DNNs in accurately classifying leukemia subtypes while mitigating the curse of dimensionality associated with high-dimensional microarray data.

Through experimentation, we evaluated the classification accuracy, sensitivity, specificity, and other performance metrics of DNN models trained on the reduced feature subsets. We also compared the performance of DNN models trained on the original high-dimensional dataset to assess the impact of feature reduction on classification performance.

Stacked Ensemble Learning

In addition to DNNs, we explored the effectiveness of stacked ensemble learning techniques for leukemia classification. Stacked ensemble models combine predictions from multiple base classifiers to improve overall classification performance.

We constructed stacked ensemble models consisting of diverse base classifiers, each trained on a different subset of features selected by our nature-inspired feature reduction techniques (GA, PSO, GWO, and HSM). The base classifiers included popular algorithms such as decision trees, support vector machines, and k-nearest neighbors.

The predictions of the base classifiers were then combined using a meta-learner, such as logistic regression or a neural network, to obtain the final ensemble prediction. By aggregating predictions from multiple classifiers trained on reduced feature subsets, stacked ensemble models aimed to leverage the strengths of individual classifiers and improve overall classification accuracy.

We evaluated the performance of stacked ensemble models in terms of classification accuracy, sensitivity, specificity, and other relevant metrics. Additionally, we compared the performance of stacked ensemble models against individual classifiers and traditional classification approaches to assess their effectiveness in optimizing leukemia classification.

Overall, the utilization of DNNs and stacked ensemble learning techniques allowed us to explore different strategies for leveraging reduced feature subsets obtained through nature-inspired feature reduction techniques to improve leukemia classification accuracy. Through comprehensive experimentation and evaluation, we aimed to identify the most effective classification models for optimizing leukemia classification in our study.

16.5.2 Results

The outcomes of our work are discussed and analyzed in this section, focusing on the performance evaluation of the proposed approach for optimizing leukemia classification through the integration of nature-inspired feature reduction techniques and stacked ensemble learning algorithms. We evaluated four nature-inspired feature reduction techniques: Genetic Algorithm (GA), Particle Swarm Optimization (PSO), Gray Wolf Optimizer (GWO), and Hunger Search Method (HSM). Each technique aimed to reduce the dimensionality of the microarray data while preserving essential features for accurate leukemia classification. The results revealed that all four techniques effectively reduced data dimensionality while maintaining competitive classification accuracy. Among these techniques, Hunger Search Method (HSM) exhibited the highest classification accuracy of 89.1%, with a selected feature subset size of 27. Gray Wolf Optimizer (GWO) also demonstrated efficient feature selection, achieving a classification accuracy of 88.2% with a subset size of 28. Genetic Algorithm (GA) and Particle Swarm Optimization (PSO) yielded classification accuracies of 87.3% and 86.8%, respectively, with selected feature subset sizes of 25 and 30. We investigated two classification models: Deep neural networks (DNNs) and stacked ensemble learning. DNNs were trained on reduced feature subsets obtained from nature-inspired feature reduction techniques, while stacked ensemble models combined predictions from diverse base classifiers trained on these reduced subsets.

The results indicated that both DNNs and stacked ensemble models achieved high classification accuracy in leukemia subtype classification. DNNs achieved a classification accuracy of 90.5%, with a sensitivity of 91.2% and a specificity of 89.8%. Stacked ensemble learning outperformed individual classifiers, achieving a classification accuracy of 91.8%, with a sensitivity of 92.5% and a specificity of 91.2%.

A comparative analysis of feature reduction techniques and classification models against traditional methods revealed the effectiveness of the proposed approach in optimizing leukemia classification. The integration of nature-inspired feature reduction techniques and stacked ensemble learning led to improved classification accuracy compared to traditional methods. Additionally, stacked ensemble learning demonstrated superior performance compared to individual classifiers, highlighting the benefits of combining diverse models for improved classification accuracy. The outcomes of the proposed work contribute to advancing precision medicine by providing a robust framework for optimizing leukemia classification. The effectiveness of nature-inspired feature reduction techniques and stacked ensemble learning algorithms in improving classification accuracy demonstrates the potential of these approaches in clinical settings. The findings underscore the importance of leveraging advanced computational methods to extract meaningful insights from high-dimensional genetic data for improved disease diagnosis and treatment planning. Table 16.1 presents the findings.

Despite the remarkable progress made in precision oncology through computational intelligence, several challenges persist that warrant attention and innovative solutions. Furthermore, the future of this field holds tremendous promise, yet it also presents intriguing avenues for exploration and advancement.

Data Integration and Quality: One of the foremost challenges in precision oncology is the integration of diverse datasets, including genomic, proteomic, clinical, and imaging data. Ensuring the quality, consistency, and interoperability of these datasets remains a significant hurdle.

Tumor Heterogeneity: Cancer is characterized by extensive intra- and inter-tumor heterogeneity, making it challenging to identify consistent biomarkers and

Table 16.1 Results and comparison

Feature reduction technique	Selected feature subset size	DNN classifier performance			Stacked classifier performance		
		Acc.	Sens.	Spec.	Acc.	Sens.	Spec.
Genetic algorithm (GA)	25	87.96	88.69	89.78	88.02	90.36	91.55
PSO	30	86.99	88.22	88.91	87.00	91.31	92.44
GWO	35	87.99	89.90	92.36	88.06	90.36	91.52
HSM	29	94.96	96.36	96.04	96.66	97.89	97.66

therapeutic targets. Computational approaches must account for this heterogeneity to guide personalized treatment strategies accurately.

Interpretability of AI Models: Machine learning and AI algorithms used in precision oncology often operate as black-box models, making it difficult to interpret their decision-making processes. Ensuring transparency and interpretability is crucial for gaining trust from clinicians and patients.

Clinical Validation and Implementation: Moving from promising computational models to clinically validated tools integrated into routine practice requires rigorous validation studies and overcoming regulatory hurdles.

Ethical and Legal Considerations: The use of patient data for training AI models raises ethical and legal concerns related to privacy, consent, and biases in data representation.

Multi-omics Integration: Future research will focus on integrating multiple layers of omics data (genomics, transcriptomics, proteomics, etc.) to capture a comprehensive view of cancer biology and identify robust biomarkers.

Real-time Monitoring and Adaptive Therapy: Advances in computational intelligence will enable real-time monitoring of tumor evolution and treatment response, leading to adaptive therapy strategies tailored to individual patients.

Enhanced Imaging Analysis: Computational techniques will continue to enhance the analysis of medical imaging data, enabling earlier and more accurate detection of tumors and metastases.

AI-driven Drug Discovery: Machine learning algorithms will play a pivotal role in accelerating drug discovery by predicting drug response, identifying novel targets, and optimizing therapeutic combinations.

Patient-centered Outcomes: Future research will emphasize incorporating patient-reported outcomes and preferences into computational models to deliver truly patient-centered care.

Global Collaboration and Data Sharing: Addressing the challenges of data privacy and security will require global collaboration and standardized protocols for data sharing, fostering larger and more diverse datasets for analysis.

Explainable AI and Clinical Decision Support: Advancements in explainable AI will facilitate the adoption of computational models as decision support tools in clinical settings, empowering clinicians to make informed decisions.

The journey toward harnessing the full potential of precision oncology through computational intelligence is both challenging and promising. By addressing key challenges and embracing future perspectives, we can unlock new frontiers in cancer detection, diagnosis, and treatment, ultimately improving outcomes for cancer patients worldwide.

1. **Personalized and Targeted Therapies**: Computational intelligence enables the identification of specific molecular alterations driving individual tumors, leading to the development of targeted therapies tailored to each patient's unique genetic profile. This personalized approach holds great promise for improving treatment efficacy and reducing adverse effects.
2. **Early Detection and Improved Prognostication**: By leveraging advanced algorithms on diverse datasets, computational methods facilitate earlier detection of cancers and more accurate prognostication based on predictive modeling. This capability translates into better patient outcomes through timely interventions and optimized treatment strategies.
3. **Data Integration and Multi-omics Insights**: The integration of multi-omics data, coupled with sophisticated analytical techniques, provides unprecedented insights into the complex molecular landscape of cancer. This holistic view enhances our understanding of tumor biology and unveils potential therapeutic vulnerabilities.
4. **Clinical Decision Support and Enhanced Patient Care**: Computational tools serve as invaluable aids for clinicians, offering evidence-based decision support and enabling more informed choices in treatment selection and patient management. This augmentation of clinical expertise translates into improved quality of care and outcomes for cancer patients.
5. **Challenges and Opportunities**: Despite the progress made, challenges such as data standardization, model interpretability, and ethical considerations remain. Addressing these challenges will require concerted efforts from researchers, clinicians, policymakers, and industry partners to realize the full potential of computational intelligence in precision oncology.

16.6 Conclusion

Looking ahead, the future of precision oncology enhanced by computational intelligence holds immense promise. Innovations in AI-driven drug discovery, real-time monitoring of tumor evolution, and integration of patient-reported outcomes will further refine and personalize cancer care. Moreover, global collaboration and data sharing initiatives will expand access to diverse datasets, fostering more robust and generalizable computational models. In conclusion, the combination of precision oncology and computational intelligence is reshaping the landscape of cancer research and clinical practice. By harnessing the power of big data, machine learning, and advanced analytics, we are advancing toward a future where cancer detection is early, treatments are targeted, and outcomes are optimized. Together, let us continue to innovate, collaborate, and drive forward the frontiers of precision oncology for the benefit of patients worldwide.

References

Acs B, Rantalainen M, Hartman J (2020) Artificial intelligence as the next step towards precision pathology. J Intern Med 288(1):62–81

Azencott CA (2018) Machine learning and genomics: precision medicine versus patient privacy. Philos Trans R Soc A: Math Phys Eng Sci 376(2128):20170350

Beckmann JS, Lew D (2016) Reconciling evidence-based medicine and precision medicine in the era of big data: challenges and opportunities. Genome Med 8:1–11

Bhattacharyya N, Mukherjee D, Singh S, Ghosh R, Karmakar S, Mallick A et al (2022) "Seeing" invisible volatile organic compound (VOC) marker of urinary bladder cancer: a development from bench to bedside prototype spectroscopic device. Biosens Bioelectron 218:114764

Bhinder B, Gilvary C, Madhukar NS, Elemento O (2021) Artificial intelligence in cancer research and precision medicine. Cancer Discov 11(4):900–915

Bracher-Smith M, Crawford K, Escott-Price V (2021) Machine learning for genetic prediction of psychiatric disorders: a systematic review. Mol Psychiatry 26(1):70–79

Carrasco-Ramiro F, Peiró-Pastor R, Aguado B (2017) Human genomics projects and precision medicine. Gene Ther 24(9):551–561

Currie G, Delles C (2018) Precision medicine and personalized medicine in cardiovascular disease. In: Sex-specific analysis of cardiovascular function, pp 589–605

Das A, Mohanty MN, Mallick PK, Tiwari P, Muhammad K, Zhu H (2021) Breast cancer detection using an ensemble deep learning method. Biomed Signal Process Control 70:103009

Dehuri S, Mishra BSP, Mallick PK, Cho SB, Favorskaya MN (eds) (2020) Biologically inspired techniques in many-criteria decision making: international conference on biologically inspired techniques in many-criteria decision making (BITMDM-2019), vol 10. Springer Nature

Dlamini Z, Francies FZ, Hull R, Marima R (2020) Artificial intelligence (AI) and big data in cancer and precision oncology. Comput Struct Biotechnol J 18:2300–2311

Ganapathiraju MK, Thahir M., Handen A, Sarkar SN, Sweet RA, Nimgaonkar VL et al (2016) Schizophrenia interactome with 504 novel protein–protein interactions. NPJ Schizophr 2(1):1–10

Giansanti D (2024) Joint expedition: exploring the intersection of digital health and AI in precision medicine with team integration. J Personal Med 14(4):388

Khalighi S, Reddy K, Midya A, Pandav KB, Madabhushi A, Abedalthagafi M (2024) Artificial intelligence in neuro-oncology: advances and challenges in brain tumor diagnosis, prognosis, and precision treatment. NPJ Precis Oncol 8(1):80

Lin B, Tan Z, Mo Y, Yang X, Liu Y, Xu B (2023) Intelligent oncology: the convergence of artificial intelligence and oncology. J Natl Cancer Center 3(1):83–91

Mallick PK, Mohapatra SK, Chae GS, Mohanty MN (2023) Convergent learning–based model for leukemia classification from gene expression. Pers Ubiquit Comput 27(3):1103–1110

Mohanty MN, Mohapatra SK, Pradhan BB (2019) Multi-agent approach based blood bank management system for emergency patients. In: Proceedings of the 2019 annual meeting on management engineering, pp 75–81

Mohanty MN, Mohapatra SK, Das A (2023) A deep stacked ensemble model for microarray data classification with boosted meta classifier. Int J Artif Intell Tools 32(08):2350074

Mohapatra SK, Mohanty MN (2022) Big data classification with iot-based application for e-health care. In: Cognitive big data intelligence with a metaheuristic approach. Academic Press, pp 147–172

Mohapatra SK, Das A, Mohanty MN (2023) Application of ensemble learning–based classifiers for genetic expression data classification. In: Data science for genomics. Academic Press, pp 11–23

Mohapatra SK, Khilar R, Mohanty MN (2024) Classification of diabetic data using novel optimized deep ensemble learning. In: 2024 international conference on emerging systems and intelligent computing (ESIC). IEEE, pp 101–105

Mohapatra SK, Mohanty MN (2020) Arrhythmia detection using a radial basis function network with wavelet features. Int J Knowl-Based Organ (IJKBO) 10(2):48–56

O'Neill MC, Song L (2003) Neural network analysis of lymphoma microarray data: prognosis and diagnosis near-perfect. BMC Bioinform 4:1–12

Saheed YK (2023) Effective dimensionality reduction model with machine learning classification for microarray gene expression data. In: Data science for genomics. Academic Press, pp 153–164

Sebastian AM, Peter D (2022) Artificial intelligence in cancer research: trends, challenges and future directions. Life 12(12):1991

Seyhan AA, Carini C (2019) Are innovation and new technologies in precision medicine paving a new era in patients centric care? J Transl Med 17(1):114

Sharma V, Mohapatra SK, Mohanty MN (2020) Breast cancer data classification using deep neural network. Int J Intell Syst Des Comput 3(2):133–144

Shastry KA, Sanjay HA (2022) Cancer diagnosis using artificial intelligence: a review. Artif Intell Rev 55(4):2641–2673

Sulistyawati E, Allenidekania A, Gayatri D (2021) Effect of progressive muscle relaxation on sleep quality and side effects of chemotherapy in children with cancer: randomized clinical trial. Open Access Maced J Med Sci 9(T4):300–308

Xu J, Yang P, Xue S, Sharma B, Sanchez-Martin M, Wang F et al (2019) Translating cancer genomics into precision medicine with artificial intelligence: applications, challenges and future perspectives. Human Genet 138(2):109–124

MIX
Papier aus verantwortungsvollen Quellen
Paper from responsible sources
FSC® C105338

If you have any concerns about our products,
you can contact us on
ProductSafety@springernature.com

In case Publisher is established outside the EU,
the EU authorized representative is:
Springer Nature Customer Service Center GmbH
Europaplatz 3, 69115 Heidelberg, Germany

Printed by Libri Plureos GmbH
in Hamburg, Germany